Mathematical Models in Developmental Biology

Jerome K. Percus
Courant Institute of Mathematical Sciences
and Department of Physics, New York University

Stephen Childress
Courant Institute of Mathematical Sciences

26

Mathematical Models in Developmental Biology

Courant Institute of Mathematical Sciences
New York University
New York, New York

American Mathematical Society
Providence, Rhode Island

2010 *Mathematics Subject Classification.* Primary 92C10, 92C15, 92C17, 92C45.

For additional information and updates on this book, visit
www.ams.org/bookpages/cln-26

Library of Congress Cataloging-in-Publication Data

Percus, Jerome K. (Jerome Kenneth)
Mathematical models in developmental biology / Jerome K. Percus, Stephen Childress, Courant Institute of Mathematical Sciences, New York University, New York, NY.
pages cm. — (Courant lecture notes in mathematics ; volume 26)
Includes bibliographical references and index.
ISBN 978-1-4704-1080-3 (alk. paper)
1. Developmental biology—Mathematical models. I. Childress, Stephen. II. Title.

QH581.2.P36 2015
571.8′2—dc23

2015004198

∞ The paper used in this book is acid-free and falls within the guidelines
established to ensure permanence and durability.
Visit the AMS home page at `http://www.ams.org/`

10 9 8 7 6 5 4 3 2 1 20 19 18 17 16 15

Contents

Preface

During the academic year 1977–78 the authors undertook to present to graduate students at the Courant Institute of Mathematical Sciences (CIMS) a course devoted to mathematical models in developmental biology. Much of the material introduced in this course at the time reflected the most active research areas, dealing primarily with bifurcation and catastrophes, biochemical pattern formation, and mechanical morphogenesis. Inevitably, though, the course evolved in unexpected directions and some original research was incorporated into the lectures. From time to time other researchers have requested copies of what were then mimeographed CIMS lecture notes, and recently interest arose for including an updated version in the present AMS lecture notes series.

In the intervening years there have been profound changes in the nature of and attitude toward the kind of modeling championed in these notes. The rapid advances in understanding of developmental processes at the molecular level have revealed in detail mechanisms that thirty years ago were highly speculative models. Perhaps the most dramatic changes have occurred in our understanding of the biochemical basis of pattern formation. As our knowledge of these systems has expanded, there also seems to be a new appreciation of the importance of mathematical modeling at the intermediate level between the molecular instructions for a developmental event and the description of the end result of that event.

The problem we faced in preparing this volume was how to give a sense of the original course, while also making some effort either to bring material up to date or else to offer references to later research. In fact, our choice of material in 1977, while perhaps reasonably timely then, can hardly be said to have fully anticipated the work of the subsequent decades. We also wanted to incorporate portions of lecture notes by one of us (Percus), given at the Courant Institute in 2006, which further expanded and extended the material. It seemed therefore preferable to preserve most of the content of the original lectures, grouping them appropriately for a monograph format, while adding a few notes to provide links to the more recent literature, and ignoring more polished versions presented in the intervening years. A few topics were dropped as tangential to the main interests of the notes. While far short of providing a state-of-the-art review, we hope the result will be useful to mathematicians interested in this exciting field of applied research.

Kenneth L. Ho prepared LaTeX files of the 2006 lecture series, and we thank him for permission to include portions of that work in the present volume.

CHAPTER 1

Introduction

1.1. Modeling Biological Development

A *mathematical model* in natural science is a systematization of data such that alteration of experimental conditions is reflected ideally by variation of a small number of model parameters. It goes without saying that models are rarely developed in this way. Rather, they are constructed by simplification of hypothetical underlying mechanisms, and their ultimate major utility lies in the physical, chemical, and other mechanisms that they suggest, which are then susceptible to experimental verification. A mathematical model has the advantage of being able to draw upon concepts and techniques from nominally alien disciplines, and the consequent danger that those properties that contribute to the uniqueness of the discipline at hand may be understated due to insufficient universality.

In this course, we address ourselves primarily to the development of multicellular animals from an initial single cell egg. In bygone years, the self-inconsistent concept of *preformation* was in vogue, in which each egg contained in miniature a copy of the adult to which it would eventually give rise. It was not terribly difficult to imagine ways in which this copy could be enlarged and developed—and it was not terribly interesting. We are now reasonably certain that the information used by the developing organism is initially present in a spatially unstructured form, and the problem then is to explain *epigenesis*, the development of the highly complex, highly reproducible spatial structure of the animal from this initial structureless mass. This should not be understood in its most extreme form—there *is* spatial structure in the initial egg, and it *does* play a significant role in some of the ensuing development, but the total information that it yields is small on the scale of that required to define the process of development. Choice of the level of a mathematical model is of crucial importance. Should it describe all animals simultaneously, animal type entering in some fashion as a variable parameter? This is not absurd during early development, and so we will do something like that. Should it instead be "content" with describing in a uniform way all changes in form that can occur in a developing animal? This is the aim of catastrophe theory, which we will examine in the next chapter. Should we attempt at all to describe the development of a single egg, or should we use a stochastic description of an ensemble of similar eggs? The stochastic approach will indeed be a useful tool.

On a more detailed level: Should an effort be made to at least distinguish between the biochemical properties that serve as markers for cell type—the problem of differentiation and pattern formation—and the spatial delineation of tissues formed

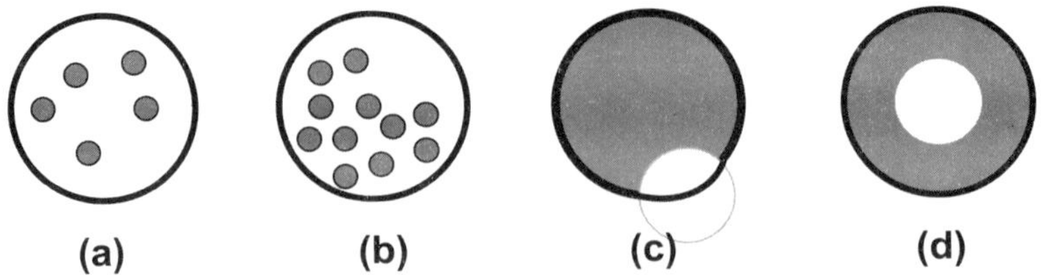

FIGURE 1.1. Yolk distribution in the egg: (a) isolecithal, (b) mesolecithal, (c) telolecithal, (d) centrolecithal.

from cells and extracellular material—the problem of change of form or morphogenesis? We will certainly see to what extent this is feasible. If we focus upon cells as units of structure, to what extent do we have to peer inwards to underlying biochemistry to have control over their overt properties, and to what extent can we deal in some continuum quasi-hydrodynamic way with cell populations, perhaps with an external biochemical field? Can we use the mechanical properties of cells as some sort of intermediate connection, mimicking cellular interaction by quasi-molecular interaction?

This list of questions and associated model types can be enlarged without difficulty, and becomes epistemological with even less difficulty. We will therefore adopt the more prudent course of following the traditional stages of developmental biology, of trying to organize as much phenomenology as possible at each stage, but of not struggling unduly when particular properties of specialized classes of organisms have to be pursued. A fairly broad swathe of mathematical techniques will perforce enter, and very few will exit.

1.2. Early Stages: A Brief Survey

The earlier, presumably more primitive stages of development show great similarities among diverse organisms, consistent of course with the gross physical form of the egg. In particular, the presence of yolk retards cleavage, producing larger cells and a general slowness of change. Eggs are thus usefully characterized according to their distribution of yolk: *isolecithal*, with little uniformly distributed yolk, as in echinoderms, on amphioxi, and mammals (Figure 1.1(a)); *mesolecithal*, with a heavier accumulation of yolk on one side, as in amphibia (Figure 1.1(b)); *telolecithal*, with only a small area bare of heavy yolk, as in birds (Figure 1.1(c)); *centrolecithal*, with yolk excluded only from the center and some channels, as in insects, and some coelenterates (Figure 1.1(d)).

The yolky side is the vegetal, the other the animal side or pole, a distinction arising on separation of egg from ovary, and valid in isolecithal eggs as well. In isolecithal eggs, cleavage is uniform straight through, or holoblastic, also true in mesolecithal eggs, but resulting in nonuniform cell size. In telelecithal eggs, cleavage is very incomplete, or meroblastic, with a cleavage area confined to a blastodisc. In centrolecithal eggs, it is generally superficial, separating nuclei but not cells proper.

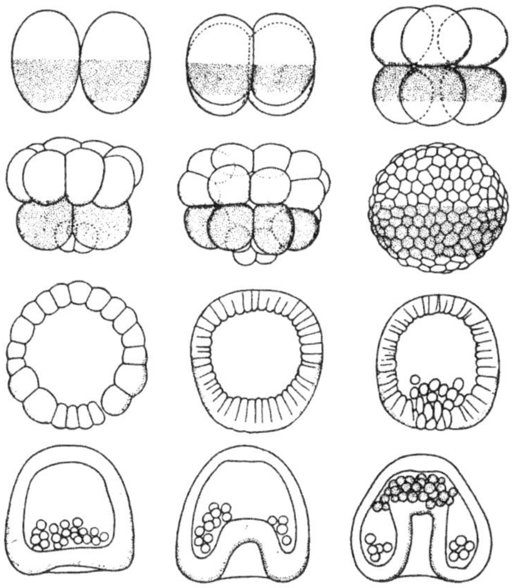

FIGURE 1.2. Cleavage and gastrulation of the sea urchin *Parcentrotus lividus*. After Boveri, 1901; drawn by Ina Mette.

The first commonly named stage of development results in a hollow ball, or *blastula*, after a number of cell cleavages. There may or may not be a discernible solid ball of cells, or morula, as an intermediate stage. There follows a complex pattern forming step of *gastrulation*, involving an invagination to produce the *endodermal* layer, the outside corresponding to *ectoderm* and the joining region to *mesoderm*. These three germ layers give rise ultimately to alimentary organs and lungs; skin and nervous system; and a support system of bone, muscle, and vascular elements. The sequences of cleavage, blastulation, and gastrulation occur in the development of diverse organisms—the sea urchin, an invertebrate echinoderm; amphioxus, a protochordate, the entering wedge to vertebrates; the frog, a vertebrate amphibian; the chick, a vertebrate bird; and the mouse, a vertebrate mammal. The form of these events, however, varies widely. We cannot attempt here detailed descriptions, and the reader is urged to consult basic texts such as [**1**]. We show in Figure 1.2 cleavage and gastrulation of the sea urchin.

1.3. An Example: Formation of the Blastocoel

We shall start by trying to model the very earliest morphogenetic change, that of blastulation, or more pointedly of *blastocoel* (the cavity in the blastula) formation; see Figure 1.3. It will turn out that several important concepts and techniques arise even at this level. But what precisely is the problem? It is to describe and at least empirically explain why a cleaving egg abandons the form of a morula or ball of cells and forms a cavity or blastocoel, a switch that occurs almost at once and gradually in sea urchins, but much later and suddenly in mammals. In fact, let us concentrate upon *isolecithal* eggs for the moment to avoid the additional parameters of yolk distribution.

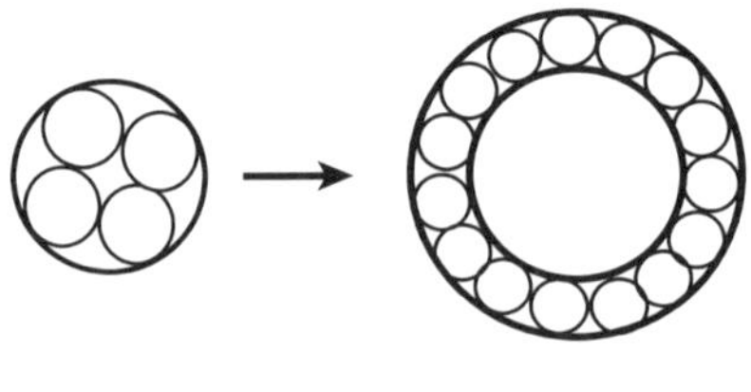

FIGURE 1.3. Formation of the blastocoel.

1.3.1. Local Behavior. Two mechanisms have been suggested for blastocoel formation, both relying on the firm attachment of cells to the inner egg surface coat or *hyaline layer*. In Dan's theory [**54**], the intercellular fluid accumulates large macromolecules, hence takes up more water osmotically and forces the cells attached to the hyaline layer outward.

According to Gustavson and Wolpert [**89**], cavity formation in the sea urchin is rather the result of hyaline layer adhesion exceeding intercellular adhesion, so that as the cells get smaller during cleavage they remain in a one-cell layer, a more or less fixed total cell volume, thus increasing the hyaline layer surface area and expanding the whole egg, with additional fluid entering passively. (The egg linear expansion need not be large—a shell from $R = 4$ to $R = 5$ radius contains roughly the same cell volume as a ball $R = 4$.) This adhesive imbalance would presumably not occur until much later in mammalian eggs.

The Gustavson-Wolpert mechanism takes advantage of cell properties that we shall meet frequently. Let us therefore investigate it in further detail. How shall we describe our system? At this stage, the major question is whether there is or is not a cell at a given point in space, so that it is reasonable to introduce a *cell density field*

$$\nu_{\mathbf{x}} = \begin{cases} 0 & \text{if there is no cell at } \mathbf{x}, \\ 1 & \text{if there is a cell at } \mathbf{x}. \end{cases}$$

We have the option of regarding $\mathbf{x}$ as continuous, as representing cell-sized boxes that may or may not be filled by cells, as mean cell occupation in a larger volume, or as something in between. We will adopt the second option but will feel free to switch as convenient; the difference is inconsequential only when there are many cells in each structural element of the organism.

Now what determines the configuration that the cells achieve in the course of time? We will start with the assumption of *quasi-static equilibrium*: a system with dynamical degrees of freedom $\{q_\alpha\}$ is described by some energy function $E(\{q_\alpha\}, t)$. This gives rise to generalized forces

$$F_\alpha = -\frac{\partial E}{\partial q_\alpha},$$

which change the dynamical variables. The change may be Newtonian, $\ddot{q}_\alpha = F_\alpha / m_\alpha$, hydrodynamic drag $\dot{q}_\alpha = F_\alpha / \gamma_\alpha$, or simply unknown. If we suppose that the time scale of the dynamics is fast compared with the explicit externally controlled time dependence of the function E, the system will always have time to equilibrate,

$$\frac{\partial E}{\partial q_\alpha} = 0,$$

and remain motionless on the large time scale, no matter what the dynamics. This is quasi-static equilibrium. Of course, we also require that the equilibrium be stable—that a fluctuation not dynamically introduce growing changes. With typical dynamics, this requires

$$\sum dF_\alpha \, dq_\alpha \leq 0$$

at equilibrium, or

$$\sum \left(\frac{\partial^2 E}{\partial q_\alpha \, \partial q_\beta} \right) dq_\alpha \, dq_\beta \geq 0.$$

This means that E is not just stationary, but must be at least at a *local minimum*.

What are the components of the system energy in the present case? To start with, there is of course the hyaline layer surface adhesion energy

$$E_{\mathrm{bd}} = -\gamma \sum_{x \in \mathrm{bd}} \nu_{\mathbf{x}}.$$

(bd $\equiv$ boundary): the more cells at the boundary, the closer to a minimum is E. Next, there is the cell-cell interaction adhesive energy

$$E_{\mathrm{int}} = -\frac{1}{2} J \sum_{\langle \mathbf{x}, \mathbf{y} \rangle} \nu_{\mathbf{x}} \nu_{\mathbf{y}},$$

where $\langle \mathbf{x}, \mathbf{y} \rangle$ indicates that the cells at $\mathbf{x}$ and $\mathbf{y}$ are nearest neighbors, in contact, and J is the mean interaction energy (a factor of $\frac{1}{2}$ occurs here because each neighboring pair occurs twice in double summation). Then there is the restriction that the cell number be fixed:

$$\sum_{\mathbf{x}} \nu_{\mathbf{x}} = N,$$

which, on insertion with a Lagrange parameter λ, masquerades as an energy contribution

$$E_N = \lambda \Big(\sum_{\mathbf{x}} \nu_{\mathbf{x}} - N \Big).$$

Finally, there is the effect of the all-pervading biological component—*fluctuation*—which enters at all levels of macroscopic description. Its simplest manifestation is via the *autonomous mechanical motion* of cells, so that external forces yield only a most likely dynamics, with a considerable spread. The same effect is of course produced by the fluctuation of coupling parameters, such as γ and J, due to variability of surface contact. And, on a different level, observation of a sample of several nominally identical organisms yields a spread of biological parameters not necessarily derivable from that of a single representative. Such unavoidable fluctuations are at the heart of traditional thermodynamics, where the unavailable energy is accounted for by inclusion of a negative entropy. This takes the form

$$E_T = T \sum_{\mathbf{x}} \big(\rho_{\mathbf{x}} \ln \rho_{\mathbf{x}} + (1 - \rho_{\mathbf{x}}) \ln(1 - \rho_{\mathbf{x}}) \big),$$

where T is an effective temperature or fluctuation level, and

$$\rho_{\mathbf{x}} \equiv \langle \nu_{\mathbf{x}} \rangle,$$

the average cell density, is no longer necessarily 0 or 1. In the absence of compelling energetic reasons, this gives rise to a nondescript soup of half cell, half not-cell. For many purposes, it is sufficient to Taylor expand E_T about its minimum at $\rho_{\mathbf{x}} = \frac{1}{2}$, yielding

$$E_T = T \sum_{\mathbf{x}} \left(\frac{1}{2}\sigma_{\mathbf{x}}^2 + \frac{1}{12}\sigma_{\mathbf{x}}^4 + \cdots \right),$$

to within an additive constant, where $\sigma_{\mathbf{x}} \equiv 2\rho_{\mathbf{x}} - 1$.

We conclude that, in terms of the field $\sigma_{\mathbf{x}}$ with limiting values

$$\sigma_{\mathbf{x}} = \begin{cases} -1 & \text{no cell at } \mathbf{x}, \\ 1 & \text{cell at } \mathbf{x}, \end{cases}$$

but intermediate values as well, the system energy can be written approximately as

$$E = \sum_{\mathbf{x}} \left(\frac{1}{12}\sigma_{\mathbf{x}}^4 + \frac{1}{2}\sigma_{\mathbf{x}}^2 + \left(\frac{\lambda}{2} - \frac{Jc}{4} \right) \sigma_{\mathbf{x}} \right) - \frac{J}{8} \sum_{\langle \mathbf{x},\mathbf{y} \rangle} \sigma_{\mathbf{x}}\sigma_{\mathbf{y}} - \frac{\gamma}{2} \sum_{\text{bd}} \sigma_{\mathbf{x}},$$

to within an additive constant. Here T has been chosen as the unit of energy, and c is the potential number of neighboring cells of a given cell, so that $\sum_{\langle \mathbf{x},\mathbf{y} \rangle} \sigma_{\mathbf{x}} = c \sum_{\mathbf{x}} \sigma_{\mathbf{x}}$, etc. Because of the fluctuation energy approximation, $\sigma_{\mathbf{x}}$ is no longer automatically bounded by -1 and 1, so that we must interpret an increase from 0 simply as the increasing likelihood of a cell, and a decrease from 0 as the increasing likelihood of no cell.

Suppose now that the energy was able to adjust itself locally. Then in any region—away from the boundary—of uniform cell density corresponding to "spin" σ (a term deriving from the use of σ in assemblies of magnetic dipoles), we would have an energy per grid point or *energy density*

$$E(\sigma) = \frac{1}{12}\sigma^4 + \left(\frac{1}{2} - \frac{Jc}{8} \right) \sigma^2 + \left(\frac{\lambda}{2} - \frac{Jc}{4} \right) \sigma$$

to within an additive constant. What does this look like? If $Jc/4 < 1$, then

$$\frac{\partial^2 E}{\partial \sigma^2} = \sigma^2 + \left(1 - \frac{Jc}{4} \right) > 0,$$

so that there is just one minimum; see Figure 1.4(a). If σ is constrained to a high value ≈ 1 at the egg boundary, it must fall rapidly to the homogeneous medium minimum, creating a thin shell. The sign of the minimum σ is precisely the reverse of that of the constant in $E(\sigma)$, i.e., $\frac{Jc}{4} - \frac{\lambda}{2}$

If $\frac{Jc}{4} > 1$, it depends. Let us write

$$E = \frac{1}{12}\sigma^4 + \frac{1}{2}u\sigma^2 + \frac{1}{3}v\sigma;$$

the uniform cell density σ is now controlled by the two parameters u and v: the stationary points are given by $E' = \frac{1}{3}\sigma^3 + u\sigma + \frac{1}{3}v = 0$. If $u \ll 0$ or $\frac{Jc}{4} \gg 1$, there are three real roots, corresponding to two minima of $E(\sigma)$; see Figure 1.4(b). If the

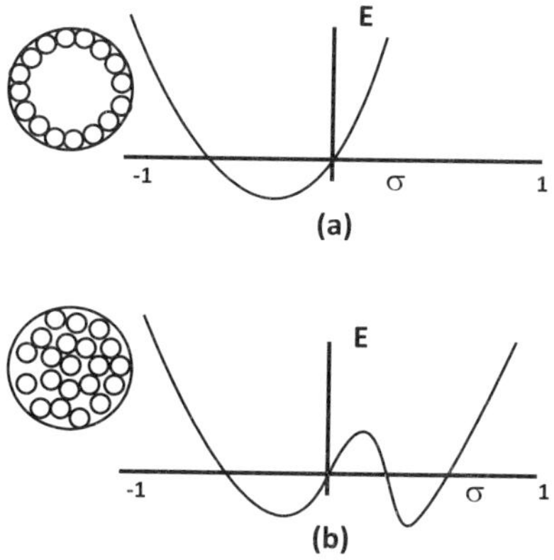

FIGURE 1.4. Stationary points of the energy function.

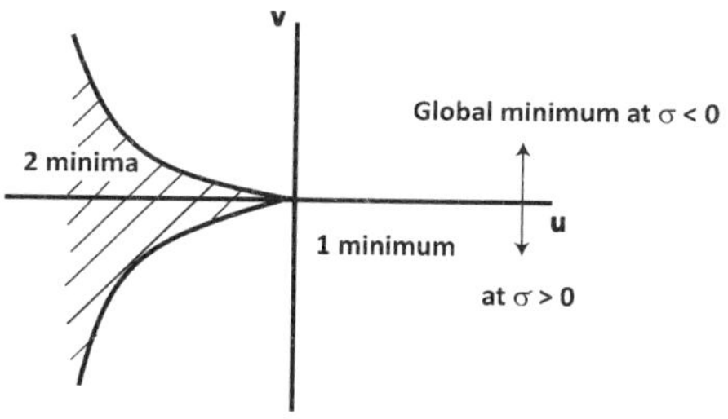

FIGURE 1.5. Regimes of blastulation.

highest one is selected by the boundary condition, we can readily have a morula structure, but the lowest root produces a blastocoel again. Once more, the lowest minimum, due to skewing by the linear term, is at $\pm$ values of σ according to the sign of $\frac{Jc}{4} - \frac{\lambda}{2}$.

The dividing line between the above regimes occurs when a maximum and minimum coincide, producing an inflection point: $E'' = \sigma^2 + u = 0$. Eliminating σ in $E' = \frac{1}{3}\sigma^3 + u\sigma + \frac{1}{3}v = 0$, the stationary character of E with respect to σ undergoes a sudden (*catastrophique* in French) change when the *control variables* u and v satisfy

$$v^2 + 4u^3 = 0.$$

We can express this diagrammatically as shown in Figure 1.5.

If it turns out that λ is increasing in time, a typical sample of developmental paths in parameter space from morula to blastula is indicated. It is clear, however, that there can be a catastrophic change in the initial state as well, as a function of evolution.

1.3.2. Global Behavior. In the presence of nonuniformity, and we have already distinguished between boundary and inside, the local analysis is insufficient. We must solve for the full three-dimensional density $\sigma_{\mathbf{x}}$, again with two control parameters. Returning to the full energy of the system,

$$E = \sum_{\mathbf{x}} \left(\frac{1}{12}\sigma_{\mathbf{x}}^4 + \frac{1}{2}\sigma_{\mathbf{x}}^2 + \left(\frac{\lambda}{2} - \frac{Jc}{4} \right) \sigma_{\mathbf{x}} \right) - \frac{J}{8} \sum_{\langle \mathbf{x},\mathbf{y} \rangle} \sigma_{\mathbf{x}}\sigma_{\mathbf{y}} - \frac{\gamma}{2} \sum_{\text{bd}} \sigma_{\mathbf{x}},$$

let us first suppose that γ is sufficiently high that it simply enforces the condition $\sigma_{\mathbf{x}} = 1$ at the boundary; the boundary cells are guaranteed to be stuck to the hyaline layer. Further, we observe that

$$\begin{aligned} - \sum_{\langle \mathbf{x},\mathbf{y} \rangle} \sigma_{\mathbf{x}}\sigma_{\mathbf{y}} &= \frac{1}{2} \sum_{\langle \mathbf{x},\mathbf{y} \rangle} (\sigma_{\mathbf{x}} - \sigma_{\mathbf{y}})^2 - \frac{1}{2} \sum_{\langle \mathbf{x},\mathbf{y} \rangle} (\sigma_{\mathbf{x}}^2 + \sigma_{\mathbf{y}}^2) \\ &= \frac{1}{2} \sum_{\langle \mathbf{x},\mathbf{y} \rangle} (\sigma_{\mathbf{x}} - \sigma_{\mathbf{y}})^2 - c \sum_{\mathbf{x}} \sigma_{\mathbf{x}}^2. \end{aligned}$$

Hence we have

$$E = \sum_{\mathbf{x}} \left(\frac{1}{12}\sigma_{\mathbf{x}}^4 + \left(\frac{1}{2} - \frac{Jc}{8} \right) \sigma_{\mathbf{x}}^2 + \left(\frac{\lambda}{2} - \frac{Jc}{4} \right) \sigma_{\mathbf{x}} \right) + \frac{J}{16} \sum_{\langle \mathbf{x},\mathbf{y} \rangle} (\sigma_{\mathbf{x}} - \sigma_{\mathbf{y}})^2,$$

where $\sigma_{\mathbf{x}} = 1$ at the boundary.

The inhomogeneity correction is a positive energy that must be supplied any time that $\sigma_{\mathbf{x}}$ changes in space; i.e., a surface tension results from breaking cell-cell adhesion contacts (see Chapter 4) and must be justified globally. Now, taking $\partial / \partial \sigma_{\mathbf{x}}$, we have at the minimum, in the previous uv-notation

$$\frac{1}{3}\sigma_{\mathbf{x}}^3 + u\sigma_{\mathbf{x}} + \frac{1}{3}v = \frac{J}{4} \sum_{\mathbf{y}}^{\langle \mathbf{x},\mathbf{y} \rangle} (\sigma_{\mathbf{y}} - \sigma_{\mathbf{x}}).$$

The right-hand side here is essentially the finite difference analogue of the Laplace operator, i.e., the mean deviation of a function on a surface surrounding the point in question. To see this, we introduce the average Av over nearest neighbors and write, for a slowly varying field $\sigma_{\mathbf{x}}$,

$$\begin{aligned} \frac{J}{4} \sum_{\mathbf{y}}^{\langle \mathbf{x},\mathbf{y} \rangle} (\sigma_{\mathbf{y}} - \sigma_{\mathbf{x}}) &= cAv_{\mathbf{y}}(\sigma_{\mathbf{y}} - \sigma_{\mathbf{x}}) \\ &= cAv_{\mathbf{y}} \left(\mathbf{y} - \mathbf{x} \cdot \nabla \sigma_{\mathbf{x}} + \frac{1}{2} (\mathbf{y} - \mathbf{x})(\mathbf{y} - \mathbf{x}) : \nabla \nabla \sigma_{\mathbf{x}} + \cdots \right). \end{aligned}$$

But if $|\mathbf{y} - \mathbf{x}| = a$, then for an arbitrary vector w, $Av(\mathbf{y} - \mathbf{x}) \cdot w = 0$ and $Av((\mathbf{y} - \mathbf{x}) \cdot w)^2 = \frac{1}{3}aw^2$, and so we may write

$$\frac{1}{3}\sigma_{\mathbf{x}}^3 + u\sigma_{\mathbf{x}} + \frac{1}{3}v = \frac{J}{12}a^2 c \nabla^2 \sigma_{\mathbf{x}},$$

or simply

$$E'(\sigma_{\mathbf{x}}) = \frac{J}{12}a^2 c \nabla^2 \sigma_{\mathbf{x}},$$

where ∇^2 is the continuous limit of the lattice Laplacian, $\nabla \cdot \nabla = \partial^2/\partial x^2 + \partial^2/\partial y^2 + \partial^2/\partial z^2$, and E denotes the non-interacting energy density.

Suppose now that σ varies only in one direction, which we can imagine as radial position r in a spherical egg; we then replace ∇^2 by $\partial^2/\partial r^2$, which causes obvious difficulties for small r. Since our nonlinear partial differential equation is thereby replaced by the ordinary differential equation

$$E'(\sigma) = \frac{J}{12} a^2 c \frac{d^2\sigma}{dr^2},$$

its solution is readily obtained. We simply multiply by $\frac{d\sigma}{dr}$ and integrate, obtaining

$$\frac{1}{24} J a^2 c \left(\frac{d\sigma}{dr}\right)^2 - E(\sigma) = K,$$

where K is a suitable constant, necessarily greater than $-E(\sigma)$ at each σ attained. This relation can be shown to be equivalent to the minimization of

$$E_1 = \int \left(E(\sigma) + \frac{1}{24} J a^2 c \left(\frac{d\sigma}{dr}\right)^2 \right) dr,$$

the cell continuum one-dimensional version of our original expression.

The value of the constant K is, of course, determined by boundary conditions. At the surface of the egg, $r = R$, we require $\sigma = 1$. Inside, we want the "motion" to stop, $\frac{d\sigma}{dr} = 0$, at $r = 0$. Since

$$dr = \frac{d\sigma}{d\sigma/dr} = \text{const} \frac{d\sigma}{\sqrt{K + E(\sigma)}},$$

K is thereby determined. It will be convenient for us to use a numerically indistinguishable but somewhat different criterion for purposes of illustration. The one-dimensional version of a three-dimensional egg makes more sense if we never have to go to $r = 0$ because the cell density has stabilized by then—has become asymptotic. Hence we shall use the condition that $\frac{d\sigma}{dr} = 0$ is reached at $r = -\infty$, but only look at what happens until $r = 0$. This requires that $K + E(\sigma)$ not only vanish at the small σ endpoint but in fact become stationary (for then $r = \int dr$ will diverge). We can now go through the two major sequences as J decreases, λ increases, or both. In each case, we draw the curve $-E(\sigma)$ (dropping the Taylor expansion approximation in σ steepens the sides) and the level line K. As λ increases to bias a portion of the egg towards $\sigma = -1$, J may start out low enough that only the single-minimum regime holds throughout, and a steady cavity formation results; see Figure 1.6(a).

On the other hand, if we start in the two-minimum region, a jump in σ_{min} can occur as soon as the left peak of $-E$ dominates the right one. There are then two regions of relatively uniform σ, but the outer one takes on a shell appearance as $-E$ degenerates to a single peak. The lesson we learn at this stage then is that the control parameters J and λ determine the global $\{\sigma_{\mathbf{x}}\}$ behavior in just the fashion that they determine the local σ-behavior—what matters is position relative to the singularities in the $J\lambda$-plane; see Figure 1.6(b).

There remains the problem of finding J and λ as functions of time. The former is in principle known (but perhaps not in units of "temperature" T—is the pulsatile sea urchin activity equivalent to high T and/or low J?), but the latter depends upon

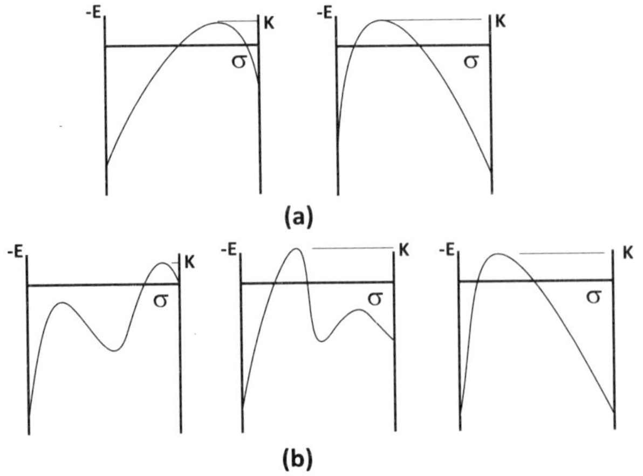

FIGURE 1.6. Cavity formation via the energy function.

the radius R of the egg, itself the result presumably of an energy minimization, contingent upon knowledge of the surface adhesivity γ.

Let us write $E(\lambda, R)$ to emphasize the unknown parameters. First fix R. Then we can write

$$E(\lambda) = \min(E_0 + \lambda g),$$

where $g = 0$ expresses the condition that the total number of cells is given. For degrees of freedom q_α, the minimizing values $q_\alpha(\lambda)$ thus satisfy $\partial(E_0+\lambda g)/\partial q_\alpha = 0$. The value of λ should be chosen such that $g = 0$ at the minimum. One way of accomplishing this is to set

$$\frac{dE(\lambda)}{d\lambda} = 0,$$

for then

$$\frac{dE(\lambda)}{d\lambda} = \left(\frac{\partial}{\partial\lambda} + \sum \frac{dq_\alpha}{d\lambda}\frac{\partial}{\partial q_\alpha}\right)E = g + \sum \frac{dq_\alpha}{d\lambda}\frac{\partial E}{\partial q_\alpha} = 0$$

implies $g = 0$. Furthermore,

$$\begin{aligned}\frac{d^2E}{d\lambda^2} &= \frac{dg}{d\lambda} + \sum \frac{d^2q_\alpha}{d\lambda^2}\frac{\partial E}{\partial q_\alpha} + \sum \frac{dq_\alpha}{d\lambda}\frac{\partial g}{\partial q_\alpha} + \sum\left(\frac{\partial^2 E}{\partial q_\alpha \partial q_\beta}\right)\frac{dq_\alpha}{d\lambda}\frac{dq_\beta}{d\lambda}\\ &= \sum \frac{\partial^2 E}{\partial q_\alpha \partial q_\beta} dq'_\alpha dq'_\beta,\end{aligned}$$

so that E is a minimum with respect to λ, too.

Now λ is a function of R, and we want $\frac{\partial}{\partial R}\, E(\lambda(R), R) = 0$ or $\frac{\partial\lambda}{\partial R}\frac{\partial E}{\partial\lambda} + \frac{\partial E}{\partial R} = 0$. We conclude that

$$\frac{\partial E(\lambda, R)}{\partial R} = \frac{\partial E(\lambda, R)}{\partial\lambda} = 0$$

and E is a minimum.

Thus the control variables J, N, and γ determine the parameters λ and R, which then serve as control variables for the cell distribution. Let us consider this piece of the problem briefly, oversimplifying greatly to clarify the point to be made. We choose to build the system energy out of surface adhesion and internal surface tension alone, regarding the volume energy at fixed total cell volume $2\pi V$ as fixed. If the spherical cavity goes from $R - d$ to R, we take $\frac{1}{d}$ as a measure of the cell density gradient, so that the surface tension energy is of the form VJ/d^2. For a thin shell, we have $2\pi V = \frac{1}{2}4\pi R^2 d$. Hence parametrizing the surface adhesion in the obvious way,

$$\Delta E = -\gamma R^2 + \frac{VJ}{d^2} \quad \text{where } R^2 d = V.$$

Eliminating d rather than using a Lagrange parameter,

$$\Delta E = \frac{J}{V}\left(R^4 - \frac{\gamma V}{J}R^2\right).$$

This is a degenerate case ($v = 0$) of the quartic expression we previously had to minimize, and always has precisely two minima at

$$|R| = \left(\frac{\gamma V}{2J}\right)^{1/2},$$

with a maximum at 0—a simple intuitively obvious $\frac{\gamma}{J}$ dependence, valid only for small d. The degeneracy is a result of radial symmetry, and if the more appropriate variable R^2, or $\frac{1}{d}$, is used, the basic quadratic minimization is recovered. The structure of the minimum never changes, and in terms of the variable $R^2 - \frac{\gamma V}{2J}$, there are no control variables at all.

CHAPTER 2

Catastrophe Theory

What should one mean by a mathematical description of biological development? The canonical form of the exact sciences—exact initial conditions determine exact future course—is less appropriate here because of innate fluctuations and ignorance of exact parameters. One solution is to adopt a stochastic or probabilistic format. But even this falls short of supplying the qualitative information that distinguishes one organism from another, one developmental path in a given organism from another. This qualitative information is presumably of a topological nature, stable against perturbation—smooth and small changes—of input and output data. Our ultimate task then is to develop such a topological dynamics, but proceeding via more narrow quantitative means, for the purpose of easier mathematical manipulation as well as that of suggesting empirical mechanisms, themselves to be analyzed at a lower biological, biochemical, or biophysical level. The example of blastulation that we have examined in Chapter 1 suggests several possible formats in which to carry out such an investigation. Let us again write the internal variables that specify the state of a space-time point as $\{q_\alpha\}$ or simply $\mathbf{q}$. Space-time will be regarded as continuous either by smoothing of discrete cell populations or by imagining that the state of each cell is determined by a continuous underlying field. Then a full quantitative nonstochastic description of the system is given by the set of functions $\mathbf{q}(\mathbf{r}, t)$. Alternative representations are also useful, e.g., the surfaces $q_\alpha(\mathbf{r}, t) = Q_\alpha$, where $Q_\alpha = 0$ might be the dividing line between having an α-type cell or not having one, resulting in a description by interfaces between cell groups of differing type. In any event, how are these functions to be characterized and categorized? Most reasonably by their literally remarkable events, i.e., the singularities of the function set which remain under small, smooth variations. How small and how smooth? Here we run the gamut from not too small mildly discontinuous changes to infinitesimal analytic ones. We in fact will have in mind suitably bounded infinitely differentiable changes, but clearly the resulting mathematics depends crucially upon the decision made here.

2.1. Basic Elements

We now have discussed the description without a hint of a mechanism. Immersion of the description into a class of mechanisms also has the mathematical advantage of reducing the set of singularity combinations to be categorized. Whatever the short- and long-range cellular interactions, one can imagine that the net effect on a given cell—or even a given spatial position—is to establish a milieu in which the state of the cell is determined. In the general quasi-static model, the determined state would occur not as an explicit function of the milieu but indirectly via minimization of some energy function $E(\mathbf{q}; \mathbf{r}, t)$ at each space-time point. In fact, we see from our model of inhomogeneous blastulation that in the absence of boundary conditions, it may be $-E$ that is minimized. Thus the merely stationary points of E and their singular behavior bear study, aside from the mathematical simplification thereby introduced.

The effect of the foregoing description is that a mechanism is introduced only at the local or biochemical level, behavior in the large remaining purely descriptive. This is certainly a useful viewpoint and is in accord with much of descriptive embryology, which is couched in terms of "*morphogenetic gradients* (amplitudes)," coordinate functions whose slow variation nonetheless induces singularities. A sufficient number (three) of gradients then supplies the "*positional information*" to determine the state of a cell at a given position. Left undiscussed are the connections between cells that establish the milieu in the first place, a matter to which we devoted some effort in our model of inhomogeneous blastulation. To establish these, it would be necessary to seek the minimum of the functional $E(\mathbf{q}(\mathbf{r}, t), t)$ at each t or perhaps even include non-quasi-static dynamics via minimization of $E(\mathbf{q}(\mathbf{r}, t), t)$ over all space-time fields $\mathbf{q}(\mathbf{r}, t)$, an extension of the Euler-Lagrange principle (= Hamilton's principle) of mechanics. In these cases, the "control space" is either one-dimensional (t) or zero-dimensional, making it inappropriate for the analysis we are about to embark on. But we have seen in the case of blastulation how internal control parameters (J and λ) may enter in a natural way, either representing changes in species or experiment, or via conservation laws, so that the control space dimension is not that well preordained. Ignoring the above disclaimers, we consider now a parametrized set of local variational conditions

$$dE(\mathbf{q}; \mathbf{u}) = 0$$

over infinitesimal variations $d\mathbf{q}$, where the control space $\mathbf{u}$ may indeed stand for space-time, but this restriction will not be needed at once. Unless specifically countermanded, all functions will be regarded as *infinitely differentiable*, i.e., belonging to C^∞, on a finite domain.

We seek to classify the ensuing stationary point regimes $\mathbf{q}(\mathbf{u})$ as functions of $\mathbf{u}$. To start, we ignore the $\mathbf{u}$-dependence and ask what E looks like near a stationary point. In the terminology to be used, such a point of vanishing gradient will be called a *singular point*. Since $\partial E/\partial q_\alpha = 0$, a Taylor expansion yields to leading order

$$E(\mathbf{q}) = E(\bar{\mathbf{q}}) + \frac{1}{2}\sum \frac{\partial^2 E(\bar{\mathbf{q}})}{\partial \bar{q}_\alpha \partial \bar{q}_\beta}(q_\alpha - \bar{q}_\alpha)(q_\beta - \bar{q}_\beta)$$

where $\bar{q}_\alpha = q_\alpha(\mathbf{u})$. The real symmetric matrix $\frac{\partial^2 E}{\partial \bar{q}_\alpha \partial \bar{q}_\beta}$ can be diagonalized by a real orthogonal transformation Ω, and so introducing the new variables ξ by

$$q_\alpha = \bar{q}_\alpha + \sum \Omega_{\alpha\beta}\xi_\beta$$

for suitable Ω, we can write

$$E(\mathbf{q}) = \bar{E} + \frac{1}{2}\sum \lambda_\alpha \xi_\alpha^2.$$

Finally, by a scaling transformation $\xi_\alpha \to k_\alpha \xi_\alpha$, we can assume that

$$\lambda_\alpha = 0, \ +1, \ \text{or} \ -1.$$

If $\lambda_\alpha \neq 0$ for all α, E is said to be *nondegenerate*, equivalent to the matrix $\partial^2 E/\partial \bar{q}_\alpha \partial \bar{q}_\beta$ being nonsingular. If it is nondegenerate at all singularities, it is called a *Morse function*. There is then a key theorem that says: *There is a Morse function lying arbitrarily close to any function with any number of degenerate singularities.* By "arbitrarily close," we refer to all derivatives simultaneously. Thus only the index (the number of $\lambda_\alpha = +1$ minus the number of $\lambda_\alpha = -1$) of a singularity has category significance, and not even that if our major interest is in minima, so that all $\lambda_\alpha = +1$. Although almost all functions are Morse, the situation changes dramatically—one might say catastrophically—when we examine parametrized classes of energy functions and the associated parametrized singular points. A hint as to how this happens is afforded by the theorem that *only nondegenerate singularities are stable—retain their index—under all small perturbations.* For example, the degenerate $E = x^3$ with one index 0 goes over to $E = x^3 - 3\epsilon^2 x$ with two nondegenerate singularities at $x = \pm\epsilon$ of indices ± 1.

A parametrized set of singularities can then have a qualitative shift in structure if it includes a degeneracy—but only then. In fact, we will characterize a parametrized function by the nature of its degenerate singular points, and this will turn out to determine its larger structure. Hence, we have to investigate the neighborhood of a degeneracy more carefully. For this purpose, it is convenient to introduce the concept of the *germ of a function*—all functions coinciding with it in a neighborhood of the point in question—and then quote the theorem: *If $\boldsymbol{\xi} \in \mathbb{R}^n$, and $\partial E/\partial \xi_\alpha(0) = 0$, but $\partial^2 E/\partial \xi_\alpha \partial \xi_\beta(0)$ has rank $n - l$, then new coordinates $\xi_1', \dots, \xi_l', \xi_1'', \dots, \xi_{n-l}''$ can be chosen such that $E(\xi_1', \dots, \xi_{n-l}'') = \phi(\xi_1', \dots, \xi_l') + \sum_1^{n-l} \lambda_\alpha (\xi_\alpha'')^2$, where $\lambda_\alpha = \pm 1$ and the germ ϕ satisfies $\partial\phi/\partial\xi_\alpha' = 0$, $\partial^2\phi/\partial\xi_\alpha'\partial\xi_\beta' = 0$.* In the neighborhood of a degeneracy, it is then ϕ, a constant plus a function starting with third-order terms, that we can concentrate on. Incidentally, we will usually not bother distinguishing between functions and germs.

We can now proceed to parametric extensions of functions, which we will term *unfoldings*. Explicitly, an unfolding $\phi(\xi_1, \dots, \xi_k; u_1, \dots, u_n) = \phi(\xi, \mathbf{u})$ of $\phi(\xi)$ is required only to satisfy the condition

$$\phi(\xi, 0) = \phi(\xi).$$

We of course are interested in stable unfoldings, which are qualitatively unchanged under small changes of internal variables and control parameters. To formalize this, we first define two unfoldings $\phi_1(\xi, \mathbf{u})$ and $\phi_2(\xi, \mathbf{v})$ as *equivalent* if we can transform one into the other by smooth transformations, i.e., if a control variable transformation $\mathbf{g}(\mathbf{u}) = \mathbf{v}$ and internal variable transformation (possibly $\mathbf{u}$-dependent) can be found identifying ϕ_1 and ϕ_2 to within a constant at each control point:

(2.1) $$\phi_1(\xi, \mathbf{u}) = \phi_2(f_{\mathbf{u}}(\xi), \mathbf{g}(\mathbf{u})) + K_{\mathbf{u}}, \quad f_{\mathbf{0}}(\xi) = \xi, \quad \mathbf{g}(\mathbf{0}) = 0,$$

and similarly for ϕ_2 in terms of ϕ_1. Then $\phi(\xi, \mathbf{u})$, an unfolding of $\phi(\xi)$, is *stable* if every nearby unfolding $\phi'(\xi, \mathbf{u})$ of $\phi(\xi)$ is equivalent to $\phi(\xi, \mathbf{u})$. For example, $\phi(x, u) = x^4 - ux^2$, with equal minima, is not stable because it is not equivalent to $\phi'(x, u) = x^4 - u(x^2 + \epsilon x)$. The restriction to stable unfoldings is made even less painful by virtue of the theorem: *"Every" unfolding ϕ_1 can be approximated arbitrarily closely by a (locally) stable unfolding ϕ_2. In fact, it can be obtained by formula* (2.1). The quotes about "every" will be explained in due course.

The crucial issue now is the fashion in which a locally stable unfolding is determined by the nature of the degenerate function $\phi(\xi)$ at each degenerate singular point. The relation is easily found. Suppose $\phi(\xi, \mathbf{u})$ is a stable unfolding of $\phi(\xi)$. Near the origin in $\mathbf{u}$-space, we can write to first order in $\mathbf{u}$,

$$\phi(\xi, \mathbf{u}) = \phi(\xi) + \sum u_\beta b_\beta(\xi).$$

A nearby unfolding of $\phi(\xi)$ to the same order in $\mathbf{u}$ must take the form (to within equivalence, we can assume $f_\beta(0) = 0$)

$$\phi_1(\xi, \mathbf{u}) = \phi(\xi) + \sum u_\beta b_\beta(\xi) + \sum u_\beta f_\beta(\xi).$$

The band of unfoldings equivalent to ϕ can be obtained by letting

$$\xi_\alpha \to \xi_\alpha + \sum a_{\alpha\beta}(\xi) u_\beta, \quad u_\alpha \to u_\alpha + \sum \lambda_{\alpha\beta} u_\beta.$$

Hence to first order in $\mathbf{u}$,

$$\phi(\xi, \mathbf{u}) \approx \phi(\xi) + \sum a_{\alpha\beta}(\xi) u_\beta \frac{\partial \phi}{\partial \xi_\alpha} + \sum u_\beta b_\beta(\xi) + \sum \lambda_{\alpha\beta} u_\beta b_\alpha(\xi).$$

This includes $\phi_1(\xi, \mathbf{u})$ if, given any $\{f_\beta\}$, the equations

$$\sum u_\beta f_\beta(\xi) = \sum a_{\alpha\beta}(\xi) u_\beta \frac{\partial \phi}{\partial \xi_\alpha} + \sum \lambda_{\alpha\beta} u_\beta b_\alpha(\xi)$$

are solvable, or taking the coefficient of u_β and dropping the β index, if

$$f(\xi) = \sum a_\alpha(\xi) \frac{\partial \phi}{\partial \xi_\alpha} + \sum \lambda_\alpha b_\alpha(\xi)$$

is solvable. We conclude, in other words, that *the $b_\alpha(\xi)$ span the space of all functions f mod* id$\{\partial\phi/\partial\xi_\alpha\}$, *with* $f(0) = 0$. Here, id$\{\partial\phi/\partial\xi_\alpha\}$ refers to the *ideal* of all functions of the form $\sum a_\alpha(\xi)\partial\phi/\partial\xi_\alpha$. It should be noted that the arbitrary control space variation $\lambda_{\alpha\beta}$ plays a crucial role in the result. If this variation is restricted by subsidiary information, then a subspace of $\{b_\alpha\}$ must do the spanning. The dimensionality of the vector space $\{b_\alpha\}$ serves a double role. On the one hand, it tells what is left after multiples of the $\partial\phi/\partial\xi_\alpha$ are removed from the function space, and is called the *codimension* of ϕ. On the other hand, it gives the smallest number of control parameters needed to span the space; a stable unfolding having this minimum number is called a *universal unfolding*.

Let us look at a few examples of universal unfoldings generated by degenerate germs. First, for dimensionality $k = 1$ for ξ-space, choose $\phi(x) = x^m$. Degeneracy requires $m \geq 3$. The ideal $\{\partial\phi/\partial x\}$ now becomes the set $\{x^{m-1}\psi(x)\}$ with complementary space vanishing at the origin simply $\text{lin}\{x^{m-2}, x^{m-3}, \dots, x\}$ (lin= "linear combination of"). We conclude that the associated universal unfolding is

$$\phi(x, u_1, \dots, u_{m-2}) = x^m + \sum_{j=1}^{m-2} u_j x^j, \quad \text{codim}(\phi) = m - 2.$$

Next, consider dimensionality $k = 2$. First, take $\phi(x, y) = x^4 + y^4$; then

$$\left\{\frac{\partial\phi}{\partial x}, \frac{\partial\phi}{\partial y}\right\} = \{x^3\psi(x, y), y^3\theta(x, y)\} = \left\{\sum_{i>2 \text{ or } j>2} a_{ij}x^i y^j\right\}$$

with complementary space

$$\left\{\sum_{i\leq 2 \text{ and } j\leq 2} a_{ij}x^i y^j\right\} = \text{lin}\left\{x, y, x^2, xy, y^2, x^2y, xy^2, x^2y^2\right\}.$$

Hence

$$\phi(x, y; u) = x^4 + y^4 + \sum_{\substack{i\leq 2, j\leq 2,\\ i+j\neq 0}} u_{ij}x^i y^j, \quad \text{codim}(\phi) = 8.$$

Simplicity does not, however, guarantee a simple unfolding. Again with $k = 2$, consider $\phi = x^2y$. Now

$$\left\{\frac{\partial\phi}{\partial x}, \frac{\partial\phi}{\partial y}\right\} = \{xy\psi(x, y), x^2\theta(x, y)\} = \left\{\sum_{i>1 \text{ or } ij>0} a_{ij}x^i y^j\right\}$$

with complementary space

$$\left\{\sum_{i\leq 1,\ ij=0} a_{ij}x^i y^j\right\} = \text{lin}\{x, y, y^2, y^3\}.$$

Thus

$$\phi(x, y; \mathbf{u}) = x^2y + u_0x + \sum_{i=1}^{\infty} u_i y^i, \quad \text{codim}(\phi) = \infty.$$

Finally, at this time, consider the lowest order $k = 3$ case. Since all second-order terms are missing, this must be $\phi(x, y, z) = x^3 + y^3 + z^3$. Hence

$$\left\{\frac{\partial\phi}{\partial x}, \frac{\partial\phi}{\partial y}, \frac{\partial\phi}{\partial z}\right\} = \left\{\sum_{\substack{i>1 \text{ or } j>1\\ \text{or } k>1}} u_{ijk}x^i y^j z^k\right\},$$

with complementary set $\text{lin}_{i,j,k\leq 1}\{x^i y^j z^k\}$. We have

$$\phi(x, y, z; u) = x^3 + y^3 + z^3 + \sum_{i,j,k\leq 1} u_{ijk}x^i y^j z^k, \quad \text{codim}(\phi) = 7.$$

Our whole point has been that unfolding from a degenerate singularity elicits qualitatively different characteristics for the resulting singularity, depending upon where in parameter space we go. It is clear that a shift occurs any time that a

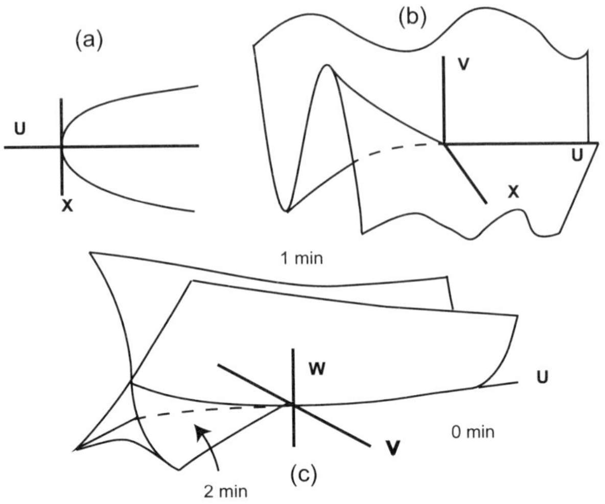

FIGURE 2.1. The (a) cusp, (b) fold, and (c) swallowtail catastrophes.

degeneracy *recurs* in a singularity of the unfolded degenerate germ. The set of control space points at which local nondegeneracy fails is called the *catastrophe* set. Formally then, for an unfolding $E(\mathbf{q},\mathbf{u})$ of an unreduced germ $E(\mathbf{q})$, we have

- singular set $\Sigma(E) = \left\{(\mathbf{q},\mathbf{u}) \mid \frac{\partial E}{\partial q_\alpha} = 0,\ \text{all } \alpha\right\}$;
- degenerate set $\Delta(E) = \left\{(\mathbf{q},\mathbf{u}) \mid \frac{\partial E}{\partial q_\alpha} = 0, \text{ all } \alpha \text{ and nullity } \frac{\partial^2 E}{\partial q_\alpha \partial q_\beta} > 0 \text{ or } \det\left(\frac{\partial^2 E}{\partial q_\alpha \partial q_\beta}\right) = 0\right\}$;
- catastrophe set $D(E) = \{(\mathbf{u}) \mid (\mathbf{q},\mathbf{u}) \in \Delta(E)\}$.

Let us return to the examples of $k = 1$, the one-dimensional reduced internal space, from this viewpoint.

FOLD: Figure 2.1(a)

$\phi(x,u) = x^3 + 3ux$; singular set $x^2 + u = 0$, so $\Sigma = (x, -x^2)$; degenerate set $x = 0$ and $x^2 + u = 0$, so $\Delta = (0,0)$; $D = (0)$. Here, two stationary points exist for $u < 0$, none for $u > 0$.

CUSP: Figure 2.1(b)

$\phi(x,u,v) = x^4 + 6ux^2 + 4vx$; singular set $x^3 + 3ux + u = 0$, so $\Sigma = (x, u, -x^3 - 3ux)$; degenerate set $x^2 + u = 0$ as well, so $\Delta = (x, -x^2, 2x^3)$; catastrophe set $D = (-x^2, 2x^3)$ or $v^2 + 4u^3 = 0$. Here, we have three stationary points for $v^2 + 4u^3 < 0$, one for $v^2 + 4u^3 > 0$.

SWALLOWTAIL: Figure 2.1(c)

$\phi(x,u,v,w) = x^5 + 10x^3u + 10x^2v + 5xw$; singular set $x^4 + 6x^2u + 4xv + w = 0$, so $\Sigma = (x, u, v, -x^4 - 6x^2u - 4xv)$; degenerate set $x^3 + 3xu + v = 0$, so $\Delta = (x, u, -x^3 - 3xu, 3x^4 + 6x^2u)$, and $D = (u, -x^3 - 3xu, 3x^4 + 6x^2u)$. The catastrophe set, nonparametrically, is the resultant of $x^4 + 6x^2u + 4xv + u = 0$, $x^3 + 3xu + v = 0$, or by Sylvester's elimination,

$$\begin{vmatrix} 0 & 0 & 1 & 0 & 6u & 4v & w \\ 0 & 1 & 0 & 6u & 4v & w & 0 \\ 1 & 0 & 6u & 4v & w & 0 & 0 \\ 0 & 0 & 0 & 1 & 0 & 3u & v \\ 0 & 0 & 1 & 0 & 3u & v & 0 \\ 0 & 1 & 0 & 3u & v & 0 & 0 \\ 1 & 0 & 3u & v & 0 & 0 & 0 \end{vmatrix} = 0,$$

reducing to

$$\frac{\text{disc}}{256} = -w^3 + 18w^2u^2 - 81wu^4 - 36wuv^2 + 108u^3v^2 + 27v^4 = 0.$$

General m:

$\phi(x, \{u_i\}) = x^m + \sum_{p=1}^{m-2} \binom{m}{p} u_p x^p$, with the singular set specified by $x^{m-1} + \sum_{p=1}^{m-2} \binom{m-1}{p-1} u_p x^{p-1} = 0$, and the degenerate set by $x^{m-2} + \sum_{p=2}^{m-2} \binom{m-2}{p-2} u_p x^{p-2} = 0$. The resultant is again obtained by Sylvester's method, and leads to a symmetric function of the roots $\alpha_1, \ldots, \alpha_{m-1}$ of the singular set; the discriminant disc $= (\prod_{i<j<m} (\alpha_1 - \alpha_j))^2$, which does indeed vanish if two singular set roots coincide. All real roots α_i of $\Sigma(E)$ maintain disc ≥ 0, corresponding, e.g., to two minima of the swallowtail, and each pair of complex roots produces one change of sign. Higher degeneracy in roots clearly produces lower m catastrophe sets, such as the $w = 0$ cusp in the swallowtail. A note on terminology: $m = 6$ is the "butterfly."

2.2. Categorization and Applications

Recapitulating, we have shown that if $\phi(\xi, \mathbf{u})$ is to be a stable unfolding, then for infinitesimal u_α, we must have

$$\phi(\xi, \mathbf{u}) = \phi(\xi) + \sum u_\alpha b_\alpha(\xi) \tag{2.2}$$

where

$$\text{lin}\{b_\alpha(\xi)\} + \text{id}\left\{\frac{\partial \phi}{\partial \xi_\alpha}\right\} = C^\infty(\xi)_0,$$

the 0 subscript restricting functions to those vanishing at $\xi = 0$. In fact, precisely this form is then a stable unfolding for some finite neighborhood of $\mathbf{u} = 0$ as well. It is really just a matter of seeing if we can iterate the infinitesimal argument. We thus ask when another unfolding $\phi_1(\xi, \mathbf{u})$ can be represented by a coordinate transformation of ϕ:

$$\begin{aligned} \phi_1(\xi, \mathbf{u}) &= \phi(\mathbf{f}(\xi, \mathbf{u}), \mathbf{g}(\mathbf{u})) + K(\mathbf{u}) \\ &= \phi(\mathbf{f}(\xi, \mathbf{u})) + \sum g_\alpha(\mathbf{u}) b_\alpha(\mathbf{f}(\xi, \mathbf{u})) + K(\mathbf{u}) \end{aligned} \tag{2.3}$$

where $\mathbf{f}(\xi, 0) = \xi$, $\mathbf{g}(0) = 0$. To check, differentiate with respect to u_β:

$$\frac{\partial \phi_1}{\partial u_\beta} = \sum \frac{\partial f_\gamma}{\partial u_\beta} \left(\frac{\partial \phi(\mathbf{f})}{\partial f_\gamma} + \sum g_\alpha(\mathbf{u}) \frac{\partial b_\alpha(\mathbf{f})}{\partial f_\gamma} \right) + \sum \frac{\partial g_\alpha}{\partial u_\beta} b_\alpha(\mathbf{f}) + \frac{\partial K}{\partial u_\beta}.$$

At $\mathbf{u} = 0$ this becomes

$$\frac{\partial \phi_1}{\partial u_\beta} = \sum \frac{\partial f_\gamma}{\partial u_\beta}\frac{\partial \phi(\xi)}{\partial \xi_\gamma} + \sum \frac{\partial g_\alpha}{\partial u_\beta} b_\alpha(\xi) + \frac{\partial K}{\partial u_\beta}$$

and is solvable for $\partial \mathbf{f}/\partial \mathbf{u}$, $\partial \mathbf{g}/\partial \mathbf{u}$, $\partial K/\partial \mathbf{u}$ under the condition (2.2). Further, if ϕ_1 is nearby, then $\partial \phi_1/\partial u_\beta \sim b_\beta(\xi)$, so that $\partial g_\alpha/\partial u_\beta \sim \delta_{\alpha\beta}$; hence g is an invertible transformation, so is f, and $\phi_1 \approx \phi$ if (2.3) is solvable. For $\mathbf{u} \neq 0$ but small enough, we can iterate with

$$\sum \frac{\partial f_\gamma}{\partial u_\beta} g_\alpha \frac{\partial b_\alpha}{\partial f_\gamma}$$

transferred as a perturbation to the left-hand side, using the fact that

$$\operatorname{lin}\{b_\alpha(\mathbf{f})\} + \operatorname{id}\left\{\frac{\partial \phi(\mathbf{f})}{\partial f_\gamma}\right\} = C^\infty(\mathbf{f})_0,$$

and so solve for $\partial f_\gamma/\partial u_\beta$, $\partial g_\alpha/\partial u_\beta$, and $\partial K/\partial u_\beta$. These relations can then be iterated to solve for $\{f_\gamma, g_\alpha, K\}$ in some finite region about the origin in $\mathbf{u}$-space. This region may, however, shrink to 0 for an infinite set of equations, so that it is necessary to assume that $\operatorname{codim}(\phi) < \infty$. We note, as an incidental result, that any unfolding is the transform of such a stable unfolding but need not be equivalent to it.

It is now easy enough to find the unfoldings of a degenerate singular germ. We also know that two germs will be equivalent in the sense of providing the same set of unfoldings, to within equivalence, if their gradients generate the same ideal, to within a coordinate transformation. A classification of inequivalent unfoldings thus requires a classification of inequivalent germs. If we have in mind physical space-time as our control space, this means that we require $\operatorname{codim}(\phi) \leq 4$. A limit is thereby placed on the dimensionality of the space of essential internal variables ξ. To see this, let us concentrate upon terms of degree ≤ 2 in a Taylor expansion. A stable unfolding of $\phi(\xi_1, \ldots, \xi_n)$, necessarily of degree ≥ 3, supplies at most n independent $\partial\phi/\partial\xi_\alpha$ of the $\binom{n+1}{2}\xi_i\xi_j + \binom{n}{1}\xi_i$ required to span all functions. Hence the remaining linear space is of dimension at least $\binom{n+1}{2} + \binom{n}{1} - n = \binom{n+1}{2}$:

$$\operatorname{codim}(\phi) \geq \binom{n+1}{2}.$$

Thus for $\operatorname{codim}(\phi) \leq 4$, we can assume $n = 1$ or 2.

To classify germs, we will use the same kind of truncated Taylor series analysis. First, to put equivalence in a usable form,

$$\phi_1(\xi) \approx \phi_2(\xi) \quad \text{if } \phi_1(\xi) = \phi_2(\mathbf{f}(\xi)) + K$$

for some smooth invertible transformation $\mathbf{f}$ and constant K. It of course follows from

$$\frac{\partial \phi_1}{\partial \xi_\alpha} = \sum \frac{\partial f_\beta}{\partial \xi_\alpha}\frac{\partial \phi_2}{\partial f_\beta}$$

that $\operatorname{id}\{\partial\phi_1/\partial\xi_\alpha\}$ and $\operatorname{id}\{\partial\phi_2/\partial\xi_\alpha\}$ differ only by a coordinate transformation. Now define a germ is *k-determined* if any $\phi'(\xi)$ agreeing with $\phi(\xi)$ through k^{th}-order terms in its Taylor series must be equivalent to ϕ.

The associated central theorem (Mather) is then:

$\phi(\xi)$ is finitely determined if and only if it is of finite codimension; if the codimension is r, ϕ is k-determined for some $k \le r+2$; if M^k, the set of all Taylor series starting with k^{th}-order terms, is contained in $M^1 \operatorname{id}\{\partial\phi/\partial\xi_\alpha\} = \operatorname{id}\{\xi_\beta \partial\phi/\partial\xi_\alpha\}$, ϕ is k-determined.

EXAMPLE 1. $\phi(x, y) = x^3 + xy^2$. Here

$$\begin{aligned} \operatorname{id}\left\{ \xi_\alpha \frac{\partial\phi}{\partial\xi_\beta} \right\} &= \operatorname{id}\{x(3x^2 + y^2), y(3x^2 + y^2), x(2xy), y(2xy)\} \\ &= \operatorname{id}\{x^3, x^2y, xy^2, y^3\} = M^3, \end{aligned}$$

so ϕ is 3-determined.

The theorem is particularly transparent for the case of one essential internal variable, $\phi(x)$. Here, without loss of generality, we can take $\phi(x) = x^m + h_{m+1}(x)$, where the first $m+1$ derivatives of $h_{m+1}(x)$ vanish at the origin $x = 0$, so that as a Taylor series it would start with x^{m+1} terms. Since $\frac{\partial\phi}{\partial x} = mx^{m-1} + h'_{m+1}(x)$, $\operatorname{id}\{\frac{\partial\phi}{\partial x}\} = \operatorname{id}\{x^{m-1}\}$ has a linear basis $\{x^{m-1}, x^m, x^{m+1}, \dots\}$ or a complementary basis $\{x, x^2, \dots, x^{m-2}\}$ of codimension $m-2$. On the other hand, if $\phi_1(x) = x^m + \ell_{m+1}(x)$, then $\phi(x(1 + p_1(x)) = \phi_1(x)$, or

$$x^m(1 + p_1(x))^m + h_{m+1}(x + xp_1(x)) = x^m + \ell_{m+1}(x),$$

written as

$$mp_1(x) + \tbinom{m}{2} p_1(x)^2 + \cdots = x^{-m}(\ell_{m+1}(x) - h_{m+1}(x + xp_1(x))),$$

can be solved for $p_1(x)$ in the vicinity of the origin. Hence ϕ is m-determined. Further, $\operatorname{id}\{x\partial\phi/\partial x\} = \operatorname{id}\{x^m\} = M^m$, as expected. For more than one variable, with discussion and proof, see [**78**]. With this theorem, it is not hard to classify inequivalent germs and their unfoldings: if $\operatorname{codim}(\phi) \le 4$, we do not have to look at anything with $k > 6$, and we need be concerned only with $\phi(x)$ and $\phi(x, y)$, and here only with Taylor series starting at third order. We have already analyzed the set $\phi(x) = x^m$. Here codim $= m-2$, so that $m = 3, 4, 5, 6$; x^m is m-determined and equivalent to any series starting at x^m.

Now for $\phi(x, y)$ we need at most terms of degree 3 through 6. There must be some terms of order 3, for with order 4 or more, we would have to pick up x, y, x^2, xy, y^2, x^3, x^2y, xy^2, and y_3 with only two basis elements supplied by $\frac{\partial\phi}{\partial x}$ and $\frac{\partial\phi}{\partial y}$; i.e., we would have $\operatorname{codim}(\phi) \ge 7$. An ancient systematic analysis of possible sets of dominant terms is that of the Newton polygon: if $\phi(x, y) = \sum a_{pq}x^py^q$, draw the terms present in (p, q)-space. If $x^{p'}y^{q'}$ and x^py^q are of the same order at the origin, this means that $O(x^{p'}y^{q'}) = O(x^py^q)$ or $y = O(x^{(p-p')/(q'-q)}) = O(x^{-\cot\theta})$; see Figure 2.2(a). Hence $x^my^n = O(x^{m-n\cot\theta})$ will be of higher order if (m, n) is to

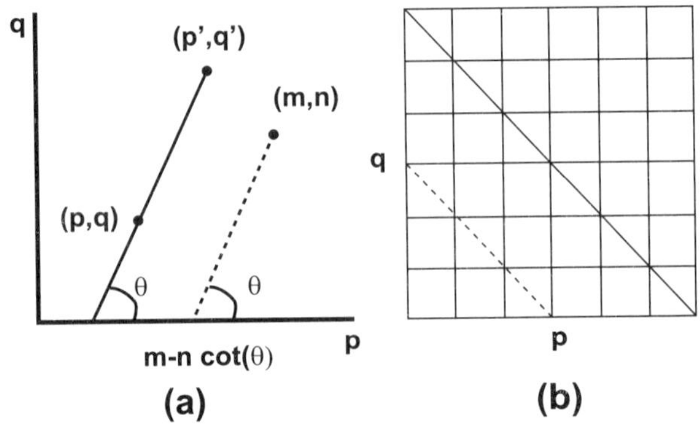

FIGURE 2.2. The Newton polygon.

the right of the (p,q) to (p',q') line. In the present case, all terms must lie between the dashed $p+q=3$ and $p+q=6$ lines; see Figure 2.2(b). If the $p+q=3$ line is occupied by two or more terms, they dominate. We can then write

$$\phi(x,y) = a_0x^3 + a_1x^2y + a_2y^3 = a_3y^3P_3\left(\frac{x}{y}\right)$$
$$= (b_1x - c_1y)(b_2x - c_2y)(b_3x - c_3y).$$

One root, say $x/y = c_1/b_1$, must be real. There are then several possibilities:

(1) All roots are real and unequal; by a change of coordinates, these can be taken as $-1, 0, 1$, so that $\phi(x,y) = x(x-y)(x+y) = x^3 - xy^2$.
(2) All roots are real, but two are equal; they can be chosen as $0, 0, \infty$ ($b_3 = 0$), so that $\phi(x,y) = x^2y$.
(3) All roots are real and equal: $\phi(x,y) = x^3$.
(4) There are two complex conjugate roots, e.g., $0, \pm i$: $\phi(x,y) = x(x+iy)(x-iy) = x^3 + xy^2$.

Case (1), and similarly case (4) are of codimension 3 (and is 3-determined): $\mathrm{id}\{3x^2 - y^2, -2xy\}$ contains, due to xy, all x_py^q with $p > 0$ and $q > 0$; then, from $x(3x^2 - y^2)$ and $y(3x^2 - y^2)$, also all x^py^q with $p > 2$ and $q > 2$. Hence the complementary space must supply the terms with $p \leq 2$ and $q \leq 2$, of which $3x^2+y^2$ and $2xy$ are already present. Hence $x^2 + y^2$, x, and y suffice for the complementary space, and $\mathrm{codim}(\phi) = 3$.

The remaining cases are in fact the most general ones with single occupancy $p+q = 3$. Both x^3 and x^2y have infinite codimension, and so must be completed by higher-degree terms. Consider $x^3 + \phi_4(x,y)$, ϕ_4 being at least degree 4, and look at the space of degree at most 3. The partial derivatives generate at most two cubic terms, one attached to $3x^2$, as well as x^3 and x^2y (from $\partial x^3/\partial x$). But $2+3+4$ monomials are needed through the third degree. Hence the complementary space has dimension of at least $2+3+4-4 = 5$ and will not enter our classification. On the other hand, $x^2y + \phi_k(x, y)$, $k \geq 4$, can be transformed to $x^2 + ay^k$.[1] A linear basis complementary to $\{xy, x^2 + y^{k-1}\}$ is $\{x, y, y^2, \dots, y^{k-1}\}$, of dimension ≤ 4 only for $k = 4$. Hence only x^2y+y^4 is available. There are then three new universal unfoldings for codim(ϕ) ≤ 4. For codimension 3, these can be written in full as

$$E(x, y, z, \dots; u, v, w) = \phi(x, y; u, v, w) + h_4(x, y; u, v, w) + m_2(z, \dots; u, v, w)$$

where

$$\phi(x, y; u, v, w) = \frac{1}{3}\alpha x^3 + xy^2 + u(y^2 - \alpha x^2) + vx + wy.$$

Here h_4 starts at fourth-order terms, and m_2 is a Morse function. Only the sign of α matters, and $\alpha = 0$ generates codimension 4. Now we have: a singular set $\alpha x^2 + y^2 - 2u\alpha x + v = 0$, $2xy + 2uy + w = 0$, or

$$\Sigma(\phi) = \{(x, y; u, -y^2 + 2\alpha ux - \alpha x^2, -2xy - 2uy)\},$$

and a degenerate set,

$$\operatorname{rank}\begin{pmatrix} 2\alpha x - 2\alpha u & 2y \\ 2y & 2x + 2u \end{pmatrix} \leq 1.$$

Rank 0 is the single point $x = y = u = 0$. For rank 1, the determinant vanishes, yielding

$$y^2 - \alpha x^2 = -\alpha u^2,$$

with $\alpha = 1$ for a *hyperbolic umbilic*, $\alpha = -1$ for an *elliptic umbilic*. Now $\Delta(\phi) = \{(x, \mp\sqrt{\alpha(x^2 - u^2)}; u, \alpha u^2 + 2\alpha ux - 2\alpha x^2, \pm 2(x+u)\sqrt{\alpha(x^2 - u^2)})\}$, resulting in the catastrophe set, parametrized by x,

$$D(\phi) = \left\{\left(u, \alpha u^2\left(-2\left(\frac{x}{u}\right)^2 + 2\left(\frac{x}{u}\right) + 1\right), \pm 2u^2\left(\frac{x}{u} + 1\right)\sqrt{\alpha\left(\left(\frac{x}{u}\right)^2 - 1\right)}\right)\right\}.$$

The general form is thus a quadratic cone $f(v/u^2, w/u^2) = 0$; see Figure 2.3(a), with the standard pattern in (v, w)-space shown in Figures 2.3(b) and (c).

From the Hessian matrix, a stable minimum requires positive definiteness, or $x + u > 0$ and $\alpha x^2 - y^2 > \alpha u^2$. Figures 2.3(b) and (c) are labeled accordingly.

The $\alpha = 0$ case is special, the *parabolic umbilic*, and must be augmented to achieve codimension 4. We can now write

$$E(x, y, z, \dots; t, u, v, w) = h_5(x, y; t, u, v, w) + m_2(z; t, u, v, w) + \phi(x, y; t, u, v, w)$$

where

$$\phi(x, y; t, u, v, w) = x^2y + \frac{1}{4}y^4 + tx^2 + uy^2 - vx - wy.$$

The singular set is given by $2xy + 2tx - v = 0$, $x^2 + y^3 + 2uy - w = 0$, or

$$\Sigma(\phi) = \{(x, y; t, u, 2(y + t)x, x^2 + y^3 + 2uy)\},$$

[1] If α and β are of degree $k - 2$, then

$$(x + \alpha)^2(y + \beta) + \phi_k(x + \alpha, y + \beta) = x^2y + x^2\beta + 2xy\alpha + \phi_k(x, y) + \cdots,$$

and $x^2\beta + 2xy\alpha$ can subtract out all k-degree terms but y^k.

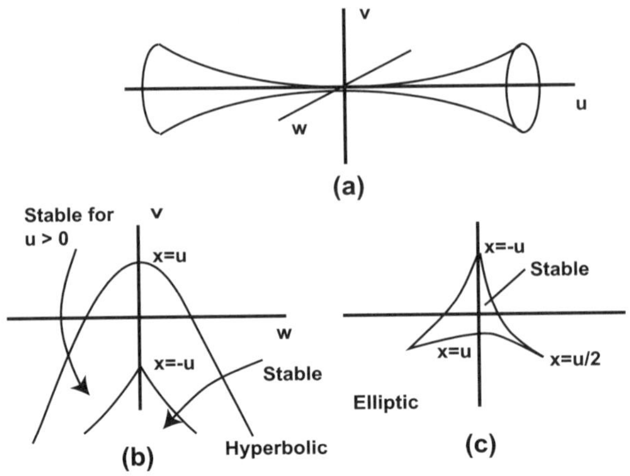

FIGURE 2.3. The hyperbolic and elliptic umbilic catastrophes.

and the degenerate set requires rank

$$\begin{pmatrix} 2(y+t) & 2x \\ 2x & 3y^2+2u \end{pmatrix} \leq 1.$$

Here there is a 0-rank section given by $2u+3t^2=0$; the full catastrophe surface is

$$(y+t)(3y^2+2u)=2x^2, \tag{2.4}$$

with $v=2(y+t)x$, $w=x^2+y^3+2uy$.

To examine the roster of possibilities, we start with the singular dividing curves of (2.4) in (t,u)-space and parametrize v and w by the corresponding xy-curve. There are two types of degenerate singularity: first, $x^2=0=(y+t)(3y^2+2u)$, $\partial x^2/\partial y=0=9y^2+6ty+2u$, yielding $(y=-t)$, $u=-\frac{3}{2}t^2$ or $(y^2=-\frac{2}{3}u)$, $u=0$, $u=-\frac{3}{2}t^2$. Second, $\partial x^2/\partial y=0=9y^2+6ty+2u$ and $\partial^2x^2/\partial y^2=0=3y+t$, or $u=\frac{1}{2}t^2$ (and $2x^2=\frac{8}{9}t^3$, so that $t\geq 0$). Further, just as the catastrophe set scaled as $v\propto u^2$, $w\propto u^2$ and $x\propto u$ in the elliptic-hyperbolic umbilics, here $u\propto t^2$, $v\propto t^{5/2}$, $w\propto t^3$, $x\propto t^{3/2}$, and $y\propto t$.

Thus, there is a typical (v,w)-section at each value of u/t^2. Filling in the degenerate points of connectivity change in u, v-space, one finds in the following table the corresponding sections shown in Figure 2.4:

Region	2, 14	4, 12	6, 10	8	16
u/t^2	0	$-3/8$	$-3/2$	$-k$	$1/2$

We thus have Thom's seven elementary catastrophes for at most four control parameters. What is their relevance to biological development? The point is that any regularly observed qualitative developmental sequence is, by definition, unaffected by small internal or external perturbations. It therefore represents a stable unfolding in space-time, in the technical sense of the term, if all differentiability criteria are satisfied. For ≤ 5 control parameters, there are in fact only a finite

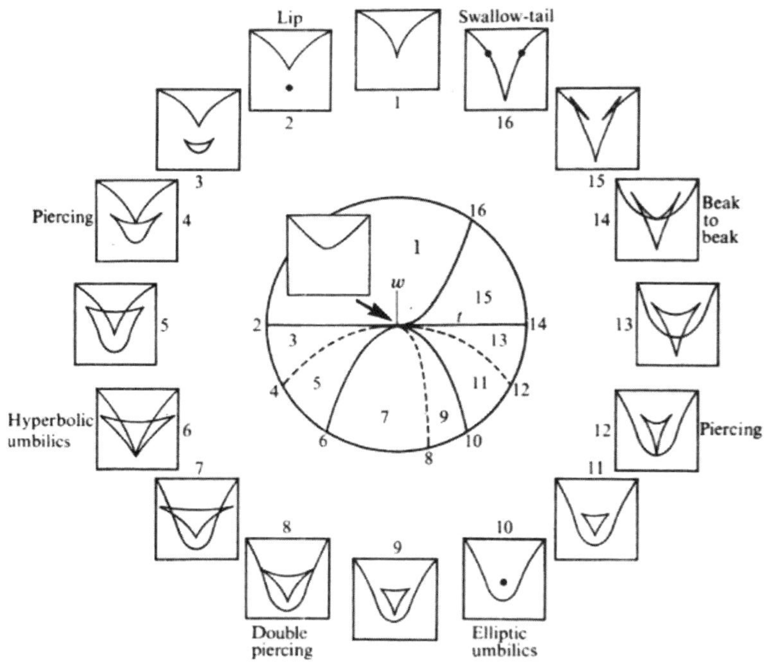

FIGURE 2.4. Sections of the parabolic umbilic. (Figure from J. F. Nye, "Optical caustics from liquid drops under gravity: observations of the parabolic and symbolic umbilics," *Philosophical Transactions A*, 1979, vol. 292, no. 1387, fig. 7. Copyright ©1979. By permission of The Royal Society.)

number of inequivalent stable unfoldings, so that a complete roster of possibilities is available. If codim ≥ 6, there are an infinite number, with two consequences. First, the classification becomes less helpful. Second, the implied opening up of control space also produces unfoldings that are not close to stable unfoldings—one may expect to find "strange" unfoldings as well. However, it is still true that all unfoldings are generated by coordinate transformations from stable ones. While high codimensionality is not strictly necessary—or even allowable—for describing a single organism, it certainly becomes so for a comparison of developments of different species or phyla.

Another point relates to the crucial concept of equivalence. The internal variables have been regarded as having an infinite domain. This is hardly true of chemical concentrations, nor is it true of the variable R^2 used to describe simplified blastulation (see Chapter 1). Thus, although we noted that $E \propto (R^2)^2 - (\gamma V/J)R^2$ can be transformed to have no control parameter, codim $= 0$, this was on the assumption that $\gamma V/J > 0$. If $\gamma V/J < 0$, the minimum does disappear from the range $0 \leq R^2 < \infty$. One conclusion (see [**25**]) is that the class of inequivalent stable unfoldings grows substantially if inequality constraints are used—in fact, becoming infinite for codim > 3. This should certainly be taken into account when a true constraint (e.g., $\sigma \leq 1$) is replaced by a merely high energy barrier. A similar warning applies to the control space: one cannot reasonably mix space and time to form equivalent observations. For consequences in classification, see [**158**].

We still have not said precisely how the stable singularity structures analyzed by catastrophe theory are to be used in biological description. For this purpose, it is only necessary to recall the original motivating quasi-static dynamics, in which the configuration was determined by energy minimization. However, there can be more than one local minimum, and we must decide how to make a choice. If the system is subject to sufficiently large dynamical changes on its way to rapid equilibration by degradation of motion, then at each potential energy it can sample all configurations of lower energy, and the *Gibbs criterion* of attaining an absolute minimum must be required. In this case, there will be primary discontinuous behavior, due to the appearance and disappearance of minima under control parameter changes, and secondary discontinuities due to the shift from one minimum to another as the value of the latter drops below that of the former. The switch-point control parameter surface, at which two minima are precisely equal, is known as the *Maxwell set*, and must be appended to the catastrophe set.

Let us examine elementary catastrophes from this viewpoint; now the overall sign of $\phi(\xi, \mathbf{u})$ matters, since we are specifically interested in minima, and any $\phi(\xi, \mathbf{u})$ that can achieve $-\infty$ is disallowed. Hence there are just two possibilities:

CUSP: $\phi(x; u, v) = x^4 + 6ux^2 + 4vx$; region $v^2 + 4u^3 < 0$; without computation, the Maxwell set is clearly $v = 0$, $u < 0$.

BUTTERFLY: $\phi(x; t, u, v, w) = x^6 + 15tx^4 + 20ux^3 + 15vx^2 + 5wx$. We must solve

$$\frac{\phi(x) - \phi(y)}{x - y} = 0, \; \phi'(x) = 0, \; \phi'(y) = 0$$

to have equal minima at two values of the argument. This is considerably more complicated. Clearly $u = w = 0$ will be a member of the Maxwell set, but it is only a line common to intersecting surfaces.

If the time scale is too rapid and the barrier too large compared to short-time dynamical changes, then a system can be stuck in a local minimum, departing only when this minimum ceases to exist. This *perfect delay* is the opposite extreme to jumping at the Maxwell set, and is regarded by Thom as being much closer to biological reality. The truth, of course, must lie in between and depend upon dynamical details. Indeed, despite immersion in a space-time framework, there should be a real asymmetry between space and time. The simplest model in this context is the *relaxation time* model. If $\xi^0(\mathbf{r}, t)$ is the relevant minimizing ξ at $(\mathbf{r}, t)$, then ξ will follow the point attractor with time constant τ if

$$\frac{\partial}{\partial t}\xi(\mathbf{r}, t) = \frac{1}{\tau}(\xi^0(\mathbf{r}, t) - \xi(\mathbf{r}, t));$$

e.g., if $\xi^0(\mathbf{r}, t)$ is time independent, ξ will home in exponentially, $\xi(t) = \xi^0 + \Delta e^{-t/\tau}$, but otherwise will lag behind. But there can be competition between attractors

$\xi^\alpha(\mathbf{r},t)$, most directly represented as

$$\frac{\partial}{\partial t}\xi(\mathbf{r},t) = \frac{1}{\tau}\sum w_\lambda(\xi^\lambda(\mathbf{r},t) - \xi(\mathbf{r},t))$$

with weight function w_λ. The weights w_λ need not be mysterious. The simplest dynamical extension of quasi-statics,

$$\frac{\partial}{\partial t}\xi_\alpha = -\kappa\frac{\partial}{\partial \xi_\alpha}E(\xi;\mathbf{r},t),$$

reduces to

$$\frac{\partial}{\partial t}\xi_\alpha = -\kappa\sum\frac{\partial^2 E}{\partial\xi^0_\alpha\partial\xi^0_\beta}\left(\xi_\beta - \xi^0_\beta\right)$$

if ξ is near ξ_0, but otherwise automatically drives ξ_α by competing attractors—not in the linear form above.

2.3. Global Implications of Local Structure

Major difficulties in application of the foregoing lie in several aspects of the same general problem. From a descriptive point of view, a compendium of local analyses says nothing about global relations and does not treat the restrictions on sets of coexistent local singularities. The mathematical problems here must be, to put it mildly, substantial. What they reflect is equally substantial—incisive description is biologically useful if it leads to at least an empirical dynamical mechanism. The only clear but implicit suggestion of catastrophe theory is that one should focus on minimization of a local energy function, and there is no compelling argument for this. These problems are not wholly mathematical, but require at least insertion of external information on possible relations between spatially and temporally separated components. We will do this now in a preliminary way in the context of catastrophe analysis, starting with phenomena of presumed lowest codimension and working our way up. But first let us see which of these difficulties are intrinsic and which are removable.

Suppose that we can indeed focus on the internal variables $\{q_\alpha\}$ at some spatial location. We have analyzed the consequences of assuming

$$E(\{q_\alpha\},t) = \text{minimum},$$

emanating, e.g., from a force opposed by a frictional drag

$$\dot{q}_\alpha = -\gamma_\alpha\frac{\partial E}{\partial q_\alpha}.$$

This is but a rather special case of the very general differential flow

$$\dot{q}_\alpha = X_\alpha(\{q_\beta\},t) \tag{2.5}$$

and its singular points defined by

$$\{X_\alpha(\{q_\beta\},t)\} = 0.$$

The intelligent way to proceed is to generalize the catastrophe theory classification to singularities of the vector maps $\{q_\alpha\} \to \{X_\alpha\}$ and their unfoldings, and this may be done. Alternatively (see [**107**]), the general case can be made to achieve a form resembling the special case. Let us see how.

One way to represent equations of motion is via the Euler-Lagrange variational principle

$$\delta \int_{t_0}^{t_1} L(\{q_\alpha(t), \dot{q}_\alpha(t)\}, t)dt = 0,$$

for then $\int_{t_0}^{t_1} L(\{q_\alpha + \delta q_\alpha, \dot{q}_\alpha + \delta\dot{q}_\alpha(t)\}, t) - L(\{q_\alpha, \dot{q}_\alpha\}, t)dt = 0$ for all perturbations $\delta q_\alpha(t)$. Expanding,

$$0 = \int_{t_0}^{t_1} \sum \left(\frac{\partial L}{\partial q_\alpha} \delta q_\alpha + \frac{\partial L}{\partial \dot{q}_\alpha} \delta \dot{q}_\alpha \right) dt,$$

and integrating by parts,

$$0 = \text{boundary terms} + \int_{t_0}^{t_1} \sum \left(\frac{\partial L}{\partial q_\alpha} - \frac{d}{dt} \frac{\partial L}{\partial \dot{q}_\alpha} \right) \delta q_\alpha \, dt$$

for all δq_α. Hence

$$\frac{\partial L}{\partial q_\alpha} = \frac{d}{dt} \frac{\partial L}{\partial \dot{q}_\alpha}.$$

To obtain (2.5), L can be taken in the form

$$L = \sum \dot{q}_\alpha U_\alpha(\{q_\beta\}, t) - U(\{q_\beta\}, t),$$

for then, dropping explicit time dependence in conformity with a quasi-static viewpoint, we have

$$\frac{d}{dt} U_\alpha = \frac{\partial}{\partial q_\alpha} \left(\sum \dot{q}_\beta U_\beta - U \right),$$

but

$$\frac{d}{dt} U_\alpha = \sum \dot{q}_\beta \frac{\partial U_\alpha}{\partial q_\beta},$$

and so

$$\sum \dot{q}_\beta \left(\frac{\partial U_\beta}{\partial q_\alpha} - \frac{\partial U_\alpha}{\partial q_\beta} \right) = \frac{\partial U}{\partial q_\alpha}.$$

In other words,

$$\dot{q}_\alpha = \sum \Gamma^{-1}_{\alpha\beta} \frac{\partial U}{\partial q_\beta} \quad \text{where } \Gamma_{\alpha\beta} = \frac{\partial U_\beta}{\partial q_\alpha} - \frac{\partial U_\alpha}{\partial q_\beta}. \tag{2.6}$$

These equations are very much of the form desired if Γ is invertible. In order to reduce precisely to $\dot{q}_\alpha = X_\alpha$, we need

$$\begin{aligned} \frac{\partial U}{\partial q_\alpha} &= \sum \left(X_\beta \frac{\partial U_\beta}{\partial q_\alpha} - X_\beta \frac{\partial U_\alpha}{\partial q_\beta} \right) \\ &= \frac{\partial}{\partial q_\alpha} \sum X_\beta U_\beta - \sum \left(U_\beta \frac{\partial X_\beta}{\partial q_\alpha} + X_\beta \frac{\partial U_\alpha}{\partial q_\beta} \right). \end{aligned}$$

Hence if

$$\left(\sum X_\beta \frac{\partial}{\partial q_\beta}\right) U_\alpha = -\sum \frac{\partial X_\beta}{\partial q_\alpha} U_\beta$$

is solvable in some neighborhood of the initial point, we can complete the description by

$$U = \sum X_\beta U_\beta.$$

Thus (2.6) is indeed a balance of force and a frictional drag with a dynamical drag matrix, while the stationary state is given by

$$\left(\frac{\partial U}{\partial q_\beta}\right) = 0,$$

precisely as desired. (Note that when stationary, $\delta \int_{t_0}^{t_1} L\,dt = -(t_1 - t_0)\delta U$, as expected.) Of course, much is hidden in the domain of existence of the U_α and invertibility of the $\Gamma_{\alpha\beta}$.

Next, what of the absence of global connections? Some are obvious and easily appended, namely those due to underlying symmetry. For example, the blastula is basically a sphere, although distorted internally and externally by cellular inhomogeneity. Thus the intrinsic blastula description should be radially symmetric, and stability of an unfolding assumed only for radially symmetric perturbations, *after which* an equivalence class can be generated by control space coordinate transformations. Discernible consequences may also arise at the level of the gastrula, which can be regarded as having underlying bilateral symmetry.

In general, a slight shift of viewpoint is convenient. If $\phi(\xi, \mathbf{u})$ is invariant under a symmetry group $G = \{g\}$ on control space, this means that $\phi(\xi, g\mathbf{u})$ can be returned to its original form by an outwardly invisible internal coordinate transformation: $\phi(\hat{g}\xi, g\mathbf{u}) = \phi(\xi, \mathbf{u})$. $\hat{G} = \{\hat{g}\}$ is thus an induced representation of G on internal space that directly affects the possible unfoldings.

For example, $x^4 + y^4$ unfolds to $x^4 + y^4 + \text{lin}\{x, y, x^2, xy, y^2, x^2y, xy^2, x^2y^2\}$. Suppose bilateral symmetry requires evenness in the unfolding parameter v, present in the universal unfolding as vy. Since $v \to -v$ induces $y \to -y$, the remaining terms must be even in y, i.e., $f(y^2)$. If we pull in from the blue a generalization to equivalence under continuous, not merely C^∞ transformations, the x^2y^2 term is not needed in the unfolding either, and so

$$\phi = \alpha(x^4 - 6x^2y^2 + y^4) + \beta(3xy^2 - x^3) + \gamma(x^2 + y^2) + ux + vy$$

suffices for the unfolding. We shall return to this example.

So, suitably modified, catastrophe theory is a superb vehicle of incisive local description. But how can indications as to mechanism be gleaned from qualitative observation in the control of catastrophe theory? By observation one means that of a discontinuity or boundary between distinct regimes. But there are two extreme interpretations of a boundary: it may be a primary catastrophe, with a shift in regime because a minimum simply disappears, or a secondary catastrophe resulting from an instantaneous shift to a new lower minimum, or of course something in between (or beyond) these perfect delay and Maxwell set extremes. To see how one might proceed, let us start with phenomena of presumed lowest codimensionality and work our way up.

EXAMPLE 2 (Late blastulation). We model this by the passage of a wave, noncell $\to$ cell, in a locally homogeneous milieu. In such a model, there is only one control parameter, $x - ct$, and the fold catastrophe

$$E(V; x, t) = V^3 + 3\gamma V(ct - x)$$

is indeed appropriate to a perfect delay mechanism if V is related to cell occupation: the minimizing

$$V^0 = (\gamma(x - ct))^{\frac{1}{2}}$$

rises from 0 when $x > ct$ and does not exist ($\to \infty$) for $x < ct$. Suppose that, at fixed time, this is to be represented in an isolecithal egg by space-independent intercell adhesion together with some cell-density weight function, i.e.,

$$E(\{V(x)\}) = \int \mathcal{E}(V, V')dx, \quad \mathcal{E}(V, V') = \frac{J}{2}(V')^2 + E(V).$$

This yields a minimizing field determined by $JV''(x) = \mathcal{E}'(V(x))$. If the structure of the homogeneous system is to be determined only by boundary conditions, there can be no $x - ct$ dependence of E. But substituting into V'', we see that

$$JV'' = -\frac{1}{4}\gamma J\big(\gamma(x - ct)\big)^{-3/2} = -\frac{\gamma^2 J}{4V^3};$$

thus

$$\mathcal{E}'(V) = -\frac{\gamma J}{4V^3} \quad \text{and} \quad \mathcal{E}(V) = \frac{\gamma^2 J}{8}\frac{1}{V^2}.$$

An energy favoring a noncell $\to$ cell transition in a special way can then reproduce the local behavior.

We are now free to insert an appropriate monotonic dependence of V on effective cell density ν,

$$V = f(\nu).$$

What are the implications, and where does the control parameter $x - ct$ enter? On comparing $\mathcal{E}(V)$ with the "thermal energy" $E_T(\nu)$ and adhesive energy $E_{\text{int}}(\nu)$ behavior, both symmetric about $\nu = \frac{1}{2}$, some antisymmetric component is needed. This suggests at once that something like a "cell number conservation" energy

$$E_N = \lambda\Big(\int \nu\, dx - N\Big)$$

with $\lambda < 0$ is a necessary part of the mechanism. The control excited by $x - ct$ must appear at the outer boundary, say $x - ct = d$. Simplest would be to specify the value of V there,

$$V(ct + d) = V_M,$$

and append the relevant conservation condition

$$\int t^{ct+d} V(x)dx = \frac{2}{3\gamma}V_M^3,$$

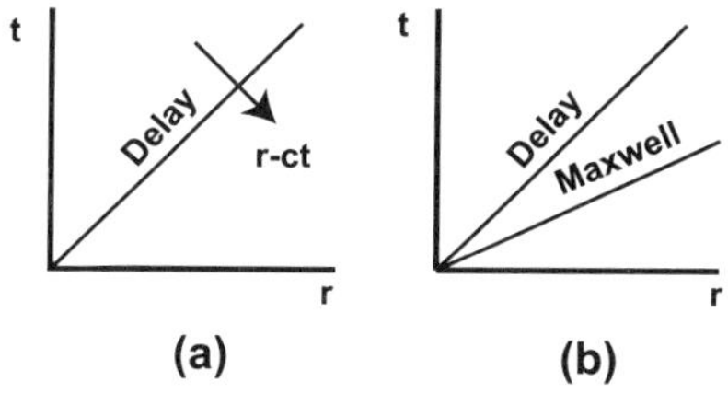

FIGURE 2.5. Discontinuity in the spherical blastula.

yielding precisely V^0 above.

EXAMPLE 3 (Spherical blastula). Taking r and t as control variables, there is no problem of symmetry, although there is a restricted range, $r \geq 0$. Now we must generalize the fold to a cusp and correspondingly the perfect delay to perhaps a Maxwell rule. Thus the cubic of Example 1 above is to be expanded to a quartic which, however, reduces back in the previous region. Such an expansion is conceptually simple enough. Suppose an $(N-1)$-degree polynomial is determined by N special points: $\phi_{N-1}(x_i) = a_i$, $i = 1, \dots, N$. The Lagrange interpolation formula then gives

$$\phi_{N-1}(x) = \sum_{i=1}^{N} a_i \prod_{j \neq i} \frac{x - x_j}{x_i - x_j}.$$

If the degree is now increased by appending a new zero, $\phi(x_0) = 0$, then

$$\phi_N(x) = \sum_{i=1}^{N} \phi_i \frac{x - x_0}{x_i - x_0} \prod_{j \neq i} \frac{x - x_j}{x_i - x_j},$$

or if $-x_0$ is large, simply

$$\phi_N(x) \sim \frac{x - x_0}{-x_0} \phi_{N-1}(x).$$

Now the picture of the discontinuity surface in Figure 2.5(a) goes over to that displayed in Figure 2.5(b), with inner interface control weakening with growth. What type of model will do this?

We can start with the general representation, inhomogeneity plus bulk energy

$$E = \int (\mathcal{E}_{\text{surf}}(V, V') + \mathcal{E}(V)) d^3 r,$$

where V is a weakly distorted cell density parameter. But the observed structure is certainly represented approximately by

$$|\nabla V| = \frac{\partial V}{\partial r} = aV(1 - V),$$

a slow change at either $V = 0$ or $V = 1$. If we insist on the fourth-degree polynomial appropriate to a cusp catastrophe, we are virtually driven to the expression

$$\mathcal{E}_{\text{surf}} = \frac{J}{2} |\nabla V|^2,$$

with precisely the basic double minimum character. The corresponding energy density

$$E(V) = \frac{Ja^2}{2}V^2(1-V)^2 + \mathcal{E}(V),$$

regarded locally, then has two minima or one, depending on the strength of a and parameters of $\mathcal{E}$. An optimal biquadratic $\mathcal{E}$ can then be picked up in the fashion of Example 1.

But the system is not homogeneous and the nominal control parameter a may have initially been closely coupled to time, but now plays the role of an internal parameter determined by minimizing with respect to a (and K)

$$E = 4\pi \int_0^{R(t)} \left(\frac{J(t)}{2} a^2 V^2(r)(1 - V(r))^2 + \mathcal{E}(V(r)) \right) r^2 \, dr$$

where

$$a\,dr = \frac{dV}{V(1-V)} = \left(\frac{1}{V} + \frac{1}{1-V}\right) dV,$$

or

$$ar + \ln K = \ln V = \frac{\ln V}{1-V},$$

so that

$$V = \frac{Ke^{ar}}{1+Ke^{ar}}.$$

The conversion of an initially spatially or temporally controlled parameter to a spatially or temporally neutral internal variable is a crucial step in the hierarchy development of catastrophe theory, setting the stage for the entry of a new control variable.

EXAMPLE 4 (Gastrulation). The nonmammalian blastula is essentially a thin sheet of cells. It appears as a sphere with suitably distorted scale of thickness. A characteristic invagination then occurs at the vegetal half, emanating from a prior *pattern formation* of a region of newly differentiated cells, and resulting in *morphogenesis*—physical motion and change of form. Descriptively, the abrupt surface spread of a point of activation, requiring control parameters r and t, is most directly represented by the penetration of a stable minimum in a hyperbolic umbilic. (See Figure 2.6, in which t_c is the penetration time.) If this is to be believed, a consequence is that two internal variables are required, consistent on the one hand with the virtual pressure of two cell-type densities V_1 and V_2 that have to have been produced by some other process or, on the other hand, directly from the balance of two biochemicals. Further, any energy associated with the latter should be readily transformable to something of the form

$$\frac{1}{3}x^3 + xy^2 + u(y^2 - x^2) + vx + wy.$$

We will later see what this suggests.

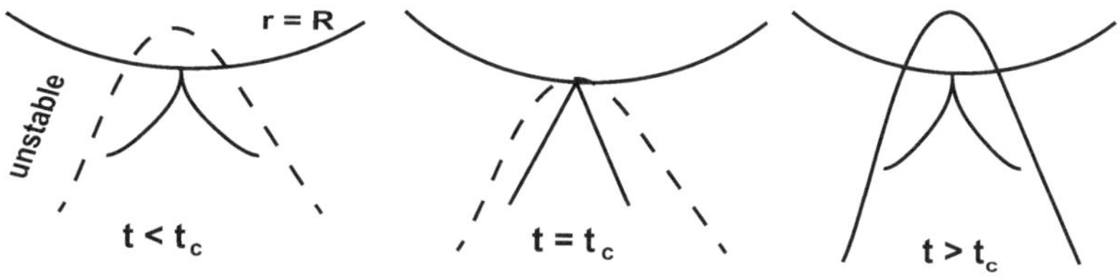

FIGURE 2.6. Gastrulation as a hyperbolic umbilic.

EXAMPLE 5 (Pentagonal symmetry). The early bilateral symmetry of the gastrula is enhanced in many cases to a more or less vague fivefold symmetry: mammalian head + limbs, starfish arms, sea urchin head + four feet, etc. Could a simple mechanism be responsible for the initial outcropping? Indeed, the bilateral extension

$$\phi = \alpha(x^4 - 6x^2y^2 + y^4) + \beta(3xy^2 - x^3) + \gamma(x^2 + y^2) + ux + vy$$

yields a catastrophe of just this form in (u, v)-space at small γ.

Here, it is important to observe that *only* two internal variables are required to generate the underlying field.

CHAPTER 3

Pattern Formation

3.1. Biological Pattern

We have observed that from the point of view of catastrophe theory, the biochemical or cellular milieu needed to evoke elementary differentiation patterns or morphogenetic movements need not have more than two components, a most conservative mechanism. We will now investigate the ways in which an underlying structure could be set up out of initial biochemical nonuniformity. Our major interest will be less in classifying the patterns and more in elucidating possible explicit mechanisms, a matter in which catastrophe theory is not of much aid. Although we shall refer to the active agents as biochemicals, they have been isolated in very few instances and could be mechanical parameters as well. The picture then is that an *underlying pattern* is formed, and thereafter *interpreted* by a cell to change its state of differentiation. Thus the biochemicals are initially internal variables, but then become control parameters. It is now to pattern formation that we pay principal attention, although observation is clearly tied to the ensuing cell interpretation. The general problem is to describe the dynamics by which an initially uniform or weakly nonuniform system autonomously develops a detailed spatial structure, and the regulatory properties which this structure possesses. Biological examples are numerous, but clear ones much less so. We have thus far focused—at least conceptually—on the milieu giving rise to the manifold forms of gastrulation, with only the animal-vegetal axis (from initial attachment point) or bilateral symmetry plane (from sperm arrival point) as directors. To assess the internal mechanism, it would be convenient to have a system in which many final forms could be achieved by varying accessible parameters. "Many" can refer either to discernible qualitative features or to accurately known quantitative features. Let us examine some of these.

When an organism is sufficiently primitive to have very few cell types in its fully developed form, a count of relative cell number becomes a significant assessment. *Cellular slime molds* fall into this category. Although a number of different lifestyles are adopted by these amoebalike (pseudopodal locomotion) organisms, let us concentrate upon that typified by *Dictyostelium discoideum*. These basically one-celled but cooperative animals start as spores—tough-shelled low-water cells—then "germinate" when moist, feed upon bacteria, and repeatedly divide. When food is low, they stop feeding and dividing, then aggregate in rivulets that combine to form an erect "sausage" which lies down and migrates by amoeboid

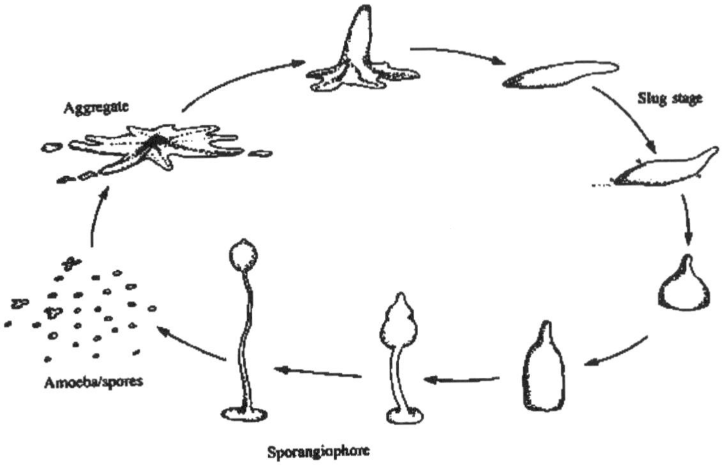

FIGURE 3.1. Life cycle of the cellular slime mold: amoeba to slug to aggregated cells to fruiting body. (Figure from Leah Edelstein-Keshet, *Mathematical Models in Biology*, 2005, p. 500, fig. 11.1. Copyright ©2005 Society for Industrial and Applied Mathematics. Reprinted with permission. All rights reserved.)

motion (towards light or heat if present) on a slime sheath which is laid down below. The *slug* then stops, builds up an internal cellulose tube, and the anterior (front) stalk cells descend by a reverse fountain to raise a bulb of originally posterior (back) spore cells to a position of easy dissemination—the "fruiting body"; see Figure 3.1.

The slug stage is the cleanest to describe. Quantitatively, the prespore/prestalk volume ratio is well regulated. It is the same at slug initiation and fruiting body initiation; since spore cells shrink, while stalk cells expand, this requires rapid last-minute conversion of stalk to spore. The regulation for different slugs or fragments of a slug is *allometric*:

$$\frac{\text{prespore volume}}{\text{prestalk volume}} = K(\text{volume})^{\lambda},$$

where λ is small but positive, ≈ 0.2. Some species shed stalk, so continuous conversion may be required. The spore $\to$ stalk conversion is very fast, stalk $\to$ spore less so, as if a new synthetic pathway had to be restarted. Also, longer migration produces higher posterior spore/stalk ratios, as if there is a progressive march towards spore in bulk. Light, fluorine ions, glucose, and above all temperature increase the spore/stalk ratio, while lithium ions and cyclic AMP decrease it. Finally, an important qualitative observation is that a polarity is maintained in the slug, with a contractile vacuole placed behind each cell nucleus.

Of course, not just cell ratio but also geometric form is well regulated in the lower organisms. One shift of form that may be related to some of our previous discussion is the extreme gastrulation-like change occurring in the single-cell colonial population of *Volvox* where, e.g., in asexual reproduction, a daughter cell population peels inward with flagellar ends inward, and then inverts through a hole in itself to form a viable locomotive colony that soon kicks itself out of its "mother." Similarly, in the next evolutionary step of multicellular sponges, cells are produced with internal flagella, and an inversion takes place to bring the flagella to the surface. The dramatic movements here are presumably the result of the formation of a pattern of novel daughter cells.

The developmental repertoire naturally becomes more extensive with more advanced organisms, the *coelenterates* (tentacular, radially symmetric, with gastrovascular cavity and three-layer body wall) being next in line. The adults take one of two forms—a sessile-attached *polyp* or free-swimming *medusa*. But much of the developmental repertoire appears here as a consequence of a single parameter—the egg size. Thus examination of numerous species of the allied classes Hydrozoa and Scyphozoa, as well as various-sized eggs within a species indicates just three developmental asymptotically medusoid forms: large eggs, e.g., *Liriope*, ≈ 0.15 mm egg, develop directly, following total equal cleavage, and gastrulation by delamination, into free-swimming medusae. Medium eggs, e.g., *Gonionemus*, ≈ 0.07 mm egg, undergo total equal cleavage, hollow ciliated blastula, delamination, planula, polyp, and then bud off what soon develops into an adult medusa. Tiny eggs, e.g., *Pegantha, Smarygdina*, eggs ≈ 0.01 mm, grow and produce many buds before functional organization. These get nourishment from a nurse cell, complete division to a morula, grow, delaminate, and repeatedly bud up to four tentacles.

On the other hand, in a single distributed-size species, *Aurela*, essentially the same thing occurs: small eggs $\rightarrow$ planula, attach and bud medusae; larger ones continue to form tentacle and mouth, a swimming actinula, and then on to a medusa; the largest go directly to medusa. Size dependence during normal egg growth is, however, the antithesis of that during bud growth or regeneration.

Another of the hydrazoa, *hydra*, morphologically a polyp that attaches, has been very extensively studied, mainly via regeneration experiments. This animal has just one form, reproducing by budding or sexually (not in the laboratory), but also even a small section of the animal head peduncle regenerates a small perfect version of the animal, and with the right polarity (direction). An implication is that a section, such as 2 3 (see Figure 3.2), will end up either as head—if part of regenerating 2 3 4 B—or foot—if part of H 1 2 3. More interesting are axial graft experiments, with various sections threaded onto a hair. Significant results are that polarity reversal takes a long time, while head determination is very fast. A small amount of head tissue induces a nearby head in the body column *if* the original head is far away; similarly, a bud appears only outside some range of inhibition of the head. The inhibition appears diffusible, taking a certain amount of time to spread from a newly created head. The static range of inhibition is assayed by observing that the graft 1 | 1 2 3 4 does not develop an intermediary head; 1 2 | 1 2 3 4 does; H 1 2 | 1 2 3 4 does not, but H 1 2 3 | 1 2 3 4 does. In summary, the head in the animal possesses short-range activation, long-range diffusible inhibition, and shows perfect *regulation* of pattern as a function of organism size.

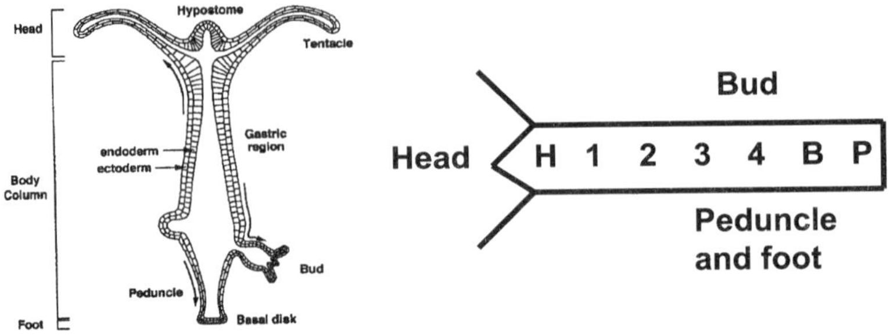

FIGURE 3.2. *Hydra*. (Figure on left from H. R. Bode, "The Role of Hox Genes in Axial Patterning in Hydra," *American Zoologist*, 2001, vol. 41, no. 3, fig. 1. By permission of Oxford University Press.)

3.2. Reaction-Diffusion Systems: Inception of Inhomogeneity

The general situation we now have is that of an active biochemical, a *morphogen*, which can change—with some delay—the internal differentiated state of a cell when present in sufficient concentration. A spatially structured pattern of cell type is then a consequence of spatially structured morphogen concentration, and we want to know how this can arise autonomously during development or regeneration from an initially uniform situation with perhaps a nondescript perturbation. Reasonable models will turn out to involve a balance between the morphogen and an *inhibitor*, or perhaps a *promoter* or *precursor* of its production. In some models, this second biochemical serves as a switch to flip to one of two or more "natural" morphogen levels, while in others there is a more graded response of the principal morphogen, or *activator*, to the inhibitor or promoter, a switching mechanism being reserved for the subsequent cell reading of activator concentration.

Let us start with a hypothetical activator X that, in some enzymatic fashion, catalyzes its own production from a precursor Y,

$$2X + Y \to 3X,$$

and is at the same time continuously degraded to B,

$$X \to B.$$

If all reaction rates are unity, and capital letters denote concentrations as well as chemical symbols, standard *homogeneous* chemical kinetics would give

$$\dot{X} = X^2Y - X.$$

At fixed Y, X can be stationary at $X^2Y - X = 0$: $X = 0$ or $X = 1/Y$. Since the enhancement rate of a perturbation $\partial\dot{X}/\partial X = 2XY - 1$, then $X = 0$ is stable with $\partial\dot{X}/\partial X = -1$, while $X = 1/Y$ with $\partial\dot{X}/\partial X = 1$ is unstable, sending X either to 0 or ∞. Indeed, the explicit solution of

$$\frac{dt}{dX} = \frac{1}{X(XY - l)} = \frac{Y}{XY - 1} - \frac{1}{X}$$

is

$$X = \frac{1}{Y + Ce^t}.$$

If Y is not at fixed concentration but is produced at constant rate from some effectively inexhaustible source, $A \to Y$, then the Y dynamics reads

$$\dot{Y} = A - X^2 Y,$$

and the balance changes. Now the only stationary point $X^2 Y = A$, $X^2 Y = X$ is given by

$$X^0 = A, \quad Y^0 = \frac{1}{A}.$$

To test its stability, we set $X = X^0 + \delta X$, $Y = Y^0 + \delta Y$ and expand:

$$\delta\dot{X} = (2X^0Y^0 - 1)\delta X + (X^0)^2 \delta Y, \quad \delta\dot{Y} = -2X^0Y^0\delta X - (X^0)^2 \delta Y.$$

These constant-coefficient linear differential equations have (except when degenerate—see below) fundamental solutions $\delta X(t) = \delta X e^{\Omega t}$, $\delta Y(t) = \delta Y e^{\Omega t}$, whence the eigenvalue relation

$$\begin{pmatrix} 1 & A^2 \\ -2 & -A^2 \end{pmatrix} \begin{pmatrix} \delta X \\ \delta Y \end{pmatrix} = \Omega \begin{pmatrix} \delta X \\ \delta Y \end{pmatrix}.$$

Stability, here desired for an isolated system, requires $\mathrm{Re}(\Omega) < 0$ for each Ω. Now

$$2\Omega = 1 - A^2 \pm \sqrt{(1 - A^2)^2 - 4A^2},$$

so that the stationary point is stable for $A^2 > 1$ with exponential oscillatory damping, and pure exponential damping for $(1 - A^2)^2 > 4A^2$ or $A > 1 + \sqrt{2}$.

Suppose now that we allow for spatial variation by considering two adjoining homogeneous systems (X_1, Y_1) and (X_2, Y_2). If the systems are physically connected, we might expect them to maintain $X_1 = X_2$ and $Y_1 = Y_2$, because any concentration difference would set up a counterbalancing transport or diffusion term, i.e.,

$$\dot{Y}_1 = \cdots + D(Y_2 - Y_1), \quad \dot{Y}_2 = \cdots + D(Y_1 - Y_2).$$

But in fact sufficiently rapid accretion of precursors might be unstable, building up more activator and hence more precursor to balance the diffusional loss. Let us see when such an inhomogeneity can occur. We assume for simplicity but not necessity that only Y diffuses:

$$\begin{aligned} \dot{X}_1 &= X_1^2 Y_1 - X_1, & \dot{X}_2 &= X_2^2 Y_2 - X_2, \\ \dot{Y}_1 &= A - X_1^2 Y_1 + D(Y_2 - Y_1), & \dot{Y}_2 &= A - X_2^2 Y_2 + D(Y_1 - Y_2). \end{aligned}$$

Since the question is that of inhomogeneity, convert to variables

$$\begin{aligned} X &= \frac{1}{2}(X_1 + X_2), \quad Y = \frac{1}{2}(Y_1 + Y_2), \\ x &= \frac{1}{2}(X_1 - X_2), \quad y = (Y_1 - Y_2). \end{aligned}$$

Hence

$$\dot{X} = X^2Y - X + 2Xxy + x^2Y, \quad \dot{Y} = A - X^2Y - (2Xxy + x^2Y),$$
$$\dot{x} = (2XY - 1)x + (X^2 + x^2)y, \quad \dot{y} = -2XYx - (X^2 + x^2 + D)y.$$

To assess stability at the uniform stationary point $X = A$, $Y = 1/A$, $x = 0$, $y = 0$, we again form the Jacobian matrix at this point:

$$\frac{\partial(\dot{X}\dot{Y}\dot{x}\dot{y})}{\partial(XYxy)} = \begin{pmatrix} 1 & A & 0 & 0 \\ -2 & -A^2 & 0 & 0 \\ 0 & 0 & 1 & A^2 \\ 0 & 0 & -2 & -A^2 - D \end{pmatrix}.$$

Now we have the eigenvalue equation from the lower 2×2 matrix:

$$(\Omega - 1)(\Omega + A^2 + D) + 2A^2 = 0 = \Omega^2 + (A^2 + D^2 - 1)\Omega + A^2 - D$$

to worry about. But the lower 2×2 determinant is

$$\Delta = A^2 - D.$$

As soon as $D > A^2$, Δ changes sign, producing one positive and one negative root Ω; thus the uniform state is no longer stable.

Generalizing, then a uniform chemically reacting system with concentrations $\xi_1, \dots, \xi_N$, say in aqueous environment, has the dynamics

$$\dot{\xi}_\alpha = F_\alpha(\dots, \xi_\beta, \dots).$$

If some concentration is nonuniform (and all concentrations are low so as not to interfere with each other), each volume will transfer molecules out by random motion at a rate proportional to the face area and rate of change of concentration. Choosing a cube L^3, then

$$L^3\dot{\xi}(\mathbf{x}) = DL^2 \sum_u \frac{\xi(\mathbf{x} + L\mathbf{u}) - \xi(\mathbf{x})}{L},$$

where $\mathbf{u}$ runs over the six neighboring cube directions. Expanding in L, this gives $L^3 D\nabla^2\xi(\mathbf{x})$, and hence the chemical kinetics extended to spatial diffusion becomes

$$\dot{\xi}_\alpha = F_\alpha(\dots, \xi_\beta, \dots) + D_\alpha \nabla^2 \xi_\alpha.$$

If the steady uniform solution is given by

$$F_\alpha(\dots, \xi_\beta^0, \dots) = 0,$$

we again assess stability by writing

$$\xi_\alpha = \xi_\alpha^0 + \delta\xi_\alpha$$

and expanding:

$$\delta\dot{\xi}_\alpha = \sum \left(\frac{\partial F_\alpha}{\partial \xi_\beta^0} \right) \delta\xi_\beta + D_\alpha \nabla^2 \delta\xi_\beta.$$

As a partial differential equation with constant coefficients, this equation has elementary solutions bounded in space given by (except with degeneracy) the exponentials

$$\delta\xi_\alpha(\mathbf{x}, t) = e^{\Omega t} e^{i\mathbf{k}\cdot\mathbf{x}} \delta\xi_\alpha,$$

so that

$$\Omega\delta\xi_\alpha = \sum\left(\frac{\partial F_\alpha}{\partial \xi_\beta^0}\right)\delta\xi_\beta - k^2 D_\alpha \delta\xi_\alpha.$$

In other words, Ω must be an eigenvalue of the matrix

$$M_{\alpha\beta}(k^2) = \frac{\partial F_\alpha}{\partial \xi_\beta^0} - k^2 D_\alpha \delta_{\alpha\beta}, \tag{3.1}$$

and the stationary uniform point will be unstable to growth of a nonuniformity if

$$\text{Re}(\Omega(k^2)) > 0$$

for some eigenvalue Ω at some k.

When may a stable homogeneous system, $\text{Re}(\Omega(0)) < 0$ for all Ω, become unstable on subtracting the positive diagonal matrix $k^2 D_\alpha \delta_{\alpha\beta}$? Certainly not if D_α is a constant D, for then all eigenvalues decrease by $k^2 D$. Certainly not if $M_{\alpha\beta}(0)$ is symmetric, for then too its eigenvalues can only be lowered. Certainly not if $M_{\alpha\beta}(0)$ is "qualitatively stable," i.e., $M_{\alpha\beta}(0) f_{\alpha\beta}$ has all $\text{Re}(\Omega) < 0$ for any set of *positive* $f_{\alpha\beta}$, for this is equivalent (see, e.g., [**157**]) to $M_{\alpha\alpha}(0) \le 0$, $M_{\alpha\beta}(0)M_{\beta\alpha}(0) \le 0$ for $\alpha \ne \beta$, $M_{\alpha\beta}(0)M_{\beta\gamma}(0)\cdots M_{\zeta\alpha}(0) = 0$ for $\alpha \ne \beta \ne \gamma \ne \delta \ne \zeta$, and $\det M(0) \operatorname{tr} M(0) \ne 0$, conditions that are unaltered by subtracting $k^2 D_\alpha \delta_{\alpha\beta}$. But instability is provoked even in our very simple 2×2 activator-precursor system. We shall soon see what is needed.

We have tacitly assumed an infinite system. For a finite system, boundary conditions are required. One possibility might be absorbing boundaries, $\xi_\alpha = 0$ at the boundary, but then there would be no uniform system with which to compare. The usual assumption is that of an isolated system with no accumulation at the boundary. If $\mathbf{u}$ is the normal direction, this implies that $\xi_\alpha(\mathbf{x}) = \xi_\alpha(\mathbf{x} + L\mathbf{u})$, or $\frac{\partial \xi_\alpha}{\partial n} = 0$ at the boundary. In general, the spatial functions that replace $e^{i\mathbf{k}\cdot\mathbf{x}}$ are highly boundary dependent, but for a one-dimensional system of length V, a model we shall return to, only $\cos(kx)$ is required, and so (3.1) remains valid if

$$k = \frac{p\pi}{V}, \quad p > 0 \text{ an integer.}$$

Consider next just two chemical components, labeled as a and h. The reaction-diffusion equations in one dimension are

$$\frac{\partial a}{\partial t} = f(a,h) + \epsilon\frac{\partial^2 a}{\partial x^2}, \quad \frac{\partial h}{\partial t} = g(a,h) + \nu\frac{\partial^2 h}{\partial x^2}.$$

For a stationary point $f(a,h) = g(a,h) = 0$, the stability matrix becomes

$$M(k^2) = \begin{pmatrix} f_a - \epsilon k^2 & f_h \\ g_a & g_h - \nu k^2 \end{pmatrix},$$

with time exponent $\Omega(k^2)$ determined by

$$(\Omega + \epsilon k^2 - f_a)(\Omega + \nu k^2 - g_h) = f_h g_a. \tag{3.2}$$

Suppose again that the uniform system has been around for a while, so that it is stable against uniform, $k = 0$ perturbations. If $\Omega^2 - (f_a + g_h)\Omega + f_a g_h - f_h g_a = 0$ is to have only negative or negative real part roots, we of course require

$$f_a + g_h < 0, \quad f_a g_h - f_h g_a > 0.$$

For large k, $\Omega = -\epsilon k^2$ or $-\nu k^2$, so the system is certainly stable. Since the linear term in (3.2), $f_a + g_h - (\epsilon + \nu)k^2$, must now remain negative, the only way for a root Ω of (3.2) of positive real part to appear for some intermediate range of k is by the constant term of (3.2), $\epsilon\nu k^4 - (\epsilon g_h + \nu f_a)k^2 + f_a g_h - f_h g_a$, changing sign. This expression is positive for $k = 0$ and $k = \infty$, and hence the sign change demands that it have two real positive roots k^2. Thus the discriminant must be positive and the linear term negative, yielding the single condition

$$\left(\frac{\epsilon}{\nu}\right)^{\frac{1}{2}} g_h + \left(\frac{\nu}{\epsilon}\right)^{\frac{1}{2}} f_a > 2(f_a g_h - f_h g_a)^{\frac{1}{2}}.$$

For autocatalytic activation, we will have $f_a > 0$, and so $\frac{\nu}{\epsilon}$ can always be found large enough to satisfy this condition—e.g., h as an "inhibitor" diffuses away fast enough.

EXAMPLE 6. Turing's model [**156**], in suitable units, takes the form

$$f = \frac{1}{16}a(h-1) - \frac{3}{4}, \quad g = 1 - \frac{1}{16}ah,$$

so that $a^0 = h^0 = 4$, and

$$f_a = \frac{h^0 - 1}{16} = \frac{3}{16}, \quad f_h = \frac{a^0}{16} = \frac{1}{4},$$
$$g_a = -\frac{h^0}{16} = -\frac{1}{4}, \quad g_h = -\frac{a^0}{16} = -\frac{1}{4}.$$

Now

$$f_a + g_h = -\frac{1}{16}, \quad f_a g_h - f_h g_a = \frac{1}{64},$$

as desired for $k = 0$ stability, while $k^2 > 0$ instability requires

$$\frac{3}{16}\left(\frac{\nu}{\epsilon}\right)^{\frac{1}{2}} - \frac{1}{4}\left(\frac{\epsilon}{\nu}\right)^{\frac{1}{2}} > \frac{1}{4} \quad \text{or} \quad \frac{\nu}{\epsilon} > 4.$$

Even if reaction-diffusion models turn out to effectively reproduce facets of development, the identification of the model variables is far from obvious. However, there is evidence in some cases that bona fide medium-size biomolecules are involved. If this is the case, we should investigate what possible models are implied by a chemical kinetic underpinning. First, it pays to examine the general properties of static uniform equilibrium of a solution of chemically reacting components. If temperature T and pressure P are fixed, thermodynamic equilibrium is controlled by the Gibbs free energy, say per solvent molecule, $\phi(A, B, C, \dots)$, which decreases monotonically with time, reaching minimum at equilibrium. Here, $A, B, C, \dots$ are also taken as concentrations in molecules per solvent molecule. For low concentrations, ideal solution theory is valid—the solutes behave like ideal gases—and the chemical potential or Gibbs free energy per molecule is given by (κ = Boltzmann's constant)

$$\mu_A = \kappa T \ln\left(\frac{A}{A_0}\right),$$

where $A_0(P,T)$ is characteristic of species A. But if the Gibbs free energy per unit volume of pure solvent is $\rho\mu_0(P,T)$, its value in the solution is reduced by the partial pressure (at fixed volume) $\kappa T \rho A$ of each solute. Hence for the system

$$\phi(A,B,C,\dots) = \mu_0 + \kappa T \sum_X X\left(\ln\left(\frac{X}{X_0}\right) - 1\right).$$

Now in equilibrium, each chemical reaction imposes a relation between equilibrium concentrations $\bar{A}, \bar{B}, \bar{C}, \dots$ by virtue of the equilibrium condition that

$$0 = d\phi = \kappa T \sum \ln\left(\frac{X}{X_0}\right) dX = \sum \bar{\mu}_X dX$$

for any set $\{dX\}$ consistent with the reaction. For example, if

$$X + 2A \rightleftharpoons B,$$

the concentration changes are restricted by $(dX, dA, dB) = (1, 2, -1)dX$, so that $(\bar{\mu}_X + 2\bar{\mu}_A - \bar{\mu}_B)dX = 0$ for all dX, or

$$\bar{\mu}_X + 2\bar{\mu}_A = \bar{\mu}_B,$$

a relation that mimics the reaction, as is always the case. Here it follows that

$$\frac{\bar{X}\,\bar{A}^2}{\bar{B}} = \frac{X_0 A_0^2}{B_0},$$

a *mass action* law for this reaction. There are two controllable concentrations here, but if this reaction were part of a chain of reactions, one could only say that each reaction enforces one relation between concentrations, determining in this way an independent set of concentrations for the whole chain.

The classical kinetics associated with the above reaction, even as part of a more complex system, is

$$\begin{aligned}
\dot{A} &= \cdots - 2kA^2X + 2k'B + \cdots, \\
\dot{X} &= \cdots - kA^2X + k'B + \cdots, \\
\dot{B} &= \cdots + 2kA^2X - k'B + \cdots,
\end{aligned}$$

where k is the rate of the forward reaction $X + 2A \to B$, and k' is the rate of the reverse reaction $B \to X + 2A$; i.e., the reaction probability goes as the product of the concentrations involved. When equilibrium for this reaction is reached,

$$k\bar{A}^2\bar{X} = k'\bar{B},$$

so that for correspondence with thermodynamic equilibrium, we have the relation

$$\frac{\bar{X}\,\bar{A}^2}{\bar{B}} = \frac{k'}{k} = \frac{X_0 A_0^2}{B_0}.$$

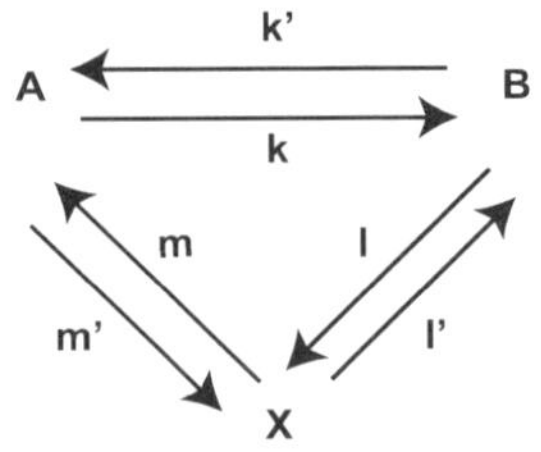

FIGURE 3.3. A reaction loop.

Identification of thermodynamics and kinetics is nailed down by observing now that away from equilibrium, reaction kinetics implies that

$$\begin{aligned}\dot{\phi} &= \cdots + \mu_A \dot{A} + \mu_B \dot{B} + \mu_X \dot{X} + \cdots \\ &= \dot{X}(\mu_X + 2\mu_A - \mu_B) + \cdots \\ &= -k' B_0 \kappa T \left(\frac{A^2 X}{A_0^2 X_0} - \frac{B}{B_0} \right) \left(\ln \frac{A^2 X}{A_0^2 X_0} - \ln \frac{B}{B_0} \right) \\ &\le 0,\end{aligned}$$

where we have used the inequality $(y - z)(\ln y - \ln z) > 0$. Thus, the contribution due to each reaction gives a monotonically decreasing ϕ. ϕ is then a *Lyapunov function* for the kinetics, which was suggested by thermodynamic arguments but has mathematical content without them: its steady decrease away from equilibrium can readily be used to show that the system approaches equilibrium asymptotically in time whenever there is a unique equilibrium. How then is anything interesting supposed to happen in time? The mere presence of spatial diffusion is insufficient, since the basic diffusion kinetics,

$$\dot{A}_1 = D(A_2 - A_1) + \cdots, \quad \dot{A}_2 = D(A_1 - A_2) + \cdots,$$

just represents the monomolecular chemical reaction

$$A_1 \rightleftharpoons A_2$$

with $k = k' = D$. Perhaps something happens by the failure of the low concentration classical chemical kinetics. But this is not in fact needed. All that is needed is for simultaneous equilibrium for all separate reactions to be impossible. This can be accomplished by having at least as many reactions as controllable variables, so that no nontrivial set of concentrations exists, or by setting some rate constants equal to 0 or ∞ to prevent local equilibrium. Technically, we then have a failure of *detailed balance*: if a system is microscopically time reversible, then in an ensemble of systems (or the equivalent thereof) in statistical equilibrium, every process will occur at the same rate as the time-reversed process. Let us consider elementary examples of the failure of detailed balance.

EXAMPLE 7 (A loop from a reactant to itself [Figure 3.3]). For local equilibrium, we would have $\bar{\mu}_A = \bar{\mu}_X = \bar{\mu}_B$ thermodynamically: $A/A_0 = B/B_0 =$

X/X_0, or from the separate kinetics,

$$\frac{\bar{A}}{\bar{B}} = \frac{k'}{k}, \quad \frac{\bar{B}}{\bar{X}} = \frac{\ell'}{\ell}, \quad \frac{\bar{X}}{\bar{A}} = \frac{m'}{m}.$$

The corresponding requirement

$$k'\ell' m' = k\ell m$$

need not be true (quite generally, the rate product about any loop is the same in both directions for detailed balance); see Figure 3.3. But although individual reactions need not balance, the loop currents certainly can; i.e.,

$$\dot{A} = -(k+m')A + k'B + mX, \quad \dot{B} = kA - (\ell + k')B + \ell' X,$$
$$\dot{X} = m'A + \ell B - (m + \ell')X$$

do have a stationary solution, $X = A = B = 0$, $\dot{X} = \dot{A} = \dot{B} = 0$, given by

$$\bar{B} = \frac{k(m+\ell') + \ell' m}{(\ell + k')(\ell' + m) - \ell\ell'}\bar{A}, \quad \bar{X} = \frac{m'(\ell + k) + k'\ell}{(\ell + k')(\ell' + m) - \ell\ell'}\bar{A}.$$

This is a stable equilibrium, e.g., at fixed A, but since A_0, B_0, and X_0 are irrelevant, it does not arise from a standard Gibbs free energy.

EXAMPLE 8 (Vanishing rate constants, reaction driven by source-sink). Here

$$A \xrightarrow{k} X \underset{\ell'}{\overset{\ell}{\rightleftharpoons}} Y \xrightarrow{m} B,$$

in which k' and m' have been set equal to 0, and the controlled concentrations A and B are fixed. This can also be automatic, e.g., let $A \to \infty$ and $k \to 0$ at fixed kA. Since $dA = dB = 0$, separate thermodynamic equilibria would require $0 = \bar{\mu}_X = \bar{\mu}_Y = 0$ or $\bar{X} = X_0, \bar{Y} = Y_0$. Kinetic equilibrium for individual steps is of course impossible, e.g., $k\bar{A} = 0\,\bar{X}$ requires $X = \infty$, but the system

$$\dot{X} = kA + \ell' Y - \ell X, \quad \dot{Y} = -(\ell' + m)Y + \ell X,$$

stably equilibrates to $\bar{X} = \frac{\ell'+m}{\ell}\frac{k}{m}A$ and $\bar{Y} = \frac{k}{m}A$.

EXAMPLE 9 (Infinite rate constant). Here

$$A \underset{k'}{\overset{k}{\rightleftharpoons}} B + C, \quad C \underset{\ell'}{\overset{\ell}{\rightleftharpoons}} X + Y,$$

where $\ell \to \infty$. Since C is instantly depleted, $C \to 0$ as $\ell \to \infty$, and the kinetic equation $\dot{C} = kA - k'BC - \ell C + \ell' XY$ tells us that $\ell C = kA + \ell' XY$. Thus, due to the second reaction, $\dot{X} = \ell C - \ell' XY = kA = \dot{Y}$, and due to the first, $\dot{A} = -kA + k'BC = -kA$. The net effect is the reaction

$$A \xrightarrow{k} B + X + Y,$$

in which effectively $k' = 0$ and the intermediate C has disappeared.

This last example also indicates how compound reactions can be built up from simple ones via intermediates. Such limiting compound reactions will be of particular utility. On the one hand, they mean that in principle only the elementary reactions

$$A \to B + C, \quad B + C \to A,$$

are required, since, e.g.,

$$B + C \xrightarrow{k} D + E$$

is the result of

$$B + C \xrightarrow{k} F, \quad F \xrightarrow{\ell} D + E,$$

as $\ell \to \infty$, a process that can be iterated. On the other hand, it means that nonstandard elementary reactants are easily constructed. Basic in this regard is

$$X + E \underset{k'}{\overset{k}{\rightleftharpoons}} E^*$$

as $k \to \infty$ and $k' \to \infty$ at fixed $K = k'/k$. Now one has instant true local equilibration (and associated Gibbs function) with $kXE = k'E^*$ or $EX = KE^*$. But suppose that chemical E occurs only in the forms E or E^*; hence $E + E^* = E_0$, a constant. Eliminating E,

$$E^* = \frac{E_0 X}{K + X}$$

is a new combination available as a reactant.

EXAMPLE 10 (Michaelis-Menten enzyme kinetics). Here

$$X + E \underset{k'}{\overset{k}{\rightleftharpoons}} E^* \underset{\ell'}{\overset{\ell}{\rightleftharpoons}} Y + E,$$

in which E is an enzyme, E^* an enzyme complex, $E = KE^*/X$, and

$$\dot{Y} = \frac{\ell E_0}{K + X} X - \frac{K\ell' E_0}{K + X} Y,$$

and since $\dot{X} + \dot{Y} = -\dot{E}^*$, then

$$\frac{d}{dt}\left(X + \frac{E_0 X}{K + X}\right) = -\frac{\ell E_0}{K + X} X + \frac{K\ell' E_0}{K + X} Y.$$

EXAMPLE 11 (Enzyme kinetics with inhibitor I). Here

$$X + E \underset{k'}{\overset{k}{\rightleftharpoons}} E^* \underset{\ell'}{\overset{\ell}{\rightleftharpoons}} Y + E, \quad I + E \underset{k^{*\prime}}{\overset{k^*}{\rightleftharpoons}} E^{**}.$$

Now enzyme conservation yields $E + E^* + E^{**} = E_0$, so that

$$E^* = \frac{E_0}{1 + \frac{I}{K^*} + \frac{X}{K}} \frac{X}{K},$$

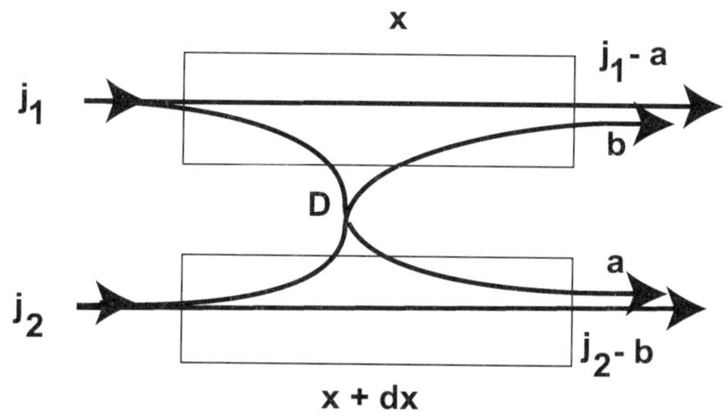

FIGURE 3.4. Breaking of detailed balance of currents $j_{1,2}$ by accompanying diffusion.

and the combined kinetics appears as

$$\dot{Y} = \frac{\ell \frac{X}{K} - \ell' Y}{1 + \frac{I}{K^*} + \frac{X}{K}},$$

$$\frac{d}{dt}\left(1 + \frac{E_0/K}{1 + \frac{I}{K^*} + \frac{X}{K}}\right) X = -\frac{\ell \frac{Y}{K} - \ell' Y}{1 + \frac{I}{K^*} + \frac{X}{K}} E_0.$$

3.3. Inhomogeneous Steady State

Returning now to our original problem, how can a nonuniform equilibrium state arise in the presence of diffusion? This requires that separate equilibrium, by which we mean detailed balance, for the concomitant diffusion "reaction" $A(x) \rightleftharpoons A(x + dx)$ (see Figure 3.4) to not be maintained.

Breaking thermodynamic equilibrium is most often accomplished by the presence, in an idealized limit, of fixed sources and sinks, like the examples we have just quoted. They prevent full equilibrium not by being fixed (this does not disturb establishment of a Gibbs function), but by the implied absence of a return path in $A \rightarrow$ or $\rightarrow B$, or by the imposition of too many relations between concentrations. In the former case, an implication is that the time scale is that during which the return path can be neglected. A concomitant effect with diffusive coupling is that there can also exist more than one stationary nonthermodynamic equilibrium, and if the uniform case is unstable, a nonuniform equilibrium can be achieved.

Suppose then that the instability of a uniform reaction-diffusion system has resulted in the exponential time growth of a perturbation. For a physically realizable system, the growth is finite and results asymptotically in either a stationary time-independent equilibrium state, an oscillatory state, or one with apparently random character. Without firm evidence to the contrary, we will assume that stationary states, with their consequent unvarying message to the differentiation machinery of the cell, are the ones of interest in development. We would now like to find the class of stationary patterns available and assess their regulatory properties with respect to perturbations, parameter changes, and boundary changes. For the last mentioned, there are two extremes: (i) the shape of activation is determined mainly by diffusion and degradation, resulting in an absolute or constant activation pattern a independent of the size of the full region, or (ii) there is a saturating production of activator, so that one needs a large activated region to produce, e.g., minimum inhibitory h over the larger region—so that relative size pattern regulation results.

Let us start with a two-component chemical system, a and h in a one-dimensional region of length L. We have seen that an unstable uniform pattern requires a large enough ratio D_h/D_a of diffusion constants. Carrying this to the extreme by way of introduction, we will assume $D_a = 0$, so that the system takes the stationary form

$$\begin{gathered} \dot{a} = 0 = f(a, h), \\ \dot{h} = 0 = g(a, h) + \nu h'', \\ a'(x) = h'(x) = 0 \quad \text{at } x = 0 \text{ and } L. \end{gathered}$$

Suppose first that $f(a, h) = 0$ has a unique solution $a = A(h)$. We then have

$$\begin{gathered} \nu h''(x) + g(A(h), h) = 0, \\ h'(0) = h'(L) = 0, \end{gathered} \tag{3.3}$$

a standard one-dimensional mechanical problem that we have seen before. Again, it is solved by setting up an "energy" integral: multiply by $h'(x)$ and integrate, obtaining

$$\frac{\nu}{2} h'^2 + V(h) = E, \tag{3.4}$$

where $V(h) = \int^h g(A(h), h) dh$, $h'(0) = h'(L) = 0$, and E is a suitable constant.

The usual analogue representation is convenient, where h represents the position of a mechanical particle with mass ν in a potential $V(h)$. The total energy is E, and x denotes "time," so that h starts at 0 velocity at $x = 0$ and stops again at $x = L$. Thus, if the pattern is not repeated, the path must go once between h_0 and h_1. Since

$$h'(x) = \left(\frac{2}{\nu}\right)^{\frac{1}{2}} \sqrt{E - V(h)},$$

then unless $h' \equiv 0$, the value of E is determined by

$$L = x_1 - x_0 = \int_{x_0}^{x_1} dx = \int_{h_0}^{h_1} \frac{1}{h'(x)} dh = \left(\frac{\nu}{2}\right)^{\frac{1}{2}} \int_{h_0}^{h_1} \frac{1}{\sqrt{E - V(h)}} dh.$$

Indeed, let us define

$$\chi(E) = \left(\frac{\nu}{2}\right)^{\frac{1}{2}} \int_{h_0(E)}^{h_1(E)} \frac{1}{\sqrt{E - V(h)}} dh,$$

the position change during a single traversal. Then, if $L < \chi_{\min}$, the minimum of χ with respect to E, then the L, E relation can never be satisfied. Hence only the uniform case can result, with the constant value of h given by $g(A(h), h) = 0$ or by $V'(h) = 0$. If $\chi_{\min} < L < 2\chi_{\min}$, the above one-passage solution is the only possibility, but if $L > 2\chi_{\min}$, the "trajectory" can cover the interval $h_0 - h_1$ more than once. p half-cycles will result if

$$L = p\chi(E),$$

but the pair p, E is not unique—the value of p will depend on the form of the initiating perturbation.

EXAMPLE 12 (The [second] Turing system). Chemically, this results from autocatalysis of X and degradation by Y. In mixed order, then,

$$X + Y \to B, \quad C + X + Y \to 2X + Y, \quad A \to Y, \quad X \to M,$$

together with the more special

$$X + D \overset{k}{\longrightarrow} F, \quad F \to G + D,$$

in the limit $D \to 0$, $k \to \infty$. From the last set, assuming unit rates when unspecified, $\dot{D} = 0 = -kXD + F$; but then $\dot{F} = kXD - F = 0$, so F is a constant, as is $\dot{X} = -kXD = -F$. All together, then,

$$\dot{X} = (C - 1)XY - X - F, \quad \dot{Y} = A - XY.$$

Rescaling and writing X as a and Y as h, we have the previously used

$$f(a, h) = \frac{1}{16}a(h - 1) - \frac{3}{4}, \quad g(a, h) = 1 - \frac{1}{16}ah.$$

Now, from $f(A, h) = 0$, we have $A(h) = \frac{12}{h-1}$, so

$$V(h) = \int_4^h \left(1 - \frac{3}{4}\frac{h}{h-1}\right)dh = \frac{1}{4}(h - 4) - \frac{3}{4}\ln\frac{h-1}{3}.$$

For small E, we have $V(4) = 0$, $V'(4) = 0$, and $V''(4) = \frac{3}{4}\frac{1}{(h-1)^2}\Big|_{h=4} = \frac{1}{12}$ so

$$V(h) = \frac{1}{24}(h - 4)^2 + \cdots.$$

In this region, therefore,

$$\chi(E) = \left(\frac{\nu}{2}\right)^{\frac{1}{2}} \int \left[E - \frac{(h-4)^2}{24}\right]^{-\frac{1}{2}} dh = \left(\frac{\nu}{2}\right)^{\frac{1}{2}} \int_{-1}^{1} \frac{\sqrt{24}}{\sqrt{1 - y^2}}\,dy = \pi\sqrt{12\nu},$$

independent of E. This indeed is $\chi_{\min}$, and so in the example of [**38**], with $\nu = \frac{1}{4}$ (and $\epsilon = \frac{1}{16}$, which we replace by 0), $L = 60$, one could then have of the order of 11 half-cycles.

But χ increases with E, and the only chance of selecting a given member p of half-cycles for a random initial perturbation would be by using a very narrow k-instability band for the initial perturbation; in fact, this does not work. For larger E, the spatial region covered is best decomposed into the portion arising from h_0 to h_M at the potential minimum V_M, and similarly h_M to h_1. The major contributions are due to the low "velocity" or high $\frac{dx}{dh}$ regions about h_0 and h_1, and so we can expand appropriately.

With just one minimum V_M in the region of interest, it is convenient to switch to V as an independent variable, and measure x_0 leftward from $V_{\min}$ and x_1 rightward from $V_{\min}$. Thus, in the first half-cycle,

$$\begin{aligned} x_1(V) &= \left(\frac{\nu}{2}\right)^{\frac{1}{2}} \int_{h_M}^{h_1} (E - V(h))^{-\frac{1}{2}}\, dh \\ &= \left(\frac{\nu}{2}\right)^{\frac{1}{2}} \int_{V_M}^{V} h_1'(V)(E - V)^{-\frac{1}{2}}\, dV \\ &= \left(\frac{\nu}{2}\right)^{\frac{1}{2}} \int_{E-V}^{E-V_M} h_1'(E - U)U^{-\frac{1}{2}}\, dU, \end{aligned}$$

or expanding about $U = 0$ ($V = E$),

$$\begin{aligned} x_1(V) &= \left(\frac{\nu}{2}\right)^{\frac{1}{2}} \int_{E-V}^{E-V_M} \left(h_1'(E) - U h_1''(E) + \cdots\right) U^{-\frac{1}{2}}\, dU \\ &= (2\nu)^{\frac{1}{2}} h_1'(E)\left((E - V_M)^{\frac{1}{2}} - (E - V)^{\frac{1}{2}} + \cdots\right). \end{aligned}$$

In particular,

$$x_1(E) = (2\nu)^{\frac{1}{2}} h_1'(E)(E - V_M)^{\frac{1}{2}} + \cdots$$

and similarly $x_0(E) = (2\nu)^{1/2}(-h_0'(E))(E - V_M)^{1/2} + \cdots$ while $\chi(E) = x_0(E) + x_1(E)$.

Pattern regulation can now be assessed in two ways. First, we may compare the size of the region $x_0(E)$ of low inhibitor concentration with that of $x_1(E)$, of high h, under change of L and hence of E. Absolute size regulation will be achieved if $h_1'(E)(E-V_M)^{1/2} = \text{const}$ or $h_1(E) = h_M + k(E-V_M)^{1/2}$, equivalent of course to the parabolic dependence $V = V_M + (h-h_M)^2/k^2$. Second, the maintenance of the pattern can be examined. The foregoing does not have absolute pattern regulation, since here

$$\begin{aligned} x_1 &= (2\nu)^{\frac{1}{2}} k\left(1 - \left(\frac{E - V}{E - V_M}\right)^{\frac{1}{2}}\right) \cdots \\ &= (2\nu)^{\frac{1}{2}}\left(k - \left(k^2 - \frac{(h - h_M)^2}{E - V_M}\right)\right)^{\frac{1}{2}} \cdots \end{aligned}$$

has substantial E-dependence, with h occurring in the combination $(h - h_M)/(E - V_M)^{1/2}$. But absolute pattern regulation is very nearly achieved if $V(h)$ reaches a maximum, $\bar{h}$, allowing an infinite V extension of the spatial domain for very small change of E.

Thus, if $h_1'(\bar{E}) = \infty$ at $h = \bar{h}$, then

$$x_0(h) = \left(\frac{\nu}{2}\right)^{1/2} \int_h^{h_M} \frac{dh}{(E - V(h))^{1/2}}$$

converges to a definite function when $E \to \bar{E}$, as does $x_1(h)$ at any fixed $h < \bar{h}$. But $x_1(\bar{h}) \to \infty$: the $h = \bar{h}$ tail expands without limit. The Turing case is less extreme, with a relatively slowly rising $V(h)$ on the right yielding only approximate absolute regulation.

Relative pattern regulationcan similarly be examined from two points of view. For relative size regulation alone, we would want

$$\frac{x_1(E)}{x_0(E)} = \text{const} = \frac{h_1'(E)}{-h_0'(E)} + \cdots .$$

Assuming leading power law dependencies, we write

$$h_0(V) = b_0 - c_0(V - V_M)^{q_0}, \quad h_1(V) = b_1 + c_1(V - V_M)^{q_1}.$$

This of course requires $q_0 = q_1$ and consequently a discontinuity in some derivative at $V = V_M$ if the region of validity extends to $b_0 = h_M = b_1$. Even then, there is not just a spatial extension of the pattern but a rescaling as well: we have, e.g.,

$$\begin{aligned} x_1(V) &= (2\nu)^{\frac{1}{2}} h_1'(E)\left((E - V_M)^{\frac{1}{2}} - (E - V)^{\frac{1}{2}}\right) \cdots \\ &= (2\nu)^{\frac{1}{2}} c_1 q_1 (E - V_M)^{q_1 - \frac{1}{2}}\left(1 - \left(1 - \frac{V - V_M}{E - V_M}\right)^{\frac{1}{2}}\right) \cdots \\ &= (2\nu)^{\frac{1}{2}} c_1 q_1 (E - V_M)^{q_1 - \frac{1}{2}}\left(1 - \left(1 - \frac{(h_1 - h_M)^{1/4}}{(E - V_M)c_1^{1/q}}\right)^{\frac{1}{2}}\right) \cdots \end{aligned}$$

so that the pattern is unchanged only in terms of the variables

$$x^* = \frac{x}{(E - V_M)^{q_1 - 1/2}}, \qquad h^* = \frac{h - h_M}{(E - V_M)^{q_1}}.$$

We have yet to examine the stability of the nonuniform pattern resulting from (3.4), with respect to the time-dependent set

$$\begin{aligned} \dot{a} &= f(a, h), & a' &= 0 \text{ at boundary}, \\ \dot{h} &= g(a, h) + \nu h'', & h' &= 0 \text{ at boundary}. \end{aligned}$$

A direct assessment, even of stability with respect to small perturbations, is not easy, but if a suitable Lyapunov function can be constructed, the task is made trivial. To prove stability with respect to any change, we need some functional of a and h that decreases monotonically in time, achieving its stationary minimum at (3.3). Since (3.3) does come from the Euler-Lagrange variational principle ,

$$\delta \int \left(\frac{\nu}{2} h'(x)^2 - V(h)\right) dx = 0, \quad h' = 0 \text{ at boundaries},$$

we are tempted to examine the functional

$$L = \int \left(\frac{\nu}{2} h'(x)^2 - W(a, h)\right) dx.$$

Indeed, we then have

$$\begin{aligned}\frac{dL}{dt} &= \int (\nu h'\dot{h}' - \dot{a}W_a - \dot{h}W_h)dx \\ &= \int [\nu h'(g' + \nu h''') - fW_a - (g + \nu h'')W_h]dx \\ &= \int [-\nu h''(g + \nu h'') - fW_a - (g + \nu h'')W_h]dx \\ &= -\int [(g + \nu h'')^2 + (W_h - g)(g + \nu h'') + fW_a]dx,\end{aligned}$$

or

$$\frac{dL}{dt} = -\int \left[g + \nu h'' + \frac{1}{2}(W_h - g)\right]^2 dx - \int \left[fW_a - \frac{1}{4}(W_h - g)^2\right]dx.$$

Hence if W can be chosen such that

$$4fW_a \geq (W_h - g)^2,$$

reaching equality only when

$$f = 0 = W_h - g,$$

stable equilibrium will occur at $f = 0$, $g + \nu h'' = 0$, as desired: L will decrease until (3.3) is satisfied. The simplest possibility is that in which $W_h = g(a, h)$, and we require $fW_a \geq 0$. Since W_a must pass through 0 when f does, then $W = w(a) + \int^h g(a, h')dh'$ implies

$$W_a = \int_{h(a)}^{h} g_a(a, h')dh',$$

where $h(a)$ is the solution to $f(a, h) = 0$, assumed single-valued. Thus if

$$fW_a > 0 \quad \text{when} \quad h \neq h(a),$$

then (3.4) is stable.

EXAMPLE 13 (The Edelstein switch [**61**]). Here

$$\begin{aligned}f(a, h) &= \frac{1 + \alpha h^2}{1 + h^2} - a, \qquad \alpha > 1, \\ g(a, h) &= \lambda a - \mu h,\end{aligned}$$

in suitable units. In uniform equilibrium, $f = g = 0$; this has one solution for small or large values of μ/λ and three solutions for an intermediate band, of which upper and lower are stable to uniform perturbations. A small change of λ can shift the uniform operating point from very low a to very high a, providing the switching characteristic. One readily finds $A(h) = \frac{1+\alpha h}{1+h^2}$, $g(h) = \lambda\alpha - \lambda\frac{\alpha-1}{1+h^2} - \mu h$, leading to

$$V(h) = \lambda\alpha h - \frac{\mu}{2}h^2 - \lambda(\alpha - 1)\tan^{-1} h.$$

The two maxima of V show that by parameter adjustment, an infinite amount of space can be allotted to either extreme h-value, but periodic patterns arise unless carefully avoided. This system does show nonuniform stability. Since

$$h(a) = \left(\frac{a-1}{\alpha - a}\right)^{\frac{1}{2}},$$

then

$$\lambda^{-1} W_a = \int_{\left(\frac{a-1}{\alpha-a}\right)^{\frac{1}{2}}}^{h} dh' = h - \left(\frac{a-1}{\alpha - a}\right)^{\frac{1}{2}}$$

or

$$f W_a = \left(\frac{1+\alpha h^2}{1+h^2} - a\right)\left(h - \left(\frac{a-1}{\alpha - a}\right)^{\frac{1}{2}}\right)$$

(where $\alpha \geq a \geq 1$ in the vicinity of $f = 0$). But

$$h \gtrless \left(\frac{a-1}{\alpha - a}\right)^{\frac{1}{2}} \quad \text{implies} \quad \frac{1+\alpha h^2}{1+h^2} - a = \alpha - a - \frac{\alpha - 1}{1+h^2} \gtrless 0,$$

so that $fWa \geq 0$, as required.

3.4. Multistable Regimes

Let us now generalize. One important assumption we have made thus far is that stationary activator production, $f(a, h) = 0$, has only one solution $a = A(h)$ for each h. This means that we have precluded the possibility of an intrinsic switching mechanism leading to two or more activator states. As an example, we consider the basic Gierer-Meinhardt [**68**] activator-inhibitor model, in suitable units,

$$f(a, h) = \frac{a^2}{h(1+a^2)} - \frac{1}{2}a + \mu, \quad g(a, h) = a^2 - \lambda h. \tag{3.5}$$

Here the production of a is autocatalytic, and has limiting inhibition by h as well as a model enzyme saturation. Now there are either one or three values of a associated with each h. Simplifying by choosing the activator source μ as $\mu = 0$, the roots are given explicitly by

$$A_0(h) = 0, \quad A_1(h) = A_2(h) = \frac{1}{h} \pm \sqrt{\frac{1}{h^2} - 1} \quad \text{for } h \leq 1.$$

Correspondingly, there are three curves of $g(A(h), h)$; see Figure 3.5(a). Which one will be followed by the system? Let us suppose that a switch can take place from the lowest to the highest curve, and see what the consequences would be. The resulting potential curve $V = \int g\, dh$ is shown in Figure 3.5(b). In the vicinity of the switch point h_s, V is a piecewise linear function of small negative slope to the left and high positive slope to the right. We thus have, as previously seen, perfect

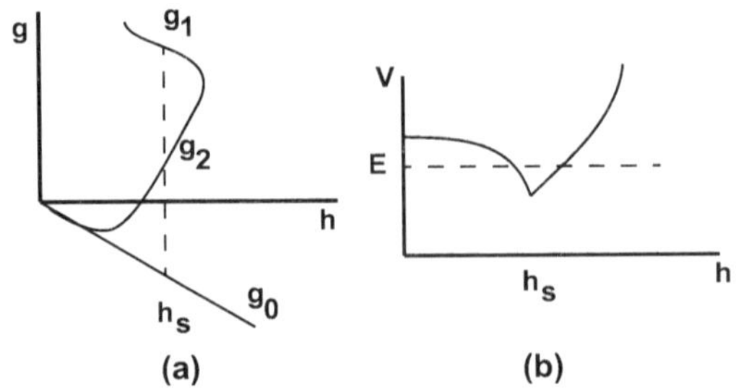

FIGURE 3.5. The Gierer-Meinhardt model.

relative regulation of the size of the smaller, high-h, high-a activated region with respect to the full region. Explicitly, if

$$V = \begin{cases} V_s - |g_0|(h - h_s), & h < h_s, \\ V_s + g_1(h - h_s), & h > h_s, \end{cases} \tag{3.6}$$

then $x_s - x_0 = \left(\frac{\nu}{2}\right)^{1/2} \int_{h_0}^{h_s} (E - V_s + |g_0|(h - h_s))^{1/2} \, dh$, yielding

$$x_s - x_0 = \left(\frac{2\nu}{g_0^2}\right)^{\frac{1}{2}} (E - V_s)^{\frac{1}{2}}, \quad x_1 - x_s = \left(\frac{2\nu}{g_1^2}\right)^{\frac{1}{2}} (E - V_s)^{\frac{1}{2}},$$

and the $h(x)$ pattern scales as well.

To account for a discontinuity in $A(h)$, we must discard another assumption that was made: the vanishing of the activator diffusion coefficient. In fact, if ϵ in the more accurate

$$f(a, h) + \epsilon a'' = 0$$

is small, the possibility arises of a very rapid change in a, leading from one small value of $f(a, h)$ to another.

In the limit as $\epsilon \to 0$, we can assume that $h = h_s$ does not change during this event ($g(a, h) + h'' = 0$ shows that even h' is continuous under a discontinuous change in a), and we may introduce the *stretched variable* X:

$$x = x_s + \epsilon^{\frac{1}{2}} X.$$

Then the *singular perturbation* $\epsilon a''$ (one that raises the differential order of the system) yields, in terms of X, the ordinary equation

$$f(a, h_s) + \frac{d^2 a}{dX^2} = 0.$$

As for the boundary conditions, we of course want to join smoothly onto $a + A_0(h_s)$, and $a = A_1(h_s)$:

$$\frac{da}{dX} = 0 \quad \text{at } a = A_0(h_s), A_1(h_s).$$

The solution is by now routine. In terms of the potential

$$U(a) = \int^a f(a, h_s) da,$$

we have

$$\frac{1}{2}\left(\frac{da}{dX}\right)^2 + U(a) = E',$$

yielding the a-transition profile. For the boundary conditions to be satisfied, we further need $U(A_0(h_s)) = E' = U(A_1(h_s))$, whence

$$\int_{A_0(h_s)}^{A_1(h_s)} f(a, h_s)da = 0.$$

In other words, h_s must be chosen such that the unsigned area between A_0 and A_1 bounded by $f(a)$ and the a-axis vanishes—necessarily a jump between the two outer solutions in the three-solution case. Alternatively, we have the (anti-) Maxwell rule: a jump can only occur when the potential $U(a)$ has two equal maxima. But must the jump occur? Yes, because the A_2-(equivalent to a g_2)-region is dynamically unstable on the local scale $\epsilon^{1/2}$ and so cannot be followed by h: $\dot{a} = \epsilon a'' + f(a, h_s)$ implies $\delta\dot{a} = \epsilon\delta a'' + f_a(a, h_s)\delta a$, or $\Omega = f_a - \epsilon k^2$, which is unstable in the limit if $f_a > 0$.

In the above activator-inhibitor model, for example, at $\mu = 0$, we need

$$0 = \int_0^{a_s}\left(\frac{a^2}{h_s(1+a^2)} - \frac{1}{2}a\right)da = \frac{a_s}{h_s} - \frac{a_s^2}{4} - \frac{1}{h_s}\tan^{-1} a_s,$$

or since $h_s = \frac{2a_s}{(1+a_s)^2}$,

$$\tan^{-1} a_s = a_s \frac{1 + \frac{1}{2}a_s^2}{1 + a_s^2},$$

with a solution $a_s = 0$ or

$$a_s = 1.525, \quad h_s = 0.9171.$$

This is very close to the point $h = 1$, at which the three solutions are reduced to one, suggesting that the system may be controllable by small external signals—we return to this below. A quick estimate of h may also be made by noting that if $f(a)$ is modeled by its generic cubic form, then the condition $\int_{A_0}^{A_1} f(a)da = 0$ is equivalent to the approximation

$$A_2(h_s) \approx \frac{1}{2}(A_0 h_s) + A_1(h_s);$$

here, e.g., this would predict $1/h - \sqrt{1/h^2 - 1} = \frac{1}{2}\left(0 + 1/h + \sqrt{1/h^2 - 1}\right)$, or $h = (8/9)^{1/2} = 0.94$, which is not bad.

Summing up, then, to the extent that $\epsilon \approx 0$, all three-state systems show very much the same asymptotic behavior in the vicinity of the switch point h_s. Taking V_s and x_s in (3.6) conveniently as 0 and integrating, we readily obtain the local solution

$$x > 0: \quad h = h_s + \frac{E}{g_1} - \frac{1}{g_1}\left(x - \frac{(2\nu E)^{\frac{1}{2}}}{g_1}\right)^{\frac{1}{2}},$$

$$x < 0: \quad h = h_s - \frac{E}{g_0} + \frac{1}{g_0}\left(x + \frac{(2\nu E)^{\frac{1}{2}}}{|g_0|}\right)^{\frac{1}{2}},$$

with $L = (2\nu E)^{1/2}(1/g_0 + 1/|g_0|)$, with its obvious relative size regulation. In the large, V will depart from linearity. For the model system at hand, as E rises, V first has a horizontal tangent, and later a vertical tangent. Thus the pattern first shifts to absolute regulation. If the vertical tangent were to occur first, there would be a maximum period for the pattern, which would require repetition at any larger L. The activator pattern is of course the constant $A_1(h_s)$ to the right of the shift, $A_s(h_s)$ to the left, going over to the corresponding functions $A(h)$ for larger excursions.

We have thus far acted as if there can only be one half-cycle at given parameter values, and that the orientation is preordained: $x < 0$ is on the left, low-h side of the h-axis, and $x > 0$ is on the right. Neither statement is correct in the absence of *some* sort of external instructions. Indeed, the point of Gierer and Meinhardt was that a diffusion-reaction system can serve as a very clever amplifier, converting a weak, erratic, quantitative signal to a strong, ordered qualitative pattern. The signal here is to be taken as the activator source μ in (3.5), or the equivalent in analogous systems. The positional dependence $\mu(x)$ is now to supply the instructions for the orientation and period of the developing pattern.

Consider first the initiation of one-dimensional pattern formation in the presence of a simple activator source:

$$\dot{a} = f(a,h) + \epsilon a'' + \mu(x),$$
$$\dot{h} = g(a,h) + \nu h'',$$

or expanding around an initial source-free homogeneous state,

$$\dot{\delta}_a = f_a \delta a + f_h \delta h + \epsilon \delta a'' + \mu(x),$$
$$\dot{\delta}_h = g_a \delta a + g_h \delta h + \nu \delta h'',$$

with

$$\delta a(x,0) = \delta h(x,0) = 0, \qquad \delta a'(0,t) = \delta a'(L,t) = \delta h'(0,t) = \delta h'(L,t) = 0.$$

Time dependence is eliminated via the Laplace transform:

$$\tilde{a}(x,\Omega) = \int_0^\infty e^{-\Omega t} \delta a(x,t)\, dt,$$

leading at once to

$$\Omega \tilde{a} = f_a \tilde{a} + f_h \tilde{h} + \epsilon \tilde{a}'' + \frac{\mu(x)}{\Omega},$$
$$\Omega \tilde{h} = g_a \tilde{a} + g_h \tilde{h} + \nu \tilde{h}'',$$

with

$$\tilde{a}'(0,\Omega) = \tilde{a}'(L,\Omega) = \tilde{h}'(0,\Omega) = \tilde{h}'(L,\Omega) = 0.$$

Then eliminate position dependence via the cosine transform:

$$\hat{a}(k,\Omega) = \int_0^L \cos(kx)\tilde{a}(x,\Omega)dx$$

where $k = p\pi/L$, and p is a nonnegative integer. Thus

$$(\Omega I - Mk^2)\begin{pmatrix}\hat{a}(k,\Omega)\\ \hat{h}(k,\Omega)\end{pmatrix} = \frac{\hat{\mu}(k)}{\Omega}\begin{pmatrix}1\\0\end{pmatrix},$$

where

$$M(k^2) = \begin{pmatrix} f_a - \epsilon k^2 & f_h \\ g_a & g_h - \nu k^2\end{pmatrix}$$

in the notation of Section 3.2.

We now no longer dwell on instability per se, but rather on the time dependence of the reaction to the external source $\mu(x)$. Since

$$\begin{pmatrix}\hat{a}\\ \hat{h}\end{pmatrix} = [\Omega\ \det(\Omega I - M(k^2))]^{-1}\hat{\mu}(k)\begin{pmatrix}\Omega - g_k + \nu k^2\\ g_a\end{pmatrix}$$

with simple poles in Ω at $\Omega = 0$ and at the eigenvalues of $M(k^2)$, the time dependence is dominated by the eigenvalue of $M(k^2)$ of maximum real part (the Ω-Laplace transform of $e^{\Omega' t}(\Omega) = \int_0^\infty e^{(\Omega'-\Omega))t}\,dt = 1/(\Omega' - \Omega)$ gives rise to all poles). As we have previously made use of, the root of maximum real part is real and positive between the values of k^2 for which $\det M(k^2) = 0$: $\Omega_{\max} \geq 0$ for

$$(3.7)\quad \frac{1}{2}\left(\frac{g_h}{\nu} + \frac{f_a}{\epsilon}\right) + \left[\frac{1}{4}\left(\frac{g_h}{\nu} + \frac{f_a}{\epsilon}\right)^2 - \frac{f_a g_h - f_h g_a}{\epsilon\nu}\right]^{\frac{1}{2}} \geq k^2$$

$$\geq \frac{1}{2}\left(\frac{g_h}{\nu} + \frac{f_a}{\epsilon}\right) - \left[\frac{1}{4}\left(\frac{g_h}{\nu} + \frac{f_a}{\epsilon}\right)^2 - \frac{f_a g_h - f_h g_a}{\epsilon\nu}\right]^{\frac{1}{2}}$$

provided the square root is nonnegative. In this region, $\Omega_{\max}$ will have a maximum for k^2 determined by $\frac{\partial}{\partial k^2}\det(\Omega I - M(k^2)) = 0$, or after a little algebra,

$$\max \Omega_{\max} = \frac{\nu}{\nu - \epsilon} f_a - \frac{\epsilon}{\nu - \epsilon} g_h - 2\frac{\sqrt{\epsilon\nu}}{\nu - \epsilon}\sqrt{-f_h g_a}$$

at

$$k^2 = \frac{\nu + \epsilon}{\nu - \epsilon}\sqrt{\frac{-f_h g_a}{\epsilon\nu}} + \frac{g_h - f_a}{\nu - \epsilon}.$$

In any event, expanding about $\Omega_{\max}$, the dominant behavior is given by

$$\begin{pmatrix}\hat{a}\\ \hat{h}\end{pmatrix} \simeq \frac{1}{\Omega - \Omega_{\max}(k^2)}\frac{\hat{\mu}(k)}{\Omega_{\max}(k^2)(2\Omega_{\max}(k^2) - f_a - g_h + (\epsilon + \nu)k^2)}\begin{pmatrix}\Omega - g_k + \nu k^2\\ g_a\end{pmatrix}.$$

If there is only one value of $k = p\pi/L$ in the range (3.7), then only one component of $\hat{\mu}(k)$ will be amplified, and the sign, equivalent to the *polarity* of the ensuing pattern, is fixed by this component. If $\mu(x)$ is removed or altered subsequently, the new effective source must be compared with the exponentially amplified original source, which thereby dominates at least as long as the linearization is valid. If there is more than one value of $k = p\pi/L$ in the critical range, two values can readily have equal or nearly equal $\Omega_{\max}$ leading to a short time pattern that depends upon the details of the source $\mu(x)$. The extreme occurs as $\epsilon \to 0$; then $\max \Omega_{\max}$ becomes the limiting value of a broad band of k^2. One effect is the possibility of simultaneous excitation of many Fourier components; another, associated with the trend to ∞ of the center of the band—or with the singular perturbation nature of the equations—is the creation of a boundary layer configuration at each boundary.

Let us proceed to the asymptotic form in the presence of a source, now in the $\epsilon \to 0$ idealization,

$$f(a,h) + \mu(x) = 0, \quad g(a,h) + \nu h'' = 0, \quad h'(0) = h'(L) = 0.$$

If $a = A(h, \mu)$, the solution of the first equation, is single valued, the resulting

$$\nu h'' + g(h,x) = 0$$

can be solved by guessing

$$\frac{\nu}{2} h'^2 + \int_{h_0(x)}^{h} g(h,x)dh = E(x),$$

where $g(h_0(x), x) = 0$. For this to be true, we have, on differentiating, $\nu h' h'' + h' g(h,x) + \int_{h_0(x)}^{h} g_x(h,x)dh = E'(x)$, or

$$E'(x) = \int_{h_0(x)}^{h} g_x(h,x)dh.$$

In particular, for small μ, we have by Newton-Raphson $A(h,\mu) = A(h) - \mu/f_a(a,h)$, whence

$$g(h,x) = g(a,h) = g(A(h),h) - \mu \frac{g_a(A,h)}{f_a(A,h)}$$

or

$$g_x(h,x) = -\mu'(x) \frac{g_a(A,h)}{f_a(A,h)},$$

so that

$$E'(x) = -\mu'(x) \int_{h_0}^{h} \frac{g_a(A(h),h)}{f_a(A(h),h)} dh.$$

Typically, in auto-cross catalytic systems, one has $g_a/f_a < 0$ ($g_a/f_a = -\lambda$ for the Edelstein model, and $-h/(h-1)$ for the Turing model). Then, for $\mu'(x) > 0$,

$$\frac{dE}{dh} = \frac{E'(x)}{h'(x)}$$

increases when h rises with x, and decreases when h falls. The qualitative effect is clear in both examples: recalling that (with $C = \sqrt{2(E - V_M)/V''_M}$)

$$\chi_{\min} = \left(\frac{\nu}{2}\right)^{\frac{1}{2}} \int_{-C}^{+C} \left(E - \frac{1}{2} V''_M (h - h_M)^2\right)^{-\frac{1}{2}} d(h - h_M) = \pi \left(\frac{\nu}{V''_M}\right)^{\frac{1}{2}},$$

one sees that $\chi_{\min}$ rises rapidly for an $h' > 0$ trajectory, resulting both in minimization of multicycle structure and the preferential placement of high-h regions at larger $\mu(x)$. The multistable case $A_i(h)$ is in fact easier to analyze in full, because we can confine our analysis to the piecewise linear domain of $V(h)$. We must first see how the source affects the switch point h_s and associated parameters. Suppose that the $\mu = 0$ case has been solved, leading to a switch between branches $\bar{A}_0$ and $\bar{A}_1$ at $\bar{h}_s$. Appending μ, which may be taken as a constant during the switch in the $\epsilon \to 0$ limit, we must now solve

$$f(A_0, h_s) + \mu = 0, \quad f(A_1, h_s) + \mu = 0, \quad \int_{A_0}^{A_1} (f(a, h_s) + \mu)da = 0,$$

where

$$f(\bar{A}_0, \bar{h}_s) = f(\bar{A}_1, \bar{h}_s) = \int_{\bar{A}_0}^{\bar{A}_1} f(a, \bar{h}_s)da = 0,$$

or, differentiating with respect to μ at $\mu = 0$,

$$\bar{f}_a^0 \frac{\partial \bar{A}_0}{\partial \mu} + \bar{f}_h^0 \frac{\partial \bar{h}_s}{\partial \mu} + 1 = 0, \quad \bar{f}_a^1 \frac{\partial \bar{A}_1}{\partial \mu} + \bar{f}_h^1 \frac{\partial \bar{h}_s}{\partial \mu} + 1 = 0,$$

$$\int_{\bar{A}_0}^{\bar{A}_1} \left(f_h(a, \bar{h}_s) \frac{\partial \bar{h}_s}{\partial \mu} + 1 \right) da = 0.$$

Hence, dropping the bars,

$$\frac{\partial h_s}{\partial \mu} = -\frac{A_1 - A_0}{\int_{A_0}^{A_1} f_a(a, h_s)da}, \qquad \frac{\partial A_0}{\partial \mu} = -\frac{1 + f_h^0 \frac{\partial h_s}{\partial \mu}}{f_a^0},$$

$$\frac{\partial A_1}{\partial \mu} = -\frac{1 + f_h^1 \frac{\partial h_s}{\partial \mu}}{f_a^1},$$

where superscripts on derivatives denote evaluation at A_0 or A_1. Correspondingly,

$$\begin{aligned} g(A_0, h_s) &= g_0 + \left(\frac{\partial A_0}{\partial \mu} g_a^0 + \frac{\partial h_s}{\partial \mu} g_h^0 \right) \mu(x) \\ &= g_0 + \frac{1}{f_a^0} \left(J^0 \frac{\partial h_s}{\partial \mu} - g_a^0 \right) \mu(x), \end{aligned}$$

where $J = f_a g_h - f_h g_a$, so that the equations of motion in the linear V (piecewise continuous g) regime become

$$h < h_s(\mu(x)) = h_s + \frac{\partial h_s}{\partial \mu} \mu(x) : \quad \nu h''(x) = \gamma_0(x),$$

$$h > h_s(\mu(x)) = h_s + \frac{\partial h_s}{\partial \mu} \mu(x) : \quad \nu h''(x) = \gamma_1(x),$$

$$h'(0) = h'(L) = 0, \quad h, h' \quad \text{continuous},$$

where

$$\gamma(x) = -g + \frac{g_a - J\frac{\partial h_s}{\partial \mu}}{f_a}\mu(x),$$

and $\gamma_{0,1}$ here refers to different regions of h. We note that for the typical auto-cross catalytic activator-inhibitor mechanism, $g_a/f_a < 0$, while by our convention, $-g_0 > 0$ and $-g_1 < 0$. Suppose there are p half-cycles in the pattern, starting say with low h at $x = 0$ and with switch points $x_1, x_2, \ldots, x_p$. On integrating the above from $x = 0$, we clearly have for $x_i < x < x_{i+1}$ (here the subscript on γ indicates the region):

$$\nu h'(x) = \int_0^{x_1} \gamma_0(y)dy + \int_{x_1}^{x_2} \gamma_1(y)dy + \cdots + \int_{x_i}^{x} \gamma_{[i]}(y)dy,$$

where $[i] = i \mod 2$. Since $h(x_{i+1}) - h(x_i) = (\partial h_s/\partial\mu)(\mu(x_{i+1}) - \mu(x))$, we further have on integrating from x_i to x_{i+1}, using $\int_a^b \int_a^x f(y)dy\,dx = \int_a^b (b-y)f(y)dy$, and dividing by $x_{i+1} - x_i$,

$$\int_0^{x_1} \gamma_0(y)dy + \int_{x_1}^{x_2} \gamma_1(y)dy + \cdots + \int_{x_{i-1}}^{x_i} \gamma_{[i-1]}(y)dy$$
$$+ \int_{x_i}^{x_{i+1}} \frac{y - x_i}{x_{i+1} - x_i}\gamma_{[i]}(y)dy = \nu\frac{\partial h_s}{\partial \mu}\frac{\mu(x_{i+1}) - \mu(x_i)}{x_{i+1} - x_i},$$

to which the condition $h'(L) = 0$ must be appended:

$$\int_0^{x_1} \gamma_0(y)dy + \int_{x_1}^{x_2} \gamma_1(y)dy + \cdots + \int_{x_p}^{L} \gamma_{[p]}dy = 0.$$

Successive subtraction then yields the complete set of conditions on the x_i. With $p > 1$ we have

$$\int_0^{x_1} \gamma_0(y)dy + \int_{x_1}^{x_2} \frac{x_2 - y}{x_2 - x_1}\gamma_1(y)dy = \nu\frac{\partial h_s}{\partial \mu}\frac{\mu(x_2) - \mu(x_1)}{x_2 - x_1} \sim \nu\frac{\partial h_s}{\partial \mu}\mu',$$

$$\int_{x_{i-1}}^{x_i} \frac{y - x_{i-1}}{x_i - x_{i-1}}\gamma_{|i-1|}(y)dy + \int_{x_i}^{x_{i+1}} \frac{x_{i+1} - y}{x_{i+1} - x_i}\gamma_{|i|}(y)dy$$
$$= \nu\frac{\partial h_s}{\partial \mu}\left(\frac{\mu(x_{i+1}) - \mu(x_i)}{x_{i+1} - x_i} - \frac{\mu(x_i) - \mu(x_{i-1})}{x_i - x_{i-1}}\right) \sim 0,$$

$$\vdots$$

$$\int_{x_{p-1}}^{x_p} \frac{y - x_{p-1}}{x_p - x_{p-1}}\gamma_{|p-1|}(y)dy + \int_{x_p}^{L} \gamma_{|p|}(y)dy$$
$$= -\nu\frac{\partial h_s}{\partial \mu}\frac{\mu(x_p) - \mu(x_{p-1})}{x_p - x_{p-1}} \sim -\nu\frac{\partial h_s}{\partial \mu}\mu',$$

where the form attained at constant $\mu'(x)$—linear μ—has been indicated by $\sim$. On the other hand, for $p = 1$, no repetition, we have just

$$\int_0^{x_1} \gamma_0(y)dy + \int_{x_1}^{L} \gamma_1(y)dy = 0.$$

There are now several observations to be made. For this purpose, we suppose that $f_h < 0$ so that $\partial h_s/\partial \mu > 0$, but $f_a < 0$, $g_a > 0$, and $g_h < 0$ so that $(g_a - J\ \partial h_s/\partial \mu)/f_a > 0$, while $-g_0 > 0$, $-g_1 < 0$, and $g_1/-g_0 \gg 1$. Thus, as $\mu(x)$ increases from 0, γ_1 becomes increasingly negative, while γ_0 starts positive and then goes increasingly negative. First then, for μ' large but still in the context of a first-order calculation, γ_0 and γ_1 are both negative, and neither the $p > 1$ nor $p = 1$ consistency conditions are solvable: there can be no switch point at all. Next, if μ' is smaller, so that $\gamma_0 > 0$ for small x, but $\nu\mu' > \max_x \int_0^x \gamma_0(y)dy > 0$, the first of the $p > 1$ conditions still cannot be satisfied, but we have a unique $p = 1$ solution. Finally, for smaller μ', $p > 1$ solutions become possible. Indeed, if $\mu(x)$ is so small that $\gamma_0 > 0$ throughout, one can in principle create an arbitrary number of switch points in this $\epsilon \to 0$ limit, just as in the case of pattern initiation. However, for finite ϵ, the resolution of the pattern is limited by the physical distance required to make the transition between $A_0(h_s)$ and $A_1(h_s)$. We can certainly estimate this minimum length as

$$\Delta x = \frac{A_1 - A_0}{da/dx}.$$

In the cubic approximation, $f = -\alpha(a-A_2)^3 + \beta(a-A_2)$, we have $A_1 = A_2(\beta/\alpha)^{1/2}$ and $f_a(A_1, h_s) = -2\beta$, while

$$\frac{1}{2}\epsilon\left(\frac{da}{dx}\right)^2\bigg|_{\Delta_2} = \int_{A_2}^{A_1} f(a, h_s)da = \frac{1}{2}\frac{\beta^2}{\alpha}.$$

Hence

$$\Delta x = 4\left(\frac{\epsilon}{-f_a'}\right)^{\frac{1}{2}},$$

the approximate minimum distance permitted for a half-cycle.

3.5. Some Applications

We proceed to applications of the reaction-diffusion mechanism. Ours will be developmental, but general methodological applications have a considerable history, e.g., from the viewpoint of nonequilibrium thermodynamics. In the Glansdorff-Prigogine work [**70**], a fluctuation-stability criterion was used, which is readily described. Consider a unidirectional reaction k, with reactants ξ_α. The nature of the reaction is given by the *stoichiometric coefficients* $\nu_{k\alpha}$, the number of α molecules produced in a unit reaction ($\nu_{k\alpha} < 0$ for the molecules that are used up). The free energy change of the system due to this unit reaction is then

$$\Delta\phi_k = \kappa T \sum \nu_{k\alpha} \ln\left(\frac{\xi_\alpha}{A_\alpha}\right)$$

with $A_\alpha(P,T)$ the characteristic concentration of species α. If the "velocity" of the unit reaction is given by u_k transformations for unit time, the system free energy then changes at the rate

$$\dot{\phi} = \kappa T \sum u_k \nu_{k\alpha} \ln\left(\frac{\xi_\alpha}{A_\alpha}\right).$$

As we know, diffusion can be included as a monomolecular reaction. A state is now in stable equilibrium if any nearby state gives up free energy:

$$\kappa T \sum \delta u_k \nu_{k\alpha} \delta \ln \xi_\alpha < 0.$$

Now the chemical kinetics

$$\dot{\xi}_\alpha = F_\alpha(\xi)$$

is determined by

$$F_\alpha = \sum u_k \nu_{k\alpha},$$

so that the above criterion becomes

$$\sum \delta F_\alpha \delta \ln \xi_\alpha < 0.$$

Written as $\sum \frac{\partial F_\alpha}{\partial \xi_\beta} \frac{1}{\xi_\alpha} \delta\xi_\alpha \delta\xi_\beta < 0$, this is equivalent to the condition

$$\text{(3.8)} \qquad \frac{1}{\xi_\alpha}\frac{\partial F_\alpha}{\partial \xi_\beta} + \frac{1}{\xi_\beta}\frac{\partial F_\beta}{\partial \xi_\alpha} \quad \text{is negative definite.}$$

We know directly from the chemical kinetics that stability is equivalent to the condition

$$\text{eigenvalues of } \frac{\partial F_\alpha}{\partial \xi_\beta} \text{ have negative real parts.}$$

Indeed, (3.8) is a sufficient condition. This follows from a general criterion of Lyapunov (here "+" denotes conjugate transpose):

> M is stable—all eigenvalues have negative real parts—if and only if there exists a positive semidefinite Q such that $QM + M^+Q$ is negative definite.

The "if" part is all we need and is very easy to prove. If M is not stable, it has an eigenvalue with $\mathrm{Re}(A) \geq 0$. Then $M\psi = \lambda\psi$, and $\psi^+M^+ = \lambda^*\psi^+$, where ψ is the corresponding eigenvector, so that for any positive semidefinite Q, we have $\psi^+(QM + M^+Q)\psi = (\lambda + \lambda^*)\psi^*Q\psi \geq 0$. Hence $QM + M^+Q$ cannot be negative definite. In the criterion (3.8), the special choice $Q_{\alpha\beta} = \delta_{\alpha\beta}/\xi_\alpha$ is made; while it can be used to prove stability, it cannot serve to prove instability, and counterexamples are readily found. We now proceed to developmental examples, restricting attention of course to systems for which quantitative data has been obtained. Since we have seen that boundary conditions play a crucial role in determining the structure of the time-asymptotic state, as well as in pattern initiation, we will focus on pattern-forming events in which the spatial domain is presumed to be the whole organism, suggesting that the standard 0-flux boundary conditions are appropriate.

Slime Mold: The biological problems are to explain the bistability of cells, prespore-prestalk conversion only at the boundary, the carrying of the prevailing spore/stalk ratio information to all cells, and the maintenance of polarity in the development of all stalk or all spore. An activator-inhibitor reaction-diffusion model is consistent with these observations, as well as with the further quantitative data, if (i) there is a weak source density decreasing from the anterior end, due to a weak sorting out during aggregation, with a more stalky front, (ii) activation $\Rightarrow$ prestalk, unactivated $\Rightarrow$ prespore, (iii) the activator transition is phenotypically prestalk.

Thus, for the volume dependence of cell-type ratio, we assume constant activator transition thickness d, and mean cross section $\mathcal{A}$ with a volume dependence $\mathcal{A} = V^{2/3}/\beta d$. Then for stalk regulation ratio α, we have the volume ratio,

$$\begin{aligned}\frac{\text{stalk}}{\text{spore}} &= \frac{\alpha(V - d\mathcal{A}) + d\mathcal{A}}{(1-\alpha)(V - d\mathcal{A})} \\ &= \frac{\alpha}{1-\alpha} + \frac{1}{1-\alpha}\frac{d\mathcal{A}/V}{1 - d\mathcal{A}/V} \\ &= \frac{\alpha}{1-\alpha}\left(1 + \frac{1}{\alpha}\quad\frac{1}{\beta V^{1/3} - 1}\right),\end{aligned}$$

which is of course larger at small V.

While the activator diffusion ϵ thus determines the asymptotic volume dependence, regeneration in this model is primarily controlled by inhibitor diffusion. Consider, for example, truncation of the head end, from L to ℓ. The initial effect is reduction of inhibitor, which then diffuses and moves the switch point to the rear. Explicitly, the $0 - L$ equilibrium pattern h_0 can be restricted to the region $0 - \ell$, maintaining zero flux boundary conditions $h'(0) = h'(\ell) = 0$ by writing

$$0 = g_0 + (g_1 - g_0)\epsilon(h_0 - h_s) + \nu h_0'' + \nu h_0'(\ell)\delta(x - \ell),$$

since integration from $\ell - \epsilon$ to $\ell + \epsilon$ yields $h_0'(\ell^+) - h_0'(\ell^-) + h_0'(\ell) = 0$. Regeneration then takes place from this established initial condition:

$$\dot{h} = g_0 + (g_1 - g_0)\epsilon(h_0 - h_s) + \nu h'' + \nu h_0'(\ell)\delta(x - \ell)\epsilon(-t),$$

and the difference pattern then satisfies

$$\Delta\dot{h} = \nu\Delta h'' + (g_1 - g_0)\big(\epsilon(h_0 - h_s + \Delta h) - \epsilon(h_0 - h_s)\big) - \nu h_0'(\ell)\delta(x - \ell)\epsilon(t),$$

diffusion with two sources. Consider the right-hand initiating source only, and carry out the Laplace transform. Since $\Delta h(x, 0) = 0$, then

$$\Omega\Delta h - \nu h'' - \frac{1}{\Omega}\nu h_0'(\ell)\delta(x - \ell);$$

i.e., $\nu\Delta h'' = \Omega\Delta h$ with right boundary condition $\Delta h'(\ell, \Omega) = -h'(\ell)/\Omega$. Since Δh must decay to the left, this is solved immediately as

$$\Delta h(x, \Omega) = -\left(\frac{\nu}{\Omega^3}\right)^{\frac{1}{2}} h_0'(\ell)e^{-(\Omega/\nu)^{\frac{1}{2}}(\ell - x)},$$

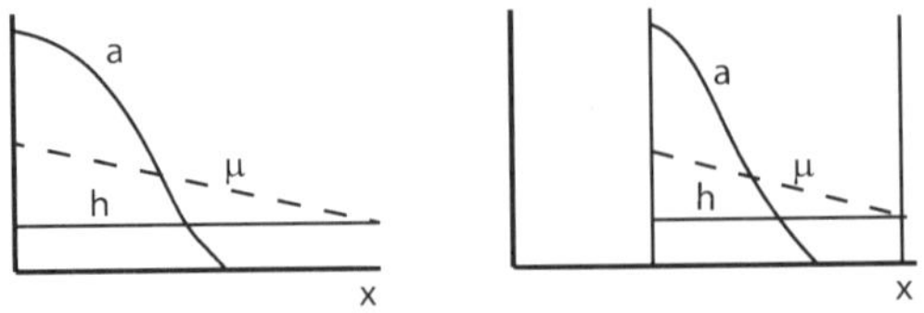

FIGURE 3.6. Regulation of hydra under surgery.

or on inverse transforming and retaining only the leading term in t,

$$\Delta h(x,t) = \frac{4}{\sqrt{\pi}} h_0'(\ell) \frac{\nu^2 t}{(\ell - x)^3} e^{-\frac{(\ell-x)^2}{4\nu t}} + \cdots .$$

It is clear that the perturbation "arrives" at the old switch point x_s at a time of the order

$$t \sim \frac{(\ell - x)^2}{\nu},$$

a characteristic diffusional dependence.

Hydra: This is undoubtedly the most heavily studied system, and results relating to head formation are consistent with a simple activator-inhibitor mechanism, with a weak activator source decreasing from head to foot. If the system is small enough, the idealized switch point behavior should be appropriate; including an a transition region, one should have the type of regulation shown in Figure 3.6, in accord with both biological regeneration experiments and numerical solution of the equations.

For large enough systems, the low a equation

$$\dot{h} = -\lambda h + \nu h''$$

would have an equilibrium solution $h \propto e^{-(\lambda/\nu)^{1/2} x}$ with an intrinsic range of inhibition $\Delta x \sim (\nu/\lambda)^{1/2}$, but this is probably out of the range of most experiments. The most interesting studies are those of axial grafts of hydra rudiments, assessed by the number of heads then appearing. The asymptotic pattern here is largely a consequence of the initial pattern of activator source, a three half-cycle asymptotic regime resulting from a sufficiently large internal peak. Typical numerical results are again very much in accord with experiment. Another series of experiments involves intermediate head removal, and these agree with the diffusional decay mechanism we discussed in slime mold.

Hydromedusae: Here we recall the phenomenon of large eggs going directly to medusa, medium ones after some interim rearrangements, and small ones becoming polyps that bud medusae. This is certainly a (two- or) three-dimensional system, and may be substantially geometry dependent. If the reaction-diffusion class of models is to be relevant, it is clear that one would want unactivated polypoid tissue, activated medusoid. In the asymptotic state, with an idealized switching mechanism,

$$\nu \nabla^2 h = \begin{cases} -|g_0| & \text{for } h < h_s, \\ g_1 & \text{for } h > h_s, \end{cases}$$

an electrostatic analogue offers an intuitive entering wedge, with $\partial h/\partial\mu = 0$ at the boundary corresponding to 0 external dielectric constant. In any event, direct activation for a large system is easy to explain, but there is no hint as to whether eventual "hot spot" activation of a small egg is a consequence of a change in geometric form via dimpling, change of boundary conditions, localized inhibitor degradation, and so on.

Vascular Patterns: An activated region, once established, serves to induce differentiation and a consequent change in cell properties. This is most easily modeled by something like the Edelstein switch. The resulting new source pattern can then induce further differentiation, resulting in the temporal buildup of a complex pattern. An example is the Meinhardt model for two-dimensional growth of the vascular pattern of a leaf [**118**]. Here a binary branching results naturally from an activator depletion system, lateral branching with activator-inhibition. Consider the former, modeled by

$$\begin{aligned} \dot{A} &= cA^2S - \mu A + D_a\nabla^2 A, \\ \dot{S} &= c_0 - cA^2S - \gamma S - \epsilon YS + D_s\nabla^2 S, \\ \dot{Y} &= dA - eY + \frac{Y^2}{1+fY^2}, \end{aligned}$$

imagined as residing on an array of cells. Y is essentially an Edelstein switch, operating at $A = 0$ between extreme values

$$Y_0 = 0, \qquad \frac{1+\sqrt{1-4e^2f}}{2ef},$$

or in the presence of A as

$$Y_A = Y_0 + A\frac{d}{Y_0 - 2e} + \cdots .$$

Here high A triggers high Y cell differentiation, which is maintained; a high Y cell now causes a local drop in S and consequently in A. Meanwhile, S, and hence A as well, move on, inhibited from behind and by any boundaries, but able to go into one or two channels, leading to a typical branching of differential cells.

Contour Control: It is also possible to construct a switch to react to a specific *level* of morphogen, so that a pattern can form on an internal contour. A simple one is given by the reaction sequence

$$P + S + SM \xrightarrow{k_1} MPS + S \xrightarrow{k_2} PSS,$$

with control exercised by the external S level (see Figure 3.7). Equilibrium concentrations of the three binding states peak at very different S concentrations, so that $[PS]$ can be used for internal contour construction. More generally, if, e.g., two morphogens S and T are bound, one or two at a time, eight rate constants are needed, but if $[PST]$ is the active complex, a surface can be created by a balance of S and T concentrations. Bone shapes with sockets have been computationally created in this way by MacWilliams [**116**] from simple sources.

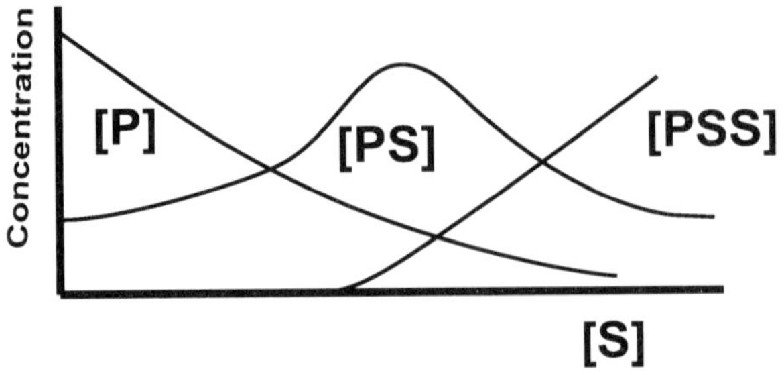

FIGURE 3.7. Multi-level control.

Meinhardt [**119**] has proposed what might be termed a biochemical analog-digital converter, in which a whole sequence of switches reacts to graded levels of, say, an inhibitor h, and used this to model the compartmentalization in insect development:

$$\dot{q}_i = c_i\left(\frac{q_i^2}{r}\right)\left(1+\frac{h}{r}\right) - \frac{\alpha}{1+\gamma q_i^2}q_i,$$
$$\dot{r} = \sum c_i q_i^2 - \beta r.$$

This has the property that as h rises to its final value, successive q_i reach their maxima and shut off, leaving just one activated q_i at each value of h.

We have given only a sample of patterns and pattern-forming mechanisms available in really quite rudimentary systems. Whether or not it would be wise to organize these under the aegis of some expanded catastrophe theory is unclear, but then of course, nature has done it in her own way.

CHAPTER 4

Differential Adhesion and Morphogenesis

We turn now to some questions in development which in one way or another involve the relative *movement* of cells or cell aggregates. There are a number of ways one might attempt to explain such dramatic developmental movements as gastrulation, but we shall begin by considering a more restricted question, which seems to have little to do with cell movement. Quite apart from the sequence of developmental events that go into the formation of a certain structure such as an organ, one may ask: what properties of the structure are in fact fully determined by the constituent cell types? Such a question implies differentiation has already occurred, as an instance of pattern formation for example, but we are also led to think about local interactions between cells, rather than in terms of the global organization implicit in models based upon reaction-diffusion equations. More important, by focusing on the properties of specific cell types one can gain an experimental tool: the artificial manipulation of dissociated cells and their aggregates.

Sponge cells, for example, can be easily obtained from sponge fragments by squeezing them through fine gauze. Two types of sponge cell in a common medium are observed to segregate and form clumps consisting of the same type of cell. Evidently there is an affinity for like cells in this experiment, although it is not obvious whether the movement is directed and cell-cell recognition is unimportant (as might be the case if there were chemotactic signaling) or rather that cell-cell affinities are paramount and movement random. In the 1940s Holtfreter [**91, 92**] studied the amphibian embryo following gastrulation into the neurula stage. Dissociation of cells was accomplished by raising the pH of the medium. The cells were observed, when mixed together in various combinations at a lower pH, to join together and form clumps as in sponge cell aggregation. However, once the clump was formed the cells began to sort themselves out in a manner consistent with their positions in the original tissue. When combined with endoderm and mesoderm, cells of ectoderm and epiderm moved out to the exterior of the clump. Mesoderm and neural plate cells moved within a clump of epidermis or endoderm. Masses of neural plate cells tended to form structures resembling brain vesicles. Evidently in this case there are cell-specific affinities involved in the aggregation process. Using enzymes to dissociate older embryos, other investigators have been able to observe sorting out into close approximations of highly developed structures. A significant self-organizing ability seems to be independent of developmental path.

4.1. Cell Sorting by Differential Adhesion

Using cells involved in organogenesis in the chick, Steinberg and others (see, e.g., [**141, 142**]) have observed selective affinities in an intriguing shish kebab experiment. Two clumps of cells, one type A, the other type B, are impaled on a glass thread. It is then found that, for example, B engulfs A; that is, B-cells form a clump which partially or entirely surrounds a clump of A-cells. We then write $B \succ A$. Experiments then reveal the transitive ordering liver $\succ$ neural tube $\succ$ heart $\succ$ epithelium $\succ$ precartilage $\succ$ epidermis. The process is very reminiscent of surface tension phenomena in fluids, and Steinberg has suggested that this analogy can be made precise through the concept of adhesive energy. As cells of one type aggregate, adhesive energy increases at the expense of free surface potential of the cells:

$$E_{\text{ad}} + V_{\text{fs}} = \text{const.}$$

Then aggregation can be viewed as a process that maximizes E_{ad} or minimizes V_{fs}. In surface tension phenomena the latter condition leads to minimal area conditions. If a sheet of cells (not a monolayer) is stretched very slowly (to eliminate dissipative losses) the work done by the surface tension is stored as new free surface potential.

For cells of more than one cell type the situation is more complicated since unlike cells may adhere and contribute to E_{ad} even though this contribution is not obviously related to the mutual adhesion and surface potential of individual cell types. For this reason it is simplest to first center the discussion on adhesive energy. Suppose that cells of types A and B have adhesive energies E_{AA}, E_{AB}, and E_{BB} per unit area, and assume all cells are the same shape and size. If two interior cells are interchanged between a clump of A-cells and a clump of B-cells, total adhesive energy will be incremented (assuming unit contact area per cell) by $2E_{AB} - E_{AA} - E_{BB}$. If this quantity is nonnegative, E_{ad} will be maximized by continuing this process until complete intermixing maximizes A-B contacts. If this quantity is negative, on the other hand, sorting out the cells into clumps of like cells should occur.

To investigate this we first relate surface tension T to adhesive energy. Consider A-cells suspended in a neutral medium (the "outside"), with the property that the adhesive energy of contact between the A-cells and the outside, E_{AO} say, vanishes. The slow extension of a sheet of A-tissue by an amount ΔX then requires work $2T_{AO}$ per unit length, the tension in the A-cell-to-outside interfaces, on each of the two sides of the sheet, being T_{AO}. The free surface increases by $2\Delta X$ and therefore ΔX bonds must be broken. Thus

$$2T_{AO}\Delta X = E_{AA}\Delta X,$$

and so $T_{AO} = \frac{1}{2}E_{AA}, T_{BO} = \frac{1}{2}E_{BB}$. To extend an A-B interface, an amount ΔX requires the breaking of $\Delta X/2$ of the A-A and B-B bonds, but results in ΔX of the A-B bonds. Therefore

$$T_{AB} = \frac{1}{2}(E_{AA} + E_{BB}) - E_{AB} = T_{AO} + T_{BO} - E_{AB}.$$

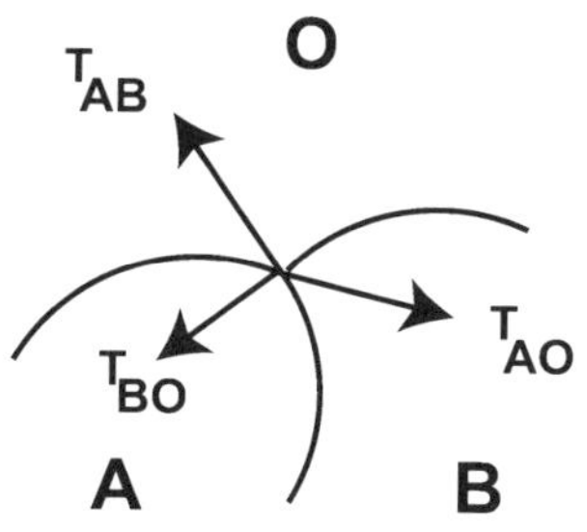

FIGURE 4.1. Force balance at a triple point.

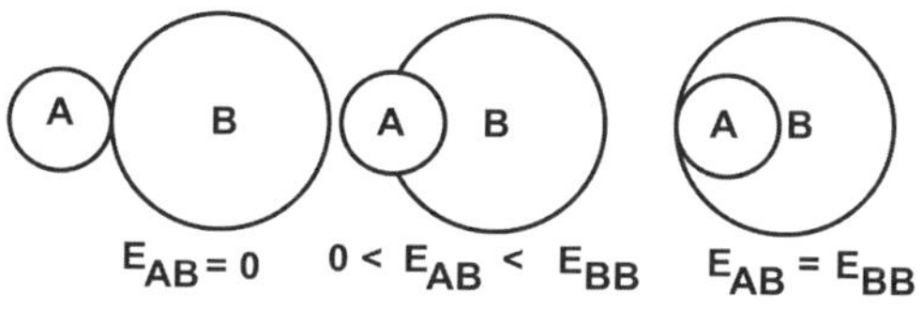

FIGURE 4.2. Engulfment.

Note that if "B" were empty and $E_{BB} = E_{AB} = 0$, this reduces to the surface tension at an A-O boundary.

Returning to the second case of sorting out into clumps of like cells, we have $T_{AB} > 0$, and we may assume that the clumps are in contact. Also, we take the cells to be so small compared to clump size that a continuum treatment is possible, so the problem is one of coalescence of fluid drops. In the manner of soap bubble equilibrium, the structure is built up of spherical surfaces of various radii. At the triple point (see Figure 4.1) the surface tension vectors form an equilibrium triad. As the clumps move apart, just at separation, vectors T_{AO} and T_{BO} rotate into coincidence so we have

$$T_{AO} + T_{BO} = T_{AB},$$

and therefore $E_{AB} = 0$.

If, on the other hand, engulfment occurs, the process may be total or partial; see Figure 4.2.

We shall now reserve the notation $(B \succ A)$ to mean that B engulfs A *totally*. Then as engulfment is reached, T_{AB} rotates into coincidence with T_{B0} and $T_{AB} + T_{BO} = T_{AO}$, so that the equilibrium implies $E_{BB} = E_{AB}$. Thereafter $E_{AB} > E_{BB}$ and *total engulfment* ensues. Note that after engulfment the position of the A clump within the B clump is arbitrary insofar as adhesive energy is concerned. There is no implied centering. Steinberg notes that eccentric positions are often observed experimentally. *Partial engulfment* of A by B occurs if $T_{AB} > 0$ and $0 < E_{AB} < E_{BB}$.

Let us consider the issue of transitivity of $\succ$: if $C \succ B$ and $B \succ A$, we now have the necessary and sufficient conditions

$$\frac{1}{2}(E_{AA} + E_{BB}) > E_{AB}, \quad E_{AB} > E_{BB}, \tag{4.1}$$

$$\frac{1}{2}(E_{BB} + E_{CC}) > E_{BC}, \quad E_{BC} > E_{CC}. \tag{4.2}$$

Note that (4.1) implies

$$0 < \frac{1}{2}(E_{AA} + E_{BB}) - E_{AB} < \frac{1}{2}(E_{AA} - E_{BB}),$$

and so $E_{AA} > E_{BB}$. In fact, $E_{AA} > E_{BB}$ and the second of (4.1) imply the first of (4.1), so this pair of of inequalities also comprises necessary and sufficient conditions for total engulfing of A by B.

Note that (4.1) and (4.2) together are seen to imply $E_{AA} > E_{CC}$ but no more, since we have absolutely no control over E_{AC}. But if it were known that engulfing *occurs* between A and C, the implication is that C engulfs A. Hence transitivity of $\succ$ is true only among cell types for which complete engulfing occurs between any pair of types.

The theory is considerably simplified if it is possible to express E_{AB} in terms of E_{AA} and E_{BB}, although this may be unrealistic biologically. Steinberg notes that if adhesion occurs between sites that are the same for all cell types but distributed with different frequencies over the membranes, then for sparse distributions η_i we shall have

$$E_{ij} = k\eta_i\eta_j, \quad i, j = A, B, C, \ldots .$$

Then necessarily

$$\frac{1}{2}(E_{AA} + E_{BB}) > E_{AA}^{\frac{1}{2}} E_{BB}^{\frac{1}{2}},$$

and so the cell type with the larger adhesion is completely engulfed and transitivity in $\succ$ occurs. Adhesion energy will then be maximized for a large number of cell types by an "onion" configuration of cell layers, all boundaries being spheres, not necessarily concentric, with cell layers ordered into the onion in ascending values of the adhesive energy as well. Other, perhaps more realistic models of adhesion will be presented below in Section 5.1.

It is helpful to see how these ideas work out when the cells are not considered small and a discrete model is used, this being the approach of most numerical simulations. Let cell types $1, 2, \ldots, M$ be given, and let

$$\begin{cases} N_i, & \text{number of } i\text{-cells}, \\ S_i, & \text{area of one } i\text{-cell}, \\ N_{ij} = N_{ji}, & \text{number of } i\text{-}j \text{ contacts}, \\ S_{ij} = S_{ji}, & \text{area of one } i\text{-}j \text{ contact}, \\ E_{ij} = E_{ji}, & \text{adhesive energy per unit area of the } i\text{-}j \text{ contact}. \end{cases}$$

Then

$$E_{\text{ad}} = \sum_{i,j\leq i} N_{ij} S_{ij} E_{ij}.$$

We also have M conservation laws for the distribution of given cell area among the various contacts:

$$N_i S_i = \text{total area of } i\text{-cells} = 2N_{ii} S_{ii} \sum_{j\neq i} N_{ij} S_{ij}.$$

Using this to eliminate

$$N_{ii} S_{ii} = \frac{1}{2} N_i S_i - \frac{1}{2} \sum_{j\neq i} N_{ij} S_{ij}$$

from the expression for E_{ad}, we have

$$\begin{aligned} E_{\text{ad}} &= \frac{1}{2} \sum_{i} N_i S_i E_{ii} - \frac{1}{2} \sum_{i} \sum_{j\neq i} N_{ij} S_{ij} E_{ij} + \sum_{i,j<i} N_{ij} S_{ij} E_{ij} \\ &= \frac{1}{2} \sum_{i} N_i S_i E_{ii} + \frac{1}{2} \sum_{i,j<i} N_{ij} S_{ij} \left(E_{ij} - \frac{1}{2} E_{ii} - \frac{1}{2} E_{jj} \right) \end{aligned}$$

and so

$$E_{\text{ad}} = K - \sum_{i,j<i} N_{ij} S_{ij} T_{ij} = K - V_{\text{fs}},$$

where

$$T_{ij} = \frac{1}{2}(E_{ii} + E_{jj}) - E_{ij}$$

is recognized as a surface tension associated with i-j contacts, and K is a constant independent of configuration. To maximize E_{ad} we therefore minimize the surface potential

$$V_s = \sum_{i,j<i} N_{ij} S_{ij} T_{ij} = \sum_{i,j<i} \int_{I_{ij}} T_{ij}\, ds$$

where I_{ij} are the interfaces between i- and j-cells. This is the discrete version of the minimization problem for fluid equilibria involving interfaces. It is definitely nontrivial in the discrete case but the general implications are clear and the fluid analogue offers a helpful guide.

So far we have described a static theory of optimal equilibria that is certainly highly simplified for even the most elementary structures. What is left out is the *path* and time evolution of formation of the structure from an initial state. But even if motility of the cells is such that E_{ad} never decreases, the pattern may reach an equilibrium that is stable locally but not globally, and therefore not optimal. Suppose, for example, that B engulfs A and we have a deformation as shown in Figure 4.3(a, b, c). The arrangement of Figure 4.3(a) is locally stable to small finite deformations and is therefore a possible end state of a sorting process. To continue toward optimality (maximum E_{ad}) the two A spheres must interact nonlocally, as in Figure 4.3(b, c). This could be the start of a hierarchical process where aggregates of cells can also move. If for the same case ($B > A$) we consider the arrangement of Figure 4.3(d), we again have a locally and finitely stable state, since $T_{AB} > 0$.

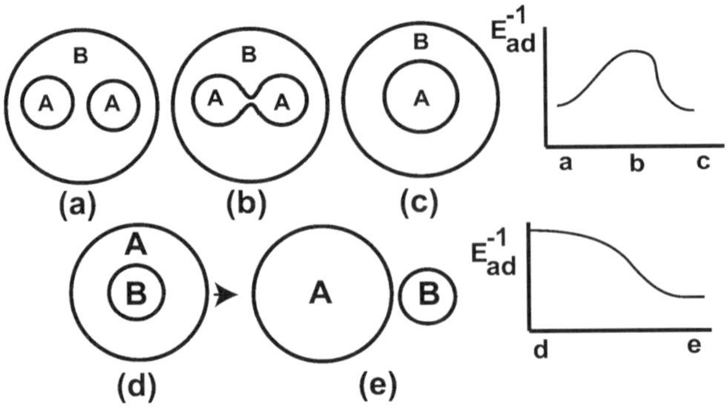

FIGURE 4.3. Failure of local sorting process.

For the deformation shown in Figure 4.3(e), however, E_{ad} is increased. Thus in this case the initial state would not have been reached by a cell-sorting process.

4.1.1. Kinematics of Sorting. These ideas bring us to a basic problem: how does the cell movement affect or determine the final structure, given local minimization of E_{ad}? The question can be studied as a kinematic problem, the sequence of states being treated without regard to the sequence of times at which they actually occur in a dynamic realization of the process.

The problem has been studied numerically by taking cells to move on a lattice and assigning certain motility rules. Antonelli, Rodgers, and Willard (see [**21**]) examine an exchange rule that interchanges nearest-neighbor cells provided that such a move does not decrease E_{ad}, a given step cycling through all cells. Working with two-dimensional square cells $B \succ A$, they find that in general only partial sorting out occurs. Eventually the only moves are 0-moves, leaving E_{ad} unchanged, with aggregates of A-cells convoluted between aggregates of B-cells.

What is apparently missing here is some long-range interaction between cells. This could be accomplished quite easily by chemotactic means (see Chapter 6) or by random aggregation. But these methods neglect a simpler fact, that resistance to compression or dilation of an incompressible deformable continuum creates long-range pressure forces. One would therefore like to allow the long-range exchange; i.e., an A-cell changes to a B-cell while simultaneously a distant B-cell changes to an A-cell. Goel and Rogers [**77**] have extended a set of motility rules developed by Leith and Goel (see [**21**]) and formulated a remarkably simple system for moving two or more cell types, which effectively embodies both moderate- and long-range interaction (the former being an attempt to model the extent of a localized stress field in a cell aggregate). Working again with two-dimensional square cells, they define, for the ith cell, an energy

$$E_i = \sum_{j=1}^{D} \sum_{k=1}^{8j} K_j E_{ijk},$$

where j refers to a series of D concentric layers of cells about i, where D is an integer determining the range of the near interactions, K is a numerical weight for the jth layer, and E_{ijk} is the energy of adhesion for the kth cell of the jth layer when

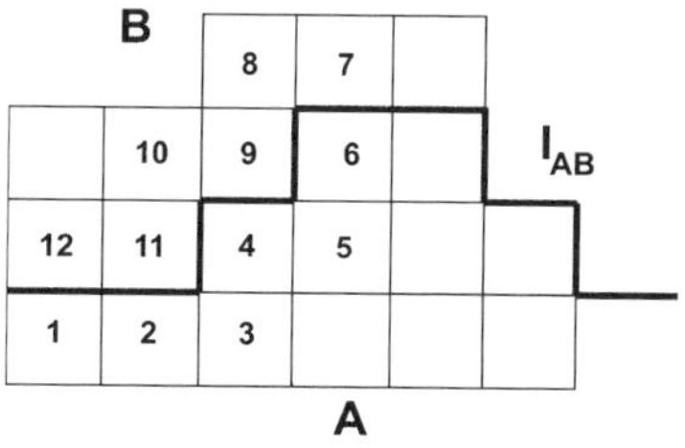

FIGURE 4.4. Effect of weak non-locality.

brought into contact with cell i (all cells are isotropic). Note that this may involve movement of a cell to make the contact.

The total energy is defined by summing over the E_i, giving an integral multiple of the usual adhesive energy owing to the multiple counting of a given edge. Cells are "moved" on the basis of the values E_i in the following way: the only cells considered are those that are within the range of influence (as determined by D) of a cell of a different type. Effectively only cells in the vicinity of the interfaces I_{ij} are considered. For all such cells one computes the increment in E_i when the given cell changes type. Thus if the interface I_{AB} is being examined, the ΔE_i for an A-cell is obtained by changing it to a B-cell and taking the difference of the two E_i's. Two lists are prepared, one for $A \to B$ conversions, the other for $B \to A$ conversions, in descending order of ΔE_i. The two lists are added and positive entries, corresponding to pairwise conversions that increase adhesive energy, are used in the move. Note that such conversions can be very long range when interpreted as "moves," since an acceptable interchange may be between distant cells along the interface. When more than one interface is present, the above procedure is repeated for each one, in descending order of the interfacial surface tension T_{ij}.

To illustrate how this works, consider the configuration of A- and B-cells shown in Figure 4.4, and suppose $D = K_1 = 1$, $E_{AA} = 3$, and $E_{AB} = 1$, $E_{BB} = 2$.

For an $A \to B$ conversion, if one starts with N_{AA} of the A-A contacts and N_{AB} of the A-B contacts there results

$$\Delta E = (N_{AA} - N_{AB})E_{AB} + N_{AB}E_{BB} - N_{AA}E_{AA}.$$

For example, for cell 3, $N_{AA} = 7$ and $N_{AB} = 1$, and therefore $E_3 = -13$. (Cell 11 is the only B-cell brought into contact.) The lists so obtained are given in the table:

$A \to B$		$B \to A$	
i	E_i	i	E_i
6	-4	11	$+4$
4	-7	9	$+1$
1	-7	12	$+1$
2	-10	7	-2
5	-13	8	-5
3	-13	10	-5

and so no moves can be made. However, if cell 7 were changed to an A-cell and cell 4 to a B-cell, reversion to the original configuration will increase energy by 9 units. Thus the procedure has some of the "smoothing" property of surface tension while at the same time allowing some "curvature" of the equilibrium boundary. Both effects are accentuated by decreasing D. Also, it should be noted that the evolution of the pattern depends very much on how the cells are sampled. If, for example, we test the two cells simultaneously, the A-cell adjacent to cell 6 in the figure, along with 6, then an allowable move would be to the symmetric figure with 6 changed to B and 11 to A. Finally, it should be kept in mind that we cannot really isolate a piece of the interface as we have done without sacrificing possible long-range interchanges.

The simulations were able to duplicate rounding up of a square clump, engulfment with eccentric inner aggregate ($D = 1$), engulfment with centering of the inner aggregate ($D = 3$), coalescence of nearby but noncontiguous aggregates ($D = 3$), and engulfment in a three-bob shish kebab experiment, where adjacent clumps engulf one over the other ($D = 3$). The sequence of states agrees quite well with observation, or at least can be made to do so by a suitable choice of adhesion parameters, and the overall behavior of the pattern gives the impression of fluid motion.

Nevertheless, the rules probably do not simulate fluid behavior precisely for any D or choice of weights K_j. While the effect of increasing D is to enhance both rounding up and centering, only the first is a fluid property. However, it should be borne in mind that a continuum fluid description may actually be no better than finite state motility rules for modeling the real behavior of cells.

4.2. Rheology of Cell Aggregates

When subjected to an applied field of stress (e.g., surface tension forces), the behavior of aggregates of animal cells bears a resemblance to that of ordinary viscous liquids, with two important differences. First, the effective viscosity of cell aggregates is enormous, comparable to that of pitch at room temperature, or about 10^4 that of glycerin. (Viscosity can be estimated by watching initially spherical cell clumps distort under gravitational or centrifugal forces, or by following the fusion of two identical spherical clumps and fitting with standard fluid dynamics of deep fusion.) Second, it is observed that the response of a tissue fragment to rapid deformation is essentially that of an elastic material. Such a transition from elastic to viscous fluid behavior under prolonged stress is typical of a class of rheological materials known as *Maxwell liquids*. A mechanical analogue of such a liquid involves an elastic spring and viscous dashpot in series, both being massless. Under rapid deformation the dashpot does not move and the spring is stretched or contracted. A constant force will, however, cause uniform motion of the dashpot.

In flow problems it is convenient to model a Maxwell liquid in the so-called Couette layer: a layer of liquid is confined between horizontal walls, the upper one moving with velocity $U(t)$. The stress exerted on a lamina of fluid obeys

$$\dot{\tau} = k\frac{du}{dy} - t_r^{-1}\tau.$$

The dot here actually means differentiation with respect to time as one follows a given fluid element, and k and t_r are constants. Under the conditions relevant here, of high viscosity and low inertia, du/dy is independent of y. Assuming also that du/dy is independent of time and that τ is initially zero we have

$$\tau = \mu\frac{du}{dy}\left(1 - e^{-\frac{t}{t_r}}\right), \quad \mu = kt_r.$$

Thus if t greatly exceeds the relaxation time t_r we obtain a stress-strain formula for a Newtonian viscous fluid,

$$\tau = \mu\frac{du}{dy},$$

where μ is the viscosity of the fluid (in units gm/cm-sec). If, on the other hand, $t \ll t_r$, then with $du/dy = \dot{\delta}$ and $\delta(0) = 0$, we obtain

$$\tau \simeq \mu\frac{\dot{\delta}(0)t}{t_r} \simeq k\delta,$$

which is Hooke's law with spring constant k.

We shall adopt the point of view that the high effective viscosity of tissue can be explained in terms of the adhesive contacts between cells and that the elastic response on short time scales is another manifestation of mechanical properties of the cell membrane. A common explanation of all these phenomena is difficult because the molecular basis of cell adhesion is not yet clear, and yet one has a range of mechanisms to model. Membrane friction, for example, introduces a dissipative process, while in the models for cell sorting adhesive energy fluxes were regarded as reversible.

Perhaps the best way to proceed is to concentrate the adhesion at discrete sites so that the number density of sites is a simultaneous measure of membrane extension and local adhesive energy. To simplify matters, we shall develop the model for a two-dimensional membrane. This is partially justified by the fact that in the layer problem just considered, applied to a cell aggregate, rows of cells in the ignorable dimension can be taken as two-dimensional cells.

Suppose then that cells of a single type are considered, and that $n(s,t) =$ number density of sites on the membrane, and let $T(s,t) =$ membrane tension, s being arc length along the membrane. We shall also need to consider the tangential velocity $u(s,t)$ with which sites move along the membrane. Conservation of sites along the membrane implies

$$\begin{aligned}\frac{\partial}{\partial t}\int_s^{s+\Delta s} n(s,t)ds &= \int_s^{s+\Delta s} \frac{\partial n}{\partial t} ds \\ &= -n(s+\Delta s,t)u(s+\Delta s,t) + u(s,t)n(s,t) + q\Delta s \\ &= -\int_s^{s+\Delta s}\left[\frac{\partial un}{\partial s} - q\right]ds,\end{aligned}$$

for arbitrary Δs, so that

$$\frac{\partial n}{\partial t} + \frac{\partial un}{\partial s} = q,$$

where q is a source density of sites. We shall take $q = 0$ in the present discussion, although it would be necessary to introduce sources if, for example, sites were extruded from within the cell.

Assuming that sites are integral features of a membrane incorporating extensible elements, we may assume that

$$T = G(n(s,t),t)$$

where the time dependence of G could reflect the hardening or softening of cell adhesion with age. A typical function G would presumably be monotone decreasing in n for fixed t. To complete the system of equations, we shall need an expression for u in terms of T, and this will usually be expressed in terms of a stress τ in the form

$$\tau = \frac{\partial T}{\partial s} = F(u,n,t).$$

The exact nature of F will depend upon details of the problem.

4.2.1. Sliding on a Substrate. As an example, let us suppose that a segment of membrane of initial length L_0 and initial site density n_0 is pulled along a plane substrate $0 \le s < \infty$. At $t = 0$ a tension corresponding to $n = n_1 < n_0$ is applied to the end $s = L_0$, where now

$$T = G(n), \quad G(n_0) = 0, \quad \frac{\partial G}{\partial n} < 0.$$

We next assume a friction law for site velocity as a function of surface stress. If sites respond independently to the presence of substrate, it is reasonable to take

$$\tau = nF(u)$$

where u is the site velocity along the substrate. Since the segment of membrane will stretch as well as translate, the problem takes the form

$$\frac{\partial n}{\partial t} + \frac{\partial un}{\partial s} = 0, \qquad \frac{\partial G}{\partial n}\frac{\partial n}{\partial s} = nF(u),$$
$$n(s_0(t),t) = n_0, \quad n(s_1(t),t) = n_1,$$

where $s_0(t)$ and $s_1(t)$ are functions to be found, determining the motion of the segment.

After an initial transient we expect to find all sites moving with the same constant velocity U. Relative to a moving observer, we have (with $s' = s - s_0(t)$), as $t \to \infty$,

$$\frac{dG(n)}{dn}\frac{dn}{ds'} = nF(u),$$

which may be solved with the condition $n(0) = n_0$. One relation between U and the ultimate segment length L is then obtained from $n(L) = n_1$. Another is obtained from the conservation of sites,

$$n_0 L_0 = \int_0^L n(s')ds'.$$

For example, in the *linearized problem* we take

$$G = -k(n - n_0), \quad F(u) = n_s f u,$$

where f is a constant friction factor, k another constant, and for a later comparison we introduce a constant n_s, which may be regarded as a fixed site concentration along the substrate. In this case, we have, as $t \to \infty$,

$$n = n_0 e^{\frac{Us'}{K}}, \quad K = k/n_s f.$$

Thus

$$\frac{UL}{K} = \ln\left(\frac{n_0}{n_1}\right), \quad n_0 L_0 = \frac{n_0 K}{U}\left(1 - \frac{n_1}{n_0}\right),$$

and therefore

$$U = \frac{K}{L_0}\left(1 - \frac{n_1}{n_0}\right), \quad L = L_0 \frac{\ln\left(\frac{n_0}{n_1}\right)}{1 - \frac{n_1}{n_0}}.$$

To examine the transients in the linearized problem, we note that the equations combine to give

$$\begin{aligned}
\frac{\partial n}{\partial t} - K\frac{\partial^2 n}{\partial s^2} &= 0, & & \text{when } s_0 \le s \le s_1,\\
-K\frac{\partial n}{\partial s} &= \dot{s}_0 n_0, & n = n_0, \quad & \text{when } s = s_0(t),\\
-K\frac{\partial n}{\partial s} &= \dot{s}_1 n_1, & n = n_1, \quad & \text{when } s = s_1(t),
\end{aligned}$$

which is a moving boundary problem for the heat equation in one dimension, often referred to as a Stefan problem. Introducing the dimensionless variables $s^* = s/L_0$ and $t^* = tK/L_0^2$ and setting $n = n_0 n^*$, we obtain a problem in which the only parameter is n_1/n_0. Thus L_0^2/K is a characteristic time for the relaxation of the segment to the stretched-sliding configuration; see Figure 4.5. Note that this time is not quite the same as the relaxation time for a Maxwell liquid. Indeed, if $s_0(t)$ also denotes elongation of the spring-dashpot device under an abrupt application of a constant force, we see that the spring is instantly elongated and simultaneously the dashpot begins to move with constant speed.

No elementary solutions of the moving boundary problem can be given, even if $|n_1 - n_0| \ll n_0$, although for small times we have, near $s = s_1(t)$,

$$n = n_0 + \frac{2(n_1 - n_0)}{\sqrt{\pi}} \int_{-\infty}^{\frac{s-s_1}{2\sqrt{Kt}}} e^{-z^2}\, dz,$$

and therefore

$$s_1 = L_0 + O\big(\sqrt{t}\big).$$

Also $\frac{\partial n}{\partial s}(s_0, t)$ is initially zero.

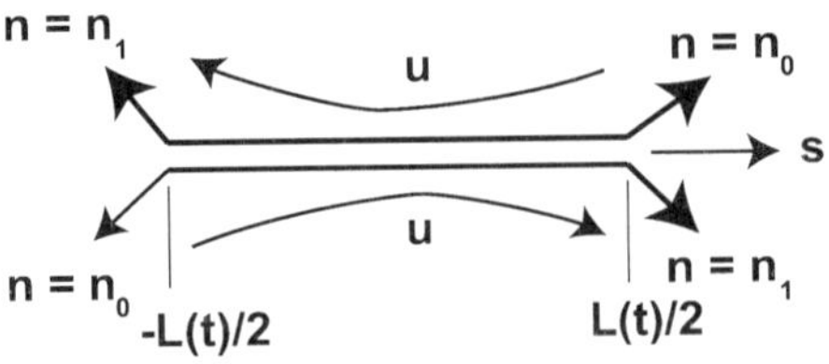

FIGURE 4.5. Relative motion of membranes in contact.

4.2.2. Interstitial Membrane. We now consider a model more directly related to the rheological properties of cell aggregates. In order to test the response of a material to stress, it is convenient to consider the Couette layer with one or both walls in motion. In our problem the material resembles a close-packed arrangement of sticky elastic bags filled with viscous fluid. We can still usefully analyze the layer problem, even though in actual morphogenetic movements the movement of the "walls," over a period of time measured in hours, may be only a fraction of the layer thickness.

Again taking a two-dimensional geometry, the cells within the layer may be visualized as rolling over each other in horizontal lamina, all cells having the same angular speed. But since each lamina moves horizontally at a different speed (because of the global shear of the material), the membranes of adjacent cells will be in relative motion. To model the stretching and sliding of any one of these regions of adhesion, we consider a one-dimensional symmetric arrangement of two membranes in contact. (In the following we consider n and u on the lower membrane.) The region of contact may be thought of as created by the pressure of one cell against another, and the relative motion will result from the enhanced tensions T_1 at diagonally opposite edges, where $T_1 = T_0 + k(n_0 - n)$ and n_0 is a constant. For the mutual friction law we take the site densities on both membranes to operate independently in creating friction (as they do in Steinberg's site model for adhesive energy), and assume a linear dependence on relative velocity. Thus

$$-k\frac{\partial n}{\partial s} = f n(s,t)n(-s,t)[u(s,t) + u(-s,t)],$$

where f is a friction factor depending only on cell type.

Given $n_1(t)$, we note first that the last equation implies that $\frac{\partial n}{\partial s}$ is even in s and therefore

$$\begin{aligned} n(-s,t) + n(s,t) &= \text{function of time} \\ &= n\left(-\frac{L}{2}, t\right) + n\left(\frac{L}{2}, t\right) = n_0 + n_1. \end{aligned}$$

Thus, counting both membranes, there is conservation of sites and $(n_0 + n_1)L(t) = 2n_0 L_0$. Thus we know $L(t)$ explicitly:

$$L(t) = \frac{2n_0}{n_0 + n_1(t)} L_0.$$

We can also easily solve for the stationary configuration ultimately attained when $n_1 = \text{const}$ for $t > 0$. In that case conservation of sites implies

$$un = u_0 n_0 = u_1 n_1,$$

where $u_0 = u\big(-\frac{L}{2}, t\big)$ and $u_1 = u\big(\frac{L}{2}, t\big)$. It follows that

$$-k\frac{\partial n}{\partial s} = f[n(-s) + n(s)]n_0 u_0 = f(n_0 + n_1)n_0 u_0.$$

But by symmetry we then have

$$n = \frac{n_0 + n_1}{2} - (n_0 - n_1)\frac{s}{L} = -\frac{f}{k}(n_0 + n_1)n_0 u_0 s + \text{const}.$$

We may thus solve for u_0 and u_1:

$$u_0 = \frac{K}{2L_0}\left(1 - \frac{n_1}{n_0}\right), \quad u_1 = \frac{n_0 u_0}{n_1},$$

where $K = k/fn_0$.

To study the transients we proceed in a similar way. We first use conservation of sites and symmetry to obtain

$$\begin{aligned}\frac{\partial n}{\partial t}(s,t) + \frac{\partial n}{\partial t}(-s,t) &= \frac{\partial}{\partial s}[u(-s,t)n(-s,t) - u(s,t)n(s,t)]\\ &= \frac{\partial}{\partial t}(n_0 + n_1) = \dot{n}_1.\end{aligned}$$

Thus

$$u(-s,t)n(-s,t) = u(s,t)n(s,t) + s\dot{n}_1,$$

$$\begin{aligned}-k\frac{\partial n}{\partial s} - f[n(s,t)u(s,t)n(-s,t) + n(s,t)u(s,t)n(s,t) + n(s,t)s\dot{n}_1]&\\ = f[(n_0 + n_1)n(s,t)u(s,t) + n(s,t)s\dot{n}_1].&\end{aligned}$$

Differentiating with respect to s, we get an equation for n involving the given function n_1:

$$f(n_0 + n_1)\frac{\partial n}{\partial t} - k\frac{\partial^2 n}{\partial s^2} = f\dot{n}_1\frac{\partial(sn)}{\partial s} = 0.$$

To proceed further, we now assume $|n_1 - n_0| \ll n_0$ and set

$$\begin{aligned} n &= n_0(1 - g(s,t)), & |g| &\ll 1,\\ n_1 &= n_0(1 - v(t)), & |v| &\ll 1,\end{aligned}$$

to obtain the following first-order approximate problem:

$$2\frac{\partial g}{\partial t} - K\frac{\partial^2 g}{\partial s^2} - \dot{v} = 0, \quad g\left(\frac{L_0}{2}, t\right) = v, \quad g\left(-\frac{L_0}{2}, t\right) = 0.$$

Assuming that v as well as g vanishes initially, the Laplace transform of this problem is easily solved to give

$$\hat{g} = \frac{\hat{v}}{2}\left[1 + \frac{\sinh\left(\sqrt{\frac{2\sigma}{K}}s\right)}{\sinh\left(\sqrt{\frac{2\sigma}{K}}\frac{L_0}{2}\right)}\right],$$

where as usual

$$\hat{g}(s,\sigma) = \int_0^\infty e^{-\sigma t} g(s,t)dt,$$

To examine behavior as a quasi-Maxwellian material, we may add the net site motion to the elastic extension to obtain the transformed total rate of deformation

$$\hat{\epsilon} = \frac{du}{dy} = \frac{1}{2}\frac{\hat{u}_0 + \hat{u}_1 + \frac{dL}{dt}}{L_0} \approx (\hat{u}_0 + \hat{u}_1)L_0^{-1} + \frac{\sigma}{2}\hat{v},$$

where now

$$\hat{u}_0 + \hat{u}_1 = \frac{K}{2}\left[\frac{\partial \hat{g}}{\partial s}\left(\frac{L_0}{2},\sigma\right) + \frac{\partial \hat{g}}{\partial s}\left(-\frac{L_0}{2},\sigma\right)\right].$$

This gives, with $\hat{\tau} = \frac{kn_0\hat{v}}{L_0}$,

$$\frac{\hat{\epsilon}}{\hat{\tau}} = \frac{L_0\sigma}{2kn_0} + \frac{1}{2n_0^2 f}\sqrt{\frac{2\sigma}{K}}\coth\left(\sqrt{\frac{\sigma L_0^2}{2K}}\right).$$

Expanding for small σ (large time), we have

$$\frac{\hat{\epsilon}}{\hat{\tau}} = \left(L_0 n_0^2 f\right)^{-1} + \frac{2}{3}\frac{L_0\sigma}{Kn_0} + O(\sigma^2).$$

From our mechanical model of a Maxwell liquid, we also have

$$\frac{\hat{\epsilon}}{\hat{\tau}} = \frac{\sigma}{k_{\text{eff}}} + \frac{1}{\mu_{\text{eff}}},$$

where k_{eff} is the effective elastic constant and μ_{eff} the effective viscosity. Comparing, we have

$$k_{\text{eff}} = \frac{3}{2}\frac{kn_0}{L_0}, \quad \mu_{\text{eff}} = L_0 n_0^2 f,$$

with relaxation time $t_r = 2L_0^2/3K$. We note that the units here are $[f]$ = dyne-sec/cm = gm/sec, $[k]$ = dyne = gm-cm/sec^2, $[\mu]$ = gm/cm-sec = poise. The viscosity of cell clusters is measured to be in the range 10^6–10^8 poise. In order to estimate relaxation time, we must know the ratio k/f, and there would seem to be no special difficulty in measuring k_{eff}. Unfortunately, it is not known how to relate either quantity to adhesive energy, although the latter is of course closely related to effective surface tension.

In summary, this model predicts Newtonian viscous behavior of cell aggregates provided that morphogenetic movements take place on a time scale that is large compared to a relaxation time $\sim L_0^2/K$, and the latter can be estimated in terms of measurable deformation and flow of tissue fragments. The model used to establish this is sufficiently simple to suggest that a corresponding stochastic treatment, allowing for correlation between sites, might be feasible.

4.3. Elements of Morphodynamics

In this section we develop those aspects of viscous flow theory that appear to be most relevant to possible models of morphogenetic movement. Our discussion will be restricted to simple Newtonian viscous fluids. As was noted above, this appears to be reasonable for sufficiently slow movements, but it is important to bear in mind that such simple substances are very crude approximations to the biological material. The latter will, in most applications, exhibit a certain internal structure that is lost in the continuum picture envisaged here. For example, we shall deal below with substances that are locally isotropic, in the sense that there are no distinguished directions in a small parcel of the material. But this is clearly not the case if a cell has a well-defined axis of symmetry, for example, and the local behavior of the aggregate could depend upon the orientation of axes. Indeed, it has been suggested that *liquid crystals* are the most "biological" of fluid substances.[1] The present models are thus intended only as a first step toward a "morphodynamics" that more faithfully reflects the structure of the constituent cells.

Consider then a fluid aggregate of isotropic cells, filling a region V in three-dimensional space, having no internal holes, and bounded by a smooth surface S. Our first task is to obtain the equations and surface forces that govern the deformation of V when surface tension acts over S.

There are two common mathematical formulations of fluid dynamics. The *Lagrangian* viewpoint essentially examines separately the motion of each fluid particle—a direct generalization, to a system with an uncountable number of degrees of freedom, of classical dynamics. The *Eulerian* viewpoint, on the other hand, is especially natural for continua, since the properties of the fluid are determined at a given point in space as a function of time, by determining globally the velocity field, the pressure field, etc., without regard to the history of a given particle. In problems of the type to be considered, involving time-dependent boundaries, both of these formulations can be useful, although we shall for the most part use Eulerian variables and develop these first.

Let $\rho(\mathbf{x}, t)$ and $u_i(\mathbf{x}, t)$, with $i = 1, 2, 3$ and $\mathbf{x} = (x_1, x_2, x_3)$, be, respectively, the continuum density and velocity fields at the point $\mathbf{x}$ at time t. Now a given point may be crossed at some time by the boundary, so in general we cannot label a point as internal. Suppose, however, that over some finite time interval there is a fixed small internal volume $V' \subset V$ with fixed boundary S'. If $q(x, t)$ is the rate at which matter is created at x (the units of q being those of density/time), the conservation

[1] That is, in general "cell state" should be a tensor rather than simply a scalar property.

of mass within V' can be written[2] (summation convention typically used)

$$\frac{d}{dt}\int \rho\, dV' = \int \frac{\partial \rho}{\partial t} dV' = -\int \rho u_i n_i dS' + \int q\, dV',$$

where **n** is the outward-pointing normal to S'. The first of the two terms on the right represents the loss (or gain) in mass due to flow out (or in) across S' of material at a density ρ, with normal velocity $u_i n_i$. The divergence (or Gauss's) theorem states in this case that (the summation convention is used)

$$\int \rho u_i n_i\, dS' = \int \frac{\partial \rho u_i}{\partial x_i} dV'.$$

Combining the last two expressions, we use the fact that V' is arbitrary to deduce that the integrand is zero pointwise. This gives us an expression for conservation of mass in the Eulerian formulation:

$$\frac{\partial \rho}{\partial t} + \frac{\partial \rho u_i}{\partial x_i} = q.$$

This is referred to in the fluid dynamics literature as the equation of *continuity*. In our applications we shall always assume that $\rho = \text{const}$, which we shall frequently refer to as the assumption of *incompressibility*. Note, however, that the two concepts are not exactly equivalent, since if

$$\frac{\partial u_i}{\partial x_i} = \frac{q}{\rho}$$

is used in the last equation, we obtain

$$\frac{\partial \rho}{\partial t} + u_i \frac{\partial \rho}{\partial x_i} = 0,$$

and this does not necessarily require that density be constant. Rather, as we shall see below, the implication of incompressibility is only that density is constant in the neighborhood of every particle but can vary from particle to particle.

The pointwise dynamical balance is expressed by Newton's law

$$\rho a_i = F_i = \frac{\partial \sigma_{ij}}{\partial x_j} + f_i,$$

where the $a_i(x,t)$ are the components of acceleration, F_i the components of the total force field, f_i a local body force, and σ_{ij} the components of the stress tensor. For a small area element dS with normal **n**, $\sigma_{ij} n_j$ is the force exerted by the fluid on the side of the element for which **n** is "outward."

In incompressible Newtonian viscous flow, the stress tensor has the relatively simple form[3]

$$\sigma_{ij} = -p(\mathbf{x},t)\delta_{ij} + \mu\left(\frac{\partial u_i}{\partial x_j} + \frac{\partial u_j}{\partial x_i}\right) - \frac{2}{3}\mu \frac{\partial u_k}{\partial x_k}\delta_{ij},$$

[2]Here and elsewhere the differentials in integrands indicate the appropriate domain of integration.

[3]We assume here that the trace of the viscous stress tensor vanishes; see [**2**], p. 144.

where $p(\mathbf{x}, t)$ is the pressure field, the second two terms on the right are referred to as the viscous stress tensor, and μ is the viscosity. We continue to adopt the usual convention that repeated indices are to be summed. The viscous stresses in a fluid are thus generated by gradients of velocity rather than by gradients of displacement as in elasticity. If this form for σ_{ij} is used in Newton's law, we obtain (with density constant)

$$a_i = -\frac{1}{\rho}\frac{\partial p}{\partial x_i} + \frac{1}{3}\frac{\nu}{\rho}\frac{\partial q}{\partial x_i} + \frac{1}{\rho}f_i + \nu\nabla^2 u_i,$$

where $\nu = \mu/\rho$ is the kinematic viscosity of the fluid. We can therefore look upon the term involving the Laplacian as arising from the diffusion of *momentum* $\rho\mathbf{u}$ with ν as coefficient of diffusion (cf. the emergence of the Laplacian in Section 1.3.2).

The form taken by the acceleration $\mathbf{a}$ in the Eulerian framework is best obtained by simultaneously introducing Lagrangian variables. To follow the position of fluid particles, let $\mathbf{X}(\mathbf{b}, t)$ be the vector position at time t of the particle that was initially located at the point $\mathbf{b}$:

$$\mathbf{X}(\mathbf{b}, 0) = \mathbf{b}.$$

If the mass source q vanishes, then in the Eulerian system the velocity is solenoidal (or divergence-free),

$$\frac{\partial u_i}{\partial x_i} = 0,$$

while in the Lagrangian system the map $M(t) : \mathbf{b} \to \mathbf{X}$ is volume preserving:

$$\det\left(\frac{\partial X_i}{\partial b_j}\right) = 1.$$

The last expression is therefore the continuity equation in Lagrangian variables for an incompressible material with no sources of mass.

The velocity of a particle is the time derivative of $\mathbf{X}$ with $\mathbf{b}$ fixed (now think of $\mathbf{b}$ as a label on each particle), which we denote by $\dot{\mathbf{X}}$. If $\mathbf{u}(\mathbf{x}, t)$ is again the Eulerian velocity field, we then have the identification

$$\dot{\mathbf{X}}(\mathbf{b}, t) = \mathbf{u}(\mathbf{X}(\mathbf{b}, t), t).$$

The acceleration of a particle is $\mathbf{a} = \ddot{\mathbf{X}}$, and therefore

$$\begin{aligned} a_i(\mathbf{X}, t) = \ddot{X}_i = \frac{\partial}{\partial t}u_i(\mathbf{X}(\mathbf{b}, t), t) &= \left[\frac{\partial u_i}{\partial t} + \dot{X}_j\frac{\partial u_i}{\partial x_j}\right]_{\mathbf{x}=\mathbf{X}} \\ &= \frac{\partial u_i}{\partial t} + u_j\frac{\partial u_i}{\partial x_j} \equiv \frac{du_i}{dt}, \end{aligned}$$

which defines the symbol d/dt applied to a Eulerian field. Thus constant density now means $d\rho/dt = 0$. Recall also that for a Maxwell fluid the viscous stress tensor τ_{ij} satisfies

$$t_r\frac{d\tau_{ij}}{dt} + \tau_{ij} = \mu\left(\frac{\partial u_i}{\partial x_j} + \frac{\partial u_j}{\partial x_i}\right) - \frac{2}{3}\mu\frac{\partial u_k}{\partial x_k}\delta_{ij}.$$

In Eulerian language d/dt is called the *material* or *total* derivative, quantities that vanish under this differentiation being constant following a particle. A scalar field vanishing under d/dt is said to be material.The Eulerian momentum equation can now be written

$$\rho\left(\frac{\partial u_i}{\partial t}+u_j\frac{\partial u_i}{\partial x_j}\right)+\frac{\partial p}{\partial x_i}-\mu\nabla^2 u_i=f_i+\frac{\nu}{3}\frac{\partial q}{\partial x_i},$$

where ρ and μ are taken to be constant whenever the material within V is of one kind.

4.3.1. Surface Forces. Suppose now that the volume V experiences a surface tension $T(\mathbf{x},t)$ with $\mathbf{x}\in S$. Since the pressure field is determined globally only up to an additive constant, and since the exterior of V is regarded as empty, the stress tensor may be assumed to vanish there, in which case we shall find that the surface force, acting on S, has the form

$$f_i^s=\sigma_{ij}^s n_j=-T(\mathbf{x},t)\left(\frac{1}{R_1(\mathbf{x},t)}+\frac{1}{R_2(\mathbf{x},t)}\right)n_i+\left(\frac{\partial T}{\partial x_i}-n_i n_j\frac{\partial T}{\partial x_j}\right), \tag{4.3}$$

where the $R_i(\mathbf{x},t)$ are defined below. The right-hand side is seen to involve a normal and a tangential force, the latter being equal to the gradient of T in the local tangent plane.

To derive (4.3), let the boundary S be subjected to a virtual displacement normal to itself determined by a displacement function $\delta\zeta(\mathbf{x},t)$ with $\mathbf{x}\in S$. The virtual work done during deformation is

$$\delta W=\int\sigma_{ij}^s n_i n_j\delta\zeta\,dS+\int T\delta\,dS.$$

Here the second term is the work done in stretching surface elements. To compute it, consider an element dS of area $d\xi_1\,d\xi_2$ where the $d\xi_i$ are elements of arc length along the principal directions (determined by the planes of symmetry of the osculating ellipsoid at dS), having corresponding principal radii of curvature $R_i(\mathbf{x},t)$. Thus

$$\begin{aligned}\delta\,dS&=d\xi_1\,d\xi_2\left(1+\frac{\delta\zeta}{R_1}\right)\left(1+\frac{\delta\zeta}{R_2}\right)-d\xi_1\,d\xi_2\\&\approx dS\delta\zeta\left(\frac{1}{R_1}+\frac{1}{R_2}\right).\end{aligned}$$

It follows that

$$\delta W=\int\delta\zeta\left[\sigma_{ij}^s n_i n_j+T\left(\frac{1}{R_1}+\frac{1}{R_2}\right)\right]dS,$$

and, since $\delta\zeta$ is arbitrary and $\delta W=0$, that

$$\sigma_{ij}^s n_i n_j=-T\left(\frac{1}{R_1}+\frac{1}{R_2}\right).$$

This determines the normal component of surface force as given above.

To find the tangential component, we consider a small, approximately planar patch S' on S, bounded by the curve C'. The tangential force acting on this path is then

$$\int_{C'} T\mathbf{t}\,ds = \int \left(\frac{\partial T}{\partial x_i} - n_i n_j \frac{\partial T}{\partial x_j}\right) dS',$$

where $\mathbf{t}$ is here a tangent vector normal to the segment C' of length ds, and we have applied Gauss's theorem to the tangent plane. Since S' is arbitrary, there is a tangential surface force as given above.

Before leaving this preliminary discussion of surface forces, we note that if the set of outer normals $\mathbf{n}$ to S is extended to a vector field $\mathbf{n}(\mathbf{x}, t)$ defined in a neighborhood of S, we have

$$\frac{1}{R_1} + \frac{1}{R_2} = \frac{\partial n_i}{\partial x_i}, \qquad \mathbf{x} \in S.$$

To see this, consider the region V' consisting of the points

$$\mathbf{x} + r\mathbf{n}(\mathbf{x}, t), \quad 0 \leq r \leq \delta\zeta, \mathbf{x} \in dS.$$

Integrating $\partial n_i/\partial x_i$ over the interior, Gauss's theorem leads to surface integrals reducing to the difference in the areas of the two surface elements that form the caps of V'. Therefore

$$\int \frac{\partial n_i}{\partial x_i} dV' \approx \left(\frac{1}{R_1} + \frac{1}{R_2}\right) dS\,\delta\zeta \approx \frac{\partial n_i}{\partial x_i} dS\,\delta\zeta,$$

which establishes the result.

4.3.2. Dimensional Considerations. Let T_0 be a typical value of surface tension, U_0 a typical speed of movement within V, and L_0 a characteristic diameter of V. The viscous stress on S will be comparable with the surface tension forces provided that $\mu U_0/L_0 \sim T_0/L_0$. Measurements on cell aggregates (see [**80**]) indicate that typically $T_0 = 15$ dynes/cm (gm/sec^2) and $\mu = 10$ poise (gm/cm-sec), giving $U_0 \sim 10^{-6}$ cm/sec. Thus appreciable deformation of an aggregate of diameter 1 mm occurs in about one day. These values have assumed the fluid analogy in reducing observational data, but it is encouraging that the timescale that emerges is typical of developmental movements.

The acceleration $\rho\, d\mathbf{u}/dt$ may be similarly estimated by $\rho U_0^2/L_0$, so its ratio to the viscous force $\sim \mu U_0/L_0^2$ is the *Reynolds number* $\mathrm{Re} = U_0 L_0/\nu$. Typically $\mathrm{Re} = 10^{-6} \times 10^{-1}/10^7 \sim 10^{-14}$. This extraordinarily small dimensionless number tells us that the inertia of the material is entirely negligible in morphodynamics. In effect, the flow is set up instantaneously by the surface forces. But of course the latter will change as V changes shape. The problem is a nonlinear one, even though in discarding the acceleration the Eulerian equations of motion become linear equations.

These estimates suggest the following set of dimensionless (starred) variables:

$$\mathbf{x}^* = \frac{\mathbf{x}}{L_0}, \quad t^* = t\frac{T_0}{\mu L_0}, \quad \mathbf{u}^* = \mathbf{u}\frac{\mu}{T_0}, \quad p^* = p\frac{L_0}{T_0},$$
$$q^* = \frac{q}{q_0}, \quad T^* = \frac{T}{T_0}, \quad R_i^* = \frac{R_i}{L_0}, \quad \mathbf{f}^* = \mathbf{f}\frac{L_0^2}{T_0}.$$

If the equations and surface conditions are expressed in these variables, the acceleration term dropped as negligible, and all stars deleted, we obtain the following system:

$$\frac{\partial p}{\partial x_i} - \nabla^2 u_i = f_i + \frac{\gamma}{3}\frac{\partial q}{\partial x_i},$$
$$\sigma_{ij}^s = -T\left(\frac{1}{R_1} + \frac{1}{R_2}\right) + \frac{\partial T}{\partial x_i} - n_i n_j \frac{\partial T}{\partial x_j},$$
$$\frac{\partial u_i}{\partial x_i} = \gamma q.$$

Whenever the dimensionless quantity $\gamma = q_0 \nu L_0 / T_0$ is small, the changes in shape of V are essentially independent of growth.

Other expressions are needed to complete this system, since we have not as yet specified how the surface tension varies with $\mathbf{x}$ and t. (We take this up in the next chapter.) Also, we need an equation relating fluid velocity on S to the rate of deformation of S. Suppose that S is determined implicitly:

$$S: \quad \Sigma(\mathbf{x}, t) = 0.$$

Since boundary points map into boundary points under the Lagrangian map M, the field must then be material on S; i.e.,

$$\frac{d\Sigma}{dt} = \frac{\partial \Sigma}{\partial t} + u_i \frac{\partial \Sigma}{\partial x_i} = 0, \quad \mathbf{x} \in S.$$

We emphasize that the neglect of inertia will have important consequences, largely due to the instantaneous dynamic response of the material to applied stresses. This basic approximation (low Reynolds number) determines what in the fluid dynamics literature is referred to as *Stokes flow* or *creeping flow*. Stokes flow, it turns out, can be characterized variationally, and this is important in its analysis and computation.

4.3.3. Dissipation. A global energy balance may be obtained from the momentum equation (Stokes flow form) by multiplying scalarly by $\mathbf{u}$ and integrating over V. We obtain

$$-\int u_i f_i \, dV = \int u_i \frac{\partial \sigma_{ij}}{\partial x_j} dV = \int u_i \sigma_{ij}^s n_j \, dS + \int \gamma q p \, dV - \Phi,$$

where

$$\Phi = \int \left[\frac{\partial u_i}{\partial x_j} \left(\frac{\partial u_i}{\partial x_j} + \frac{\partial u_j}{\partial x_i} \right) - \frac{2}{3}\gamma^2 q^2 \right] dV$$

$$= \frac{1}{2} \int \left[\left(\frac{\partial u_i}{\partial x_j} + \frac{\partial u_j}{\partial x_i} \right)^2 - \frac{4}{3}\gamma^2 q^2 \right] dV$$

is the (dimensionless) total viscous dissipation, i.e., the rate at which internal energy is converted into heat by the viscosity of the deforming material. But we also note that the time rate of change of free surface energy is

$$\frac{d}{dt} \int T\, dS = \int \frac{\partial T}{\partial t} dS + \int T \left(\frac{1}{R_1} + \frac{1}{R_2} \right) n_i u_i \, dS,$$

where, as we have seen earlier, the second term accounts for the change in area of a surface element under normal displacement. We may remove it by substituting from the energy balance, once σ^s_{ij} is expressed in terms of T. We then obtain

$$\frac{d}{dt} \int T\, dS - \int \left(\frac{\partial T}{\partial t} + u_i \frac{\partial T}{\partial x_i} - u_i n_i n_j \frac{\partial T}{\partial x_j} \right) dS = \int u_i f_i \, dV + \int \gamma p q \, dV - \Phi.$$

Thus the free surface energy is increased by the working of the body force **f**, or by a work contribution associated with growth, but is dissipated by deformation.

CHAPTER 5

The Origins of Movement

5.1. Chemistry and Geometry of Adhesion

So far the surface tension $T(\mathbf{x}, t)$ has been treated as a given function defined on the aggregate boundary, but in fact, as we have already noted, to fully determine the dynamics we must relate T to the shape and flow of the aggregate. Our next task is therefore to attempt to account for this dependence on shape as well as the distribution of the constituent cells. In so doing we shall generalize the theory to allow a large number (even an infinite number) of cell types.

5.1.1. Surfactants. Let us first consider once more a single cell type filling a region V with smooth boundary S. A surfactant will be a chemical substance whose concentration is nonzero only on S. If $T(\mathbf{x}, t) = T(c(\mathbf{x}, t))$ where c is an m-vector of surfactant concentrations, the dynamics can be completed by adding a conservation law for c. To obtain this, let S' be a patch on S, bounded by the closed curve C'; see Figure 5.1. If c were a scalar and exactly conserved in S', we would have

$$\frac{d}{dt}\int c\, dS' = F,$$

where F is the flux of c across C' into S'. Now

$$\begin{aligned}\frac{d}{dt}\int c\, dS' &= \int \frac{\partial c}{\partial t}\, dS' + \int c\left(\frac{1}{R_1} + \frac{1}{R_2}\right) u_j n_j\, dS' \\ &= \int \frac{\partial c}{\partial t}\, dS' + \int c\frac{\partial n_i}{\partial x_i} u_j n_j\, dS',\end{aligned}$$

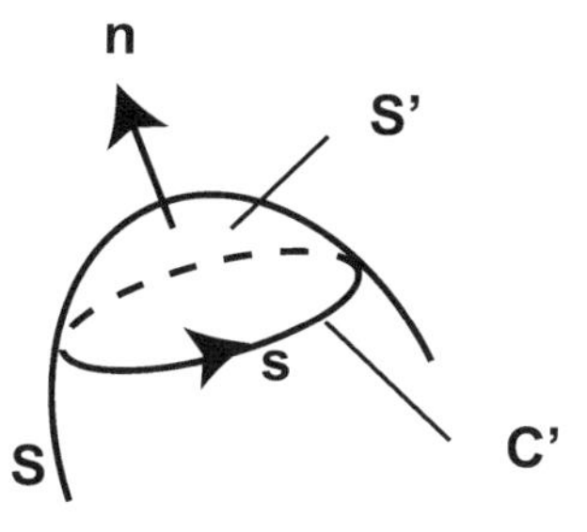

FIGURE 5.1. Surfactant on a boundary with curvature.

by a result of Section 4.3.1. To compute F, let $\mathbf{ds} = (ds_1, ds_2, ds_3)$ be the vector differential along C. Then in terms of vector products we have, by Stokes' theorem,

$$F = \int_{C'} c\mathbf{u} \cdot (\mathbf{n} \times \mathbf{ds}) = \int_{C'} c(\mathbf{u} \times \mathbf{n}) \cdot \mathbf{ds} = \int \mathbf{n} \cdot [\nabla \times c(\mathbf{u} \times \mathbf{n})] dS'.$$

Combining these expressions and using

$$\nabla \times c(\mathbf{u} \times \mathbf{n}) = c\nabla \times (\mathbf{u} \times \mathbf{n}) + \nabla c \times (\mathbf{u} \times \mathbf{n}),$$

we obtain the pointwise conservation law in vector form,

$$\frac{\partial c}{\partial t} + \mathbf{u} \cdot \nabla c - (\mathbf{u} \cdot \mathbf{n})\mathbf{n} \cdot \nabla c + c\Gamma = 0, \quad \mathbf{x} \in S,$$

where

$$\Gamma = (\mathbf{n} \cdot \mathbf{u})\nabla \cdot \mathbf{n} - \mathbf{n}\nabla \times (\mathbf{u} \times \mathbf{n}).$$

If we now take c as an m-vector, allow c to diffuse in S, and introduce surface sources q^c, then since there is an additional flux

$$F_{\text{diff}} = \int_{C'} (\mathbf{n} \times \nabla c) \cdot \mathbf{ds} = \int \mathbf{n} \cdot \nabla \times (\mathbf{n} \times \nabla c) dS'$$

into S', the final equation becomes, with $\mathbf{u}_t = \mathbf{u} - (\mathbf{u} \cdot \mathbf{n})\mathbf{n}$,

$$\frac{\partial c}{\partial t} + \mathbf{u}_t \cdot \nabla c + c\Gamma - D\nabla \times (\mathbf{n} \times \nabla c) = q^c,$$

where D is a constant diagonal matrix of surfactant diffusivities. If variables are taken as dimensionless (see Section 4.3.2), the entries of the diffusion matrix are $\mu\delta_i / T_0 L_0$, $i = 1, \dots, m$, δ_i being the diffusion coefficient for the ith surfactant. When $q^c = q(c)$ the equations become a reaction-diffusion system in S.

One application of the surfactant concept is to cell cleavage, a problem we shall consider later. There the cell membrane carries the surfactant, and something analogous to a membrane would appear to be needed if this idea is to be applied to multicellular structures; presumably the variability of the surface tension is then an instance of variable cell properties determined by a chemical field. A more obvious approach is, however, implied by the differential adhesion hypothesis: one simply regards surface tension as a property of interfaces between unlike cells, hence as a material property in the fluid analogue.

5.1.2. Surface Tension Determined by Cell Type. To study this last possibility, suppose that V is divided up into M regions $V_1, \dots, V_M$, each composed of cells of one type. Let V_0 denote the exterior of V, and let $S_{\alpha\beta}$ for $\alpha < \beta$, $\alpha > 0$, $\beta = 1, \dots, M$, be the interface between V_α and V_β. If $T_{\alpha\beta}$ is the (constant) surface tension on $S_{\alpha\beta}$, the equations of motion are completed by prescribing the $T_{\alpha\beta}$ and requiring the interfaces defined by the level sets $\Sigma_{\alpha\beta}$ to be material; i.e.,

$$\frac{d}{dt}\Sigma_{\alpha\beta} = \frac{\partial \Sigma_{\alpha\beta}}{\partial t} + u_i \frac{\partial \Sigma_{\alpha\beta}}{\partial x_i} = 0, \quad S_{\alpha\beta} : \Sigma_{\alpha\beta}(\mathbf{x}, t) = 0.$$

The interfacial conditions are now

$$(\sigma_{ij}^{\beta} - \sigma_{ij}^{\alpha})n_j = T_{\alpha\beta}\left(\frac{\partial n_j}{\partial x_j}\right)n_i, \quad \mathbf{x} \in S_{\alpha\beta},$$

and the dimensionless equations of motion, assuming viscosity varies with cell type, have the form

$$\nabla p - \mu_\alpha \nabla^2 \mathbf{u} = \mathbf{f}_\alpha + \frac{\mu_\alpha}{3}\nabla \tilde{q}_\alpha, \quad \nabla \cdot \mathbf{u} = \tilde{q}_\alpha = \frac{q_\alpha}{\rho_\alpha}, \quad \mathbf{x} \in V_\alpha.$$

In the above we take $\sigma_{ij}^{0} = 0$.

One immediate consequence of this model follows from the extension of the energy integral derived in Section 4.3.3 to the case of several cell types. One obtains

$$\frac{d}{dt}\Sigma_{\alpha<\beta}\int T_{\alpha\beta}\, dS_{\alpha\beta} = \Sigma_\alpha(\mathbf{u}\cdot \mathbf{f}_\alpha + p\tilde{q}_\alpha - \Phi_\alpha),$$

$$\Phi_\alpha = \frac{\mu_\alpha}{2}\int \left[\left(\frac{\partial u_i}{\partial x_j} + \frac{\partial u_j}{\partial x_i}\right)^2 - \frac{4}{3}\tilde{q}_\alpha^2\right] dV_\alpha.$$

Therefore if mass sources and body forces are absent we see that the *free surface energy*

$$\Sigma_{\alpha<\beta}\int T_{\alpha\beta}\, dS_{\alpha\beta}$$

decreases monotonically during movement at a rate equal to the instantaneous total viscous dissipation. This is of course completely reasonable since in creeping flow the kinetic energy is negligible and the first law must imply that the rate of working of surface tension forces, i.e., the rate of change of surface potential, will equal the rate at which heat is generated. This result is the dynamical analogue of the minimization problem encountered in cell sorting, although it does not in itself fully specify a dynamical model. We shall see later that a variational property of creeping flow may be used to show that the decrease in surface potential is the slowest possible among all flows consistent with incompressibility and the form-determined surface forces.

It is of interest to be able to treat similarly a *continuum* of cell types.We shall assume that the material is layered; i.e., sets of constant cell type are surfaces. First suppose that the layering is discrete, and consider a smooth field of virtual displacement $\delta\xi(\mathbf{x})$, vanishing everywhere except inside a small volume V' containing a large number of layers. If now ϕ is a piecewise constant scalar function with possible jumps in value at interfaces (see Figure 5.2), we have (assuming that $\partial n_j/\partial x_j$ is continuous)

$$\int \delta\xi_i \frac{\partial}{\partial x_i}\left(\phi \frac{\partial n_j}{\partial x_j}\right) dV' = \Sigma_\alpha \phi_\alpha \int \delta\xi_i \frac{\partial^2 n_j}{\partial x_i \partial x_j}\, dV'_\alpha$$

$$+ \Sigma_{\alpha,\beta=\alpha+1}\int (\phi_\beta - \phi_\alpha)\delta\xi_i n_i \frac{\partial n_j}{\partial x_j}\, dS_{\alpha\beta},$$

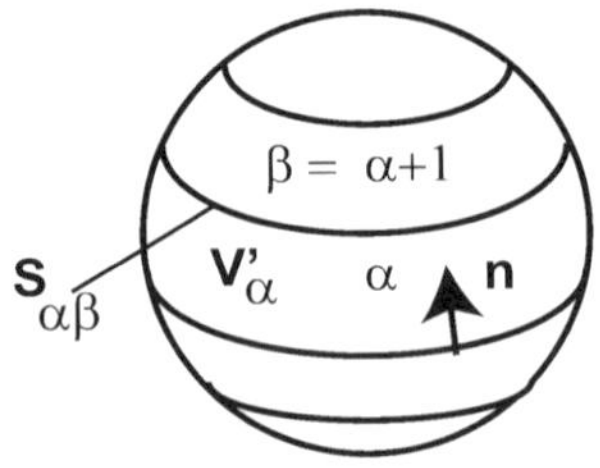

FIGURE 5.2. Surface tension with continuous cell type.

where ϕ_α is the value of ϕ in the αth layer, and we have assumed that the field of normals $\mathbf{n}(\mathbf{x}, t)$ points from α to β along the interface $S_{\alpha\beta}$; see Figure 5.2. If now we assume

$$T_{\alpha\beta} = \phi_\alpha - \phi_\beta \geq 0,$$

then the above expression can be written

$$\int \delta\xi_i \frac{\partial}{\partial x_i}\left(\phi \frac{\partial n_j}{\partial x_j}\right) dV' - \int \delta\xi_i \phi \frac{\partial^2 n_j}{\partial x_i \partial x_j}\, dV' = -\Sigma_{\alpha,\beta=\alpha+1} T_{\alpha\beta} \int \delta\xi_i n_i \frac{\partial n_j}{\partial x_j}\, dS_{\alpha\beta}$$
$$= \int \delta\xi_i \frac{\partial\phi}{\partial x_i}\frac{\partial n_j}{\partial x_j}\, dV'.$$

Since the summation is over all surface forces within V', we see that if now ϕ is replaced by a smooth function, the effective body force is just

$$f_i = \frac{\partial\phi}{\partial x_i}\frac{\partial n_j}{\partial x_j}, \quad n_j = \left|\frac{\partial\phi}{\partial x_i}\right|^{-1}\frac{\partial\phi}{\partial x_i}.$$

It should be stressed that what we have done is to define ϕ as a smooth approximation to a piecewise constant function $T_{\alpha\beta}$ determined from the adhesive energies by $2T_{\alpha\beta} = E_{\alpha\alpha} + E_{\beta\beta} - 2E_{\alpha\beta}$, as well as from the initial layering of cells. Thereafter ϕ is a cell property, hence material and satisfying (in the interior)

$$\frac{d\phi}{dt} = 0.$$

But a priori ϕ is not *universally* a cell property, in the sense that there can be found a function ϕ defined on cell types whose differences equal the surface tensions between cell types. This is not the case because the jumps in ϕ are interfacial properties and it is the jumps which, in any one layering, determine the function. To put this a different way, since $E_{\alpha\beta}$ cannot be predicted from $E_{\alpha\alpha}$ and $E_{\beta\beta}$ (in general), $T_{\alpha\beta}$ cannot be predicted solely from properties of isolated cell types.

Interestingly, Steinberg's sparse site model of (4.1), where $T = \frac{k}{2}(N_\alpha - N_\beta)^2$, with $N_{\alpha,\beta}$ equal to site densities, does predict a surface tension determined by cell properties, but not as a difference. It is therefore not obvious that a suitable ϕ can be generated by a differential adhesion model.

5.1.3. A Floating Site Model of Adhesion. Although we do not know how to treat this question systematically, we next give an example to show that such a ϕ exists. Assume that the energy of adhesion of cell A to cell A (per unit area) is $E_{AA} = 2k_A N_A$ where N_A is regarded as a site density per unit area, and all sites covering the surface of contact are used. Then k_A can be viewed as the adhesion energy, which is the potential in one A-site, with two sites involved in each contact between sites. We now define

$$E_{AB} = (k_A + k_B)\min(N_A, N_B).$$

This would be a reasonable definition if indeed all sites (on the cell with sparser sites) are used and contacts between distinct sites contribute the sum of site energies. We therefore refer to this as the *floating site model.* Then

$$\begin{aligned} T_{AB} &= k_A N_A + k_B N_B - (k_A + k_B)\min(N_A, N_B), \\ &= \begin{cases} k_A(N_A - N_B), & N_A > N_B, \\ k_B(N_B - N_A), & N_B > N_A. \end{cases} \end{aligned}$$

Unexpectedly (and in contrast to the sparse-site model considered previously), it turns out that this model is *unconditionally transitive*, i.e., transitive irrespective of how the k's are assigned. To prove this we must show (see Section 4.1) that if $k_A N_A \geq k_B N_B$ and $k_B N_B \geq k_C N_C$ and also

$$\begin{aligned} (k_A + k_B)\min(N_A, N_B) &\geq 2k_B N_B, \\ (k_B + k_C)\min(N_B, N_C) &\geq 2k_C N_C, \end{aligned}$$

it then follows that $k_A N_A \geq k_C N_C$ and also

$$(k_A + k_C)\min(N_A, N_C) \geq 2k_C N_C.$$

The first statement is immediate. Now the first of the two hypotheses involving min implies if $N_B \leq N_A$, then $k_A \geq k_B$, and if $N_B \geq N_A$, then

$$2k_B N_B \leq (k_A + k_B)N_A \leq (k_A + k_B)N_B,$$

and so in any case $k_A \geq k_B$. Similarly, $k_B \geq k_C$. The conclusion would follow immediately if $N_A \geq N_C$, for then

$$(k_A + k_C)\min(N_A, N_C) = (k_A + k_C)N_C \geq 2k_C N_C.$$

Therefore we assume that $N_A \leq N_C$. There are then three cases to consider:

(i) $N_A \leq N_B \leq N_C$: here we have, by our hypotheses,

$$(k_A + k_B)N_A \geq 2k_B N_B, \quad (k_B + k_C)N_B \geq 2k_C N_C.$$

Adding, we obtain

$$k_A N_A + k_B N_A + k_C N_B - k_B N_B \geq 2k_C N_C$$

or

$$\begin{aligned} (k_A + k_C)N_A = (k_A + k_C)\min(N_A, N_C) \\ &\geq 2k_C N_C + (k_B - k_C)(N_B - N_A) \\ &\geq 2k_C N_C. \end{aligned}$$

(ii) $N_B \leq N_A \leq N_C$: Then we have

$$\begin{aligned} 2k_C N_C &\leq (k_B + k_C)N_B \\ &\leq (k_A + k_C)N_A \\ &= (k_A + k_C)\min(N_A, N_C). \end{aligned}$$

(iii) $N_A \leq N_C \leq N_B$: Then $(k_A + k_B)N_A \geq 2k_B N_B$, and therefore

$$\begin{aligned} (k_A + k_C)N_A &\geq 2k_B N_B - k_B N_A + k_C N_A - 2k_C N_C + 2k_C N_C \\ &= 2k_C N_C + (k_B - k_C)(2N_B - N_A) + 2k_C(N_B - N_C) \\ &\geq 2k_C N_C. \end{aligned}$$

Thus in each case $(k_A + k_C)\min(N_A, N_C) \geq 2k_C N_C$, and the result is proved.

For continuous cell types we have $T_{AB} \approx k|\Delta N|$, and so we reach the desired differential identification

$$d\phi = k dN.$$

This relation suggests a thermodynamic interpretation, although we shall not pursue this here.[1]

5.1.4. A Reaction-Diffusion Model of Cell Adhesion. Of the many questions that have arisen in our discussion, perhaps the most important concerns the molecular nature of cell-cell contacts. We have seen examples of very local developmental control which presumably is indicative of sensitivity of a given cell to neighboring cells, and there are also reasons to believe that cell-specific properties can be controlled globally by diffusible substances, but it is only at the level of the structure of the cell membrane that the various phenomenological models for diffusion, differential adhesion, local control of the mitotic cycle, etc., can be tested critically.

We therefore conclude our consideration of adhesion with a brief discussion of a model due to Bell [**39**]. It is based upon fluid mosaic models of the cell membrane. There are many ways which cell-cell bonds might be formed, involving various ways of linking one membrane to another, but generally some sort of receptor site will likely be present, capable of binding with some ligand, where the latter may be some structure bound to the second membrane, or else an intermediary link in such a bond, bond formation being viewed then as a chemical reaction. An important point is that many membrane-bound proteins are more or less free to move about in the lipid bilayer, and one can view this as diffusion in the plane of the membrane. It is helpful to regard the diffusion as leading to a reaction in its own right, the idea being that a receptor-ligand "complex" is realized prior to bonding once diffusion brings the two within some critical distance. We thus take the bonding process to be in two steps:

$$R + L \underset{d_-}{\overset{d_+}{\rightleftharpoons}} C \underset{\rho_-}{\overset{\rho_+}{\rightleftharpoons}} B$$

[1] We also point out that these rules for adhesion energy can be applied to the rheological model of Section 4.2.2, with very similar results.

where C denotes complex and B a bond. The reaction equations are then

$$\frac{dr}{dt} = -d_+ r\ell + d_- c, \qquad \frac{d\ell}{dt} = -d_+ r\ell + d_- c,$$
$$\frac{dc}{dt} = d_+ r\ell - (d_- + \rho_+)c + \rho_- b, \qquad \frac{db}{dt} = \rho_+ c - \rho_- n.$$

Let us suppose that both receptor and ligand are attached to their respective membranes, the latter being separated by a distance H, and that a complex is formed where the two are within a distance $\Delta > H$. Now each reactant has dimensions of a plane number density, so that d_+ has the dimensions length^2/time, hence those of a diffusivity. If D_R and D_L are the diffusivities of receptor and ligand, and if Δ is the only other parameter determining d_+, then, since d_+ must be symmetric in the diffusivities, it will be of the form $D_R + D_L$ times an arbitrary function of $D_R D_L/(D_R + D_L)^2$. Similarly, d_- will be $\Delta^{-2}(D_R + D_L)$ times such a function. Calculations cited by Bell [**39**] yield the linear expressions

$$d_+ = 2\pi(D_R + D_L), \quad d_- = 2(D_R + D_L)\Delta^{-2}.$$

Assuming that the encounter complex density c is sufficiently small, a Michaelis-Menten approximation may be made by neglecting dc/dt and solving for c in terms of r, ℓ, and b. The resulting equations describe a reaction

$$R + L \underset{k_-}{\overset{k_+}{\rightleftharpoons}} B,$$

where

$$k_+ = \frac{d_+\rho_+}{d_- + \rho_+}, \quad k_- = \frac{d_-\rho_-}{d_- + \rho_+}.$$

Moreover, in this case $r + b = r_0$ and $\ell + b = \ell_0$ are the two number densities prior to bonding. The equation for b is therefore

$$\frac{db}{dt} = k_+(r_0 - b)(\ell_0 - b) - k_- b.$$

The rate of bond formations will be greatest when the membranes are initially brought together,

$$\left(\frac{db}{dt}\right)_{\text{max}} = k_+ r_0 \ell_0.$$

The equilibrium density of bonds is given by

$$b_{\text{eq}}(r_0, \ell_0) = \frac{1}{2}\left(r_0 + \ell_0 + \frac{1}{K}\right) - \frac{1}{2}\left[\left(r_0 + \ell_0 + \frac{1}{K}\right)^2 - 4r_0\ell_0\right]^{\frac{1}{2}},$$

where $K = \frac{k_+}{k_-}$.

The last expression has application to models for differential adhesion, and in particular we note that if $K(r_0 + \ell_0) \ll 1$ (and therefore $K^2 r_0 \ell_0 \ll 1$), we have

$$b_{\text{eq}} = K r_0 \ell_0,$$

so that bond density is approximately proportional to the product of site densities, as in the model discussed by Steinberg (see Section 4.1). But if $K(r_0 + \ell_0) \gg 1$, we get instead

$$b_{\mathrm{eq}} = \min(r_0, \ell_0),$$

which is an ingredient in the floating site model discussed earlier.

If we assume that adhesive energy in suitable units is just b_{eq} for any cell-cell contact (and so do not allow the different binding energies of the floating site model), then $T_{r\ell}$ is given here by

$$\begin{aligned}
2KT_{r\ell} = \frac{1}{2}\Big\{ &(2r_0K+1) - \big[(2r_0K+1)^2 - 4r_0^2K^2\big]^{\frac{1}{2}} \\
&+ (2\ell_0K+1) - \big[(2\ell_0K+1)^2 - \ell_0^2K^2\big]^{\frac{1}{2}} \Big\} \\
&- \Big[(r_0K+\ell_0K+1) + \big[(r_oK+\ell_0K+1)^2 - 4r_0\ell_0K^2\big]^{\frac{1}{2}}\Big] \\
= \big[(1+r_0K+\ell_0K)^2 &- 4r_o\ell_0K^2\big]^{\frac{1}{2}} - \frac{1}{2}(1+4r_0K)^{\frac{1}{2}} - \frac{1}{2}(1+4\ell_0K)^{\frac{1}{2}}.
\end{aligned}$$

To see that this is indeed nonnegative, set $R = \frac{1}{2}(1+4r_0K)^{\frac{1}{2}}$ and $L = \frac{1}{2}(1+4\ell_0K)^{\frac{1}{2}}$ to obtain

$$2KT_{r\ell} = \big[(R+L)^2 + (R-L)^2 + (R^2-L^2)^2\big]^{\frac{1}{2}} - R - L \geq 0.$$

Suppose now we try to calculate a differential volume potential. We would like to have

$$\begin{aligned}
\Delta\phi &= [b_{\mathrm{eq}}(N,N) + b_{\mathrm{eq}}(N+\Delta N, N+\Delta N) - 2b_{\mathrm{eq}}(N, N+\Delta N)] \\
&= \frac{1}{K}\Big\{ \big[(K\Delta N)^2 + 2K(2N+\Delta N) + 1\big]^{\frac{1}{2}} \\
&\qquad - \frac{1}{2}[4NK+1]^{\frac{1}{2}} - \frac{1}{2}[4(N+\Delta N)K+1]^{\frac{1}{2}} \Big\}.
\end{aligned} \tag{5.1}$$

We want gradual change of site density so that $\Delta N \ll N$. But from (5.1) we see that $\Delta\phi$ takes the form of a difference only if also we have

$$\sqrt{KN} \ll K\Delta N,$$

in which case $\Delta\phi \approx K|\Delta N|$.

In the preceding paragraph we have taken ligand and receptor as "bonding sites" in the sense used previously to interpret differential adhesion, but there is little reason to assume that this is typically the case. We list some other possibilities:

(i) Each membrane carries movable ligand and receptors. Assuming the bonding systems do not interact, there are two contributions to equilibrium bond density of the form just considered.

(ii) Two cells have similar receptors, but present on each membrane are ligands capable of bivalent bonding to two receptors, either on the same or on two nearby membranes. Monovalent ligands

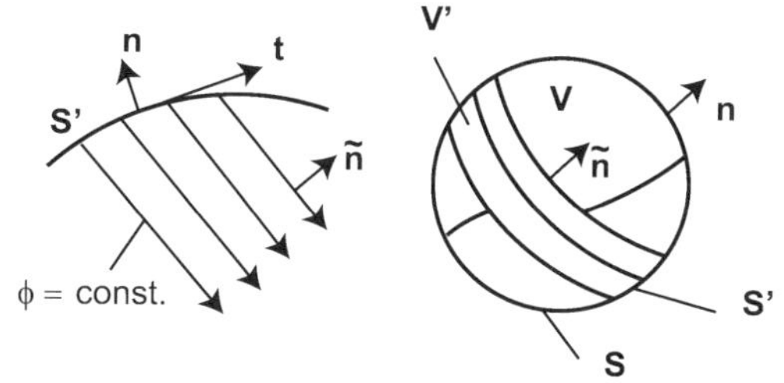

FIGURE 5.3. Tissue with distributed cell types.

capable of occupying a receptor without forming a bond might also be present.

(iii) A bridge between receptors is enzyme-activated from monovalent ligands.

In all of these there is an implied reaction scheme, and so results of the type given above can be obtained quite easily. The matter of adhesion energy is more difficult to study, and the additive energies postulated in the site models of differential adhesion are only guesses. In order to explain variability of membranes, it is important that we know why bonding structures do *not* vary from cell to cell. Without this information the underlying basis for differential adhesion remains obscure. It will be interesting to see if the answer allows global control of cell-cell adhesion.

5.1.5. Free Energy Balance for Distributed Cell Types. We may now ask how the total free surface energy should be expressed in the energy balance when cell types are distributed continuously. We suppose that a portion V' of V consists of continuous cell types, and suppose that in a small neighborhood of the internal boundary of V' the function ϕ may be taken as constant (so we will not have to worry about discontinuities in ϕ along the boundary of V'). We shall also suppose that the variability of surface tension along the part S' of S that bounds V' is already accounted for by a suitable surface variation of T. There remains, however, the variation of surface force along S' due to the layers beneath it, with tension along the lines of constant ϕ.

We will now simultaneously consider the normals to an interface and to S'. We therefore now denote the former by $\tilde{\mathbf{n}}$; see Figure 5.3. If σ^s_{ij} denotes surface stress before these terms are accounted for, the correct surface stress is now

$$\begin{aligned}\tilde{\sigma}^s_{ij} n_j &= \sigma^s_{ij} n_j - |\nabla\phi|[n_i(1-\gamma^2) + (\gamma n_i - \tilde{n}_i)\gamma], \quad \gamma = n_j \tilde{n}_j,\\ &= \sigma^s_{ij} n_j - |\nabla\phi|[n_i - \gamma\tilde{n}_i].\end{aligned}$$

This may be shown by noting that $|\nabla\phi|$ is the continuous surface tension within V' and $t_i = \tilde{n}_i - \tilde{n}_j n_j n_i$ is just a tangent vector to S'. With $q = 0$ the energy balance is now

$$\int u_i \tilde{\sigma}^s_{ij} n_j \, dS + \int u_i f_i \, dV - \Phi = 0,$$

where Φ encompasses all sources of dissipation. As V' deforms, the time derivative of the continuous version of free surface energy may be calculated as follows:

$$
\begin{aligned}
\frac{d}{dt}\int |\nabla\phi| dV' &= \int \frac{\partial|\nabla\phi|}{\partial t}\, dV' + \int \mathbf{u}\cdot\mathbf{n}|\nabla\phi| dS' \\
&= \int \frac{1}{|\nabla\phi|}\nabla\phi\cdot\nabla\phi_t\, dV' + \int \mathbf{u}\cdot\mathbf{n}|\nabla\phi| dS' \\
&= -\int \frac{1}{|\nabla\phi|}\nabla\phi\cdot\nabla(\mathbf{u}\cdot\nabla\phi) dV' + \int \mathbf{u}\cdot\mathbf{n}|\nabla\phi| dS' \\
&= -\int \frac{\nabla\phi\cdot\mathbf{n}}{|\nabla\phi|}(\mathbf{u}\cdot\nabla\phi) dS' + \int \nabla\cdot\left(\frac{\nabla\phi}{|\nabla\phi|}\right)\mathbf{u}\cdot\nabla\phi\, dV' \\
&\quad + \int \mathbf{u}\cdot\mathbf{n}|\nabla\phi| dS' \\
&= \int (\tilde{\mathbf{n}}\cdot\mathbf{n})(\mathbf{u}\cdot\tilde{\mathbf{n}})|\nabla\phi| dS' - \int \mathbf{u}\cdot\mathbf{f}\, dV' \int \mathbf{u}\cdot\mathbf{n}|\nabla\phi| dS'.
\end{aligned}
$$

Combining these results we see that

$$\frac{d}{dt}\left[\int |\nabla\phi| dV' + V_{\mathrm{fs}}\right] = -\Phi,$$

where V_{fs} accounts for all surface energies (including S') other than the contribution from continuous cell types. A remark concerning terminology: to avoid confusion with contributions to free surface energy from discrete layering, we henceforth refer to ϕ as the *volume potential* and to the contribution of continuous cell types to free energy as the *volume free energy*.

5.2. Equilibrium and Stability

Before considering specific applications, it is helpful to examine a problem which suggests both advantages and disadvantages of the fluid approach to morphogenesis. We assume the differential adhesion hypothesis and consider an aggregate of a number of cell types, one type of any pair being known to engulf the other. The question we address is: do stable or unstable configurations according to the cell-sorting picture carry over to analogous stable or unstable fluid structures when the aggregate is replaced by a continuum of cell types? This question arises because in the latter case only a continuous distribution of volume free energy intervenes, whereas in the cell-sorting context it is known that total engulfment of cell type A by cell type B is not implied solely by the positivity of T_{AB} (recall we also need $E_{AB} > E_{BB}$). Some information needed to test stability thus seems to have been lost in going over to the continuum picture. The difficulty, which is a basic one in continuum modeling, can be understood by considering two virtual deformations of a cell aggregate. Suppose that three cell types A, B, and C are arranged in an "onion" configuration, A an inner sphere of cells surrounded by annular layers B and C. As discrete cells the inner sphere can move so as to make contact with both B- and C-cells by breaking through the boundary between B- and C-cells. However, this is not possible if the cells' groups are replaced by homogeneous fluids subject only to smooth deformations. In essence, the fluid layers thin down but never break.

Consequently, even when a layering is unstable in that cell type B lies beneath A (where in fact B engulfs A), perturbations of the fluid need not uncover it. The problem is related to the inability of kinetic cell sorting to attain an absolute minimum of surface potential when cell-cell interactions are too local.

One must accept this limitation, but it is interesting to see how it is reflected in the local stability of the fluid structure, analyzed by conventional continuum perturbation theory. We will consider the continuous version of the "onion" equilibrium in which cell types occupy concentric spherical shells. For discrete fluid layers the equilibrium is only marginally stable to small perturbations, since although interfaces must be spherical they need not remain concentric. For continuous cell types the situation is less clear, but we shall show that to second order (in the displacement amplitude) any sufficiently smooth continuous layering is at most neutrally stable. This confirms the absence of $a \to b$ displacements and yet indicates that the continuous layering behaves similarly to the discrete one.

To prove this, let the volume potential ϕ be $\Phi(r)$ at equilibrium, where r is the spherical radius, with $r = 1$ on S. If $\Phi(1) \neq 0$, there is a surface tension on S, hence a positive increment to free surface energy for any deformation of S. We therefore try to prove neutral stability for deformations which do not affect S. The deformation can be carried out by a Lagrangian flow $\mathbf{x} = \mathbf{X}(\mathbf{b}, t)$, where we are interested in small displacements and therefore small times. In particular,

$$\mathbf{X} \cong \mathbf{b} + t\mathbf{u}(\mathbf{b}) \quad \text{or} \quad \delta\mathbf{X}(\mathbf{x}, t) \cong t\mathbf{u}(\mathbf{x}),$$

where $\mathbf{u}$ is the Eulerian velocity field, taken as solenoidal and with vanishing normal component on S. Since $\phi = \Phi + \delta\phi$ is material, we have

$$\frac{\partial \delta\phi}{\partial t} = -\mathbf{u} \cdot \nabla\Phi - \mathbf{u} \cdot \nabla\delta\phi \cong -\mathbf{u}(\mathbf{x}) \cdot \nabla\Phi + \mathbf{u}(\mathbf{x}) \cdot \nabla(t\mathbf{u}(\mathbf{x}) \cdot \nabla\Phi),$$

and therefore

$$\begin{aligned} \delta\phi &\cong -t\mathbf{u}(\mathbf{x}) \cdot \nabla\Phi + \frac{1}{2} t\mathbf{u} \cdot \nabla t\mathbf{u}(\mathbf{x}) \cdot \nabla\Phi) \\ &= -\delta\mathbf{X} \cdot \nabla\Phi + \frac{1}{2}\delta\mathbf{X} \cdot \nabla(\delta\mathbf{X} \cdot \nabla\Phi) + O(\delta\mathbf{X}^3). \end{aligned}$$

To compute the volume free energy we need

$$\begin{aligned} |\nabla\phi| &= |\nabla\Phi + \nabla\delta\phi| \\ &= |\nabla\phi| \left[1 + \frac{\nabla\Phi \cdot \nabla\delta\phi}{|\nabla\Phi|^2} + \frac{1}{2}\frac{|\nabla\delta\phi|^2}{|\nabla\Phi|^2} - \frac{1}{2}\frac{(\nabla\Phi \cdot \nabla\delta\phi)^2}{|\nabla\Phi|^4} \right] + O(\delta\phi^3). \end{aligned}$$

Inserting the expansion for $\delta\phi$ in terms of $\delta\mathbf{X}$ we have, after some reduction using the dependence of Φ on r alone,

$$\begin{aligned} |\nabla\phi| = |\Phi'| \left[1 - \frac{1}{\Phi'}\frac{\partial}{\partial r}(\delta X_r \Phi') + \frac{1}{2\Phi'}\frac{\partial}{\partial r}[\delta\mathbf{X} \cdot \nabla(\delta X_r \Phi')] + \frac{1}{2}|\nabla_s(\delta X_r)|^2 \right] \\ + O(\delta\mathbf{X}^3). \end{aligned}$$

Here δX_r is the radial component of $\delta\mathbf{X}$, ∇_s is the gradient in the spherical surface at r, and $\Phi' = d\Phi/dr$.

The increment in volume free energy coming from the term linear in $\delta\mathbf{X}$ is

$$-\int \operatorname{sgn}(\Phi')\frac{\partial}{\partial r}(\delta X_r \Phi')dV = \int \frac{2}{r}|\Phi'|\delta X_r\, dV = 0,$$

provided that Φ' is continuous in r. This last result follows from the fact that for solenoidal displacements the integral of the radial component over any spherical surface will vanish. The first contribution quadratic in $\delta\mathbf{X}$ is

$$\frac{1}{2}\operatorname{sgn}(\Phi')\frac{\partial}{\partial r}[\delta\mathbf{X}\cdot\nabla(\delta X_r \Phi')]dV = -\int \frac{\operatorname{sgn}(\Phi')}{r}\nabla\cdot(\delta\mathbf{X}\delta X_r\Phi')dV$$
$$= -\int\left(\frac{\delta X_r}{r}\right)^2|\Phi'|dV,$$

provided that Φ' is of one sign. Combining this with the final contribution, we have

$$\delta\int|\nabla\phi|dV = \frac{1}{2}\int|\Phi'|\left[|\nabla_s\delta X_r|^2 - 2\left(\frac{\delta X_r}{r}\right)^2\right]dV.$$

But it can be shown that, if S_r is the spherical surface of radius r and ψ is any function with zero integral over S_r,

$$\int|\nabla_s\psi|^2\, dS_r \geq 2\int\left(\frac{\psi}{r}\right)^2 dS_r,$$

where the equality is satisfied, for example, when $\psi = x_1$. Consequently,

$$\delta\int|\nabla\phi|dV \geq 0,$$

with equality possible for nontrivial choices of δX_r. This establishes neutral stability under the condition that Φ' has no zeros.

If Φ' does have zeros at points r_α, there is an added contribution

$$-\sum_\alpha|\Phi''(r_\alpha)|\int(\delta X_r)^2\, dS_{r_\alpha},$$

which is destabilizing. This instability has a clear physical meaning. In a neighborhood of any zero of ϕ' there will be layering such that local surface tension increases as r increases. In such a region the free surface energy can be decreased by interchanging two layers.

Since the fluid analogue is no better than local kinetic cell-sorting models in dealing with internalized layering, we must add a hopeful note. The early morphogenetic movements in vertebrates are to some extent concerned with the process of internalization itself. In other words, the structures that are less amenable to this kind of description tend to lie near the end rather than toward the beginning of the primary movements. (In particular, one tends to see centering of an onion structure rather than its breakup.) In any event, some property other than cell adhesion must be involved in the unique specification of onionlike structures.

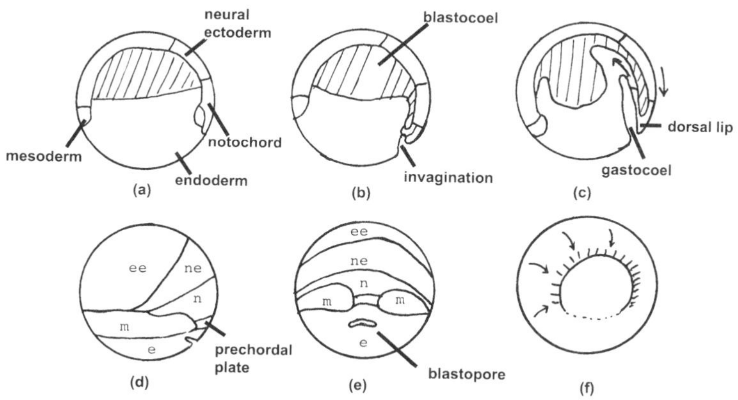

FIGURE 5.4. Gastrulation in amphibians.

5.3. Gastrulation in Amphibians

Amphibians have been extensively studied in connection with the mechanics of gastrulation as well as differential adhesion and cell sorting. Moreover, while amphibian gastrulation is far from the simplest example of the movement establishing the basic organization of the germ layers, from our point of view the large cell population of the blastula, as well as the relatively large number of cell types, offers some justification for a continuum model. Nevertheless, we must always regard any one example as only suggestive of a few basic events that we might hope to capture in a mechanical model.

5.3.1. Summary of the Process of Gastrulation. The late blastula of the amphibian may be divided into an *epiblast* composed of presumptive cells of ectoderm, mesoderm, and notochord surrounding the blastocoel, and a *hypoblast* consisting of endoderm and other organ-forming cells. Gastrulation is the most dramatic example of the morphogenetic process of *emboly* (a movement of inward thrusting or invagination), the end result being the internalization of the presumptive organ-forming cells.

The movement begins with a small invagination in the surface endoderm. As gastrulation proceeds, the endodermal material moves inward and the invagination continues, forming the blastocoel. The latter is bounded by endoderm on one side, and eventually on the other by notochordal material that has moved along the surface toward the dorsal lip, rounded the lip, and moved back along the inner wall. (In Figure 5.4, (a)–(c) are midplane sections, and (d) and (e) show the surface from the side and front, respectively.) On the surface of the blastula the initial invagination occurs more or less into a small curved segment. This segment extends as invagination continues, eventually forming the circular edge of the blastopore (see (f)), into which the surface cells converge, then move backward along the inner wall. Eventually the blastopore begins to contract with ectodermal cells forming its rim as it closes. Associated with the surface movements are the movements of *epiboly*, arising from the marked cell proliferation and spreading in the ectodermal layers. The final result of the movement (which we do not show in the figure) is a completely internalized endoderm, surrounded for the most part by a layer of mesoderm, and finally the outer layer of epidermal and neural ectoderm. Throughout this sequence the blastula maintains a spherical shape and essentially its original size.

Also of interest are the changes that occur in individual cell shape during gastrulation. Certain cells of the endodermal material near the blastopore are found to elongate dramatically during invagination and to form a bottlelike shape with the neck of the bottle attached to the surface of the blastula. Moreover, on the surface the streaming that occurs toward the blastoporal lip is accompanied by stretching of cells along an axis parallel to the direction of streaming. One cannot say that these observed changes of shape are necessarily passive responses to the movement of material; they might well be *causing* the movement. Holtfreter [**91, 92**] performed an experiment in which isolated cells were observed to form a bottle shape with the neck attached to a glass plate.

There have been a number of mechanisms put forward to explain the overall motions of invagination and streaming that accompany gastrulation. The motions do not resemble those of a buckling spherical shell, and the vegetal pole where invagination takes place is also the region where the cells are thickest (and presumably the blastula wall the strongest). Consequently, hypotheses based on osmotic pressure differentials have generally been abandoned, as have explanations based solely upon cell proliferation. As early as 1902, Rhumbler (see [**91, 92**]) proposed a model based upon effective surface tension forces derived from changes of cell shape. Holtfreter went on to explain these changes of shape in terms of the properties of a surface coat surrounding the blastula. Contemporary thinking seems to lean toward change of cell shape induced by differentials of activity among contractile fibers, possibly microtubules under the control of an activator substance.

One surprising point: the literature prior to 1960 generally dismisses surface forces as insignificant because measurements of the surface tension of individual cells yielded an insignificant value—about $1/100^{\text{th}}$ of that needed to account for movement. The fact is of course that the properties of the surface of an individual cell have no direct bearing on the effective surface tractions that might be generated by changes in shape. (Think of sticky paper bags filled with fluid but able to change shape autonomously.)

In any attempt to judge models, it is important to bear in mind the level of the attempted description. In fact, our attitude here is to regard the various plausible explanations of movement based upon cell shape as distinguishable only at the cellular level. At the aggregate level it is plausible (but not obvious) that they are indistinguishable and equivalent to a differential adhesion model, in much the same way that any number of gas kinetic theories may suitably account for the basic form of the equation of state of the gas.

To take one example, according to Holtfreter the modification of shape of a cell might be an expression of contact on one side with the surface coat. It seems likely that on close examination the attempt of cells to change shape will set up a field of stress that will be in equilibrium everywhere except at the surface of the blastula (as well as at interfaces between cells behaving differently). In that case one can simply introduce surface forces from the outset and regard cells as passive. Note, however, that one cannot say a priori whether variability of force is because of the variable response of a cell to a uniform surface coat (surface tension a cell property) or rather that it is some property of the surface coat which is variable (in which case one has a prescribed surfactant). We must admit that many of these possible processes might be used simultaneously.

5.3.2. Interpretation of the Movement in Terms of Cell Affinities. When one compares the structure of the blastula with the sorting experiments between the constituent cell types [**91, 92**], the initial disposition of cell types conforms roughly to a hierarchy based on engulfment. To idealize this for a moment, ignore the blastocoel and imagine a solid blastula whose fate map conforms to a simple polarization along some axis, with surfaces of constant cell type being orthogonal to the axis. We presume the surface tension at the epidermal or animal "pole" to be less than that at the endodermal or vegetal pole, with a continuous variation in between. To simplify the problem as much as possible, we assume that the forces are confined to the surface S. One way to justify this assumption is to write, as in Section 5.1.3, $d\phi = kdN = 0$ over the interior, so that variability of surface tension $T = kN$ comes from variation of k at constant N. Given such an initial layering, we expect the surface forces to produce invagination, spreading of epidermal cells, and generally flow toward the onion equilibrium.

In this highly idealized context, we therefore view gastrulation as a mechanical response to an initial layering that is "far" from equilibrium. This "explanation" of course raises another question, namely, what establishes the initial layering and why should it have a particular form distinct from the equilibrium one. If one relies upon Turing's model (see Section 3.2) and a diffusive instability to set up the initial fate map of the blastula, it is at least not unreasonable to suppose that the pattern of morphogen concentrations is completely independent of the nature of the mechanical equilibrium established by differential adhesion.

The above initial condition has been chosen to allow a relatively simple analysis in spherical polar coordinates. Assuming the initial shape of the blastula is a sphere of radius unity, the surface tension is initially taken to have the following general expansion in Legendre polynomials:

$$T = \sum_{n=0}^{\infty} A_n P_n(\cos\theta)$$
$$= A_0 + A_1 \cos\theta + A_2\left(\frac{3}{2}\cos^2\theta - \frac{1}{2}\right) + A_3\left(\frac{5}{2}\cos^3\theta - \frac{3}{2}\cos\theta\right) + \cdots.$$

Our aim is to compute the initial motion of the boundary, given T. The equations to be solved within the unit sphere are

$$\nabla p - \nabla^2 \mathbf{u} = 0, \quad \nabla \cdot \mathbf{u} = 0,$$

and all fields are symmetric about the z-axis. If u_r and u_θ are the components of velocity in spherical polar coordinates, the solenoidal condition takes the form

$$\frac{1}{r^2}\frac{\partial}{\partial r}\left(r^2 u_r\right) + \frac{1}{r\sin\theta}\frac{\partial}{\partial\theta}(\sin\theta u_\theta) = 0,$$

and this expression is identically satisfied if u_r and u_θ are defined in terms of the *Stokes stream function* ψ:

$$u_r = \frac{1}{r^2\sin\theta}\frac{\partial\psi}{\partial\theta}, \qquad u_\theta = -\frac{1}{r\sin\theta}\frac{\partial\psi}{\partial r}.$$

If pressure is eliminated from the two equations of motion and ψ is introduced, it is found after some reduction that ψ satisfies $L^2\psi = 0$, where L is the differential operator

$$L = \frac{\partial^2}{\partial r^2} + \frac{\sin\theta}{r^2}\frac{\partial}{\partial\theta}\left(\frac{1}{\sin\theta}\frac{\partial}{\partial\theta}\right).$$

The general solution of this equation is easily seen to have the form

$$\psi = \sum_{n=1}^{\infty} \sin\theta\, d P_n(\cos\theta)\big[(n+1)(4n+6)B_n r^{n+3} + C_n r^{n+1} + D_n r^{2-n} + E_n r^{-n}\big].$$

The corresponding pressure field is

$$p = -\sum_{n=1}^{\infty} P_n(\cos\theta)\big[(n+1)(4n+6)B_n r^n + 2n(2n-1)D_n r^{-(n+1)}\big] + K,$$

where K is a constant. In the present interior problem we must take $D_n = E_n = 0$ to ensure regularity at $r = 0$. The constants B_n and C_n are then to be chosen to match surface stresses to the prescribed surface tension on $r = 1$. The two surface conditions are

$$-p + \sigma_{rr} = -p + 2\frac{\partial u_r}{\partial r} = -\frac{2T}{r},$$

$$\sigma_{\theta r} = r\frac{\partial}{\partial r}\left(\frac{u_\theta}{r}\right) + \frac{1}{r}\frac{\partial u_r}{\partial\theta} = \frac{1}{r}\frac{\partial T}{\partial\theta}, \qquad r = 1.$$

Recalling that

$$\frac{d}{d\theta}\sin\theta\frac{dP_n}{d\theta} = -n(n+1)\sin(\theta)P_n,$$

these are seen to reduce to $K = 2A_0$ and

$$-2A_n = (n+1)(4n+6)B_n - 2n(n+1)[(n+1)B_n + (n-1)C_n],$$
$$-A_n = n(n+3)B_n + (n+1)(n-2)C_n + n(n+1)(B_n + C_n),$$

for $n \geq 1$. We obtain $B_1 = -A_1/6$ with C_1 arbitrary. This arbitrariness expresses the fact that the initial boundary motion is determined by surface movement only up to a uniform translation along the z-axis. We also find

$$B_2 = -\frac{2}{19}A_2, \quad C_2 = \frac{13}{114}A_2, \quad B_3 = -\frac{29}{870}A_3, \quad C_3 = \frac{42}{870}A_3.$$

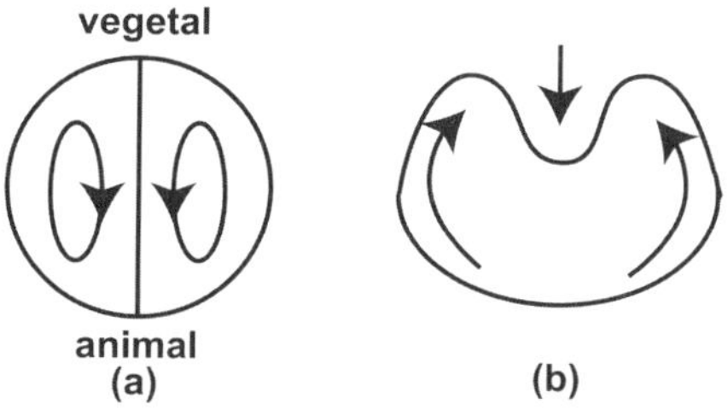

FIGURE 5.5. Gastrulation flow pattern.

The surface velocities are

$$u_r(1,\theta) = -\sum_{n=1}^{\infty} n(n+1)(B_n + C_n)P_n(\cos\theta),$$

$$u_\theta(1,\theta) = -\sum_{n=1}^{\infty} \frac{dP_n}{d\theta}(\cos\theta)[(n+3)B_n + (n+1)C_n],$$

and we wish to use these expressions to study the initial deformation of the boundary.

Suppose first that T depends upon θ only through the P_1-component. Then C_1 may be chosen to make $u_r(1,\theta) = 0$, and so the initial motion is entirely tangential with surface streaming toward the vegetal pole; see Figure 5.5(a). (The sign follows from the fact that $u(1,\theta) = -(A_1/3)\sin\theta$ and $A_1 > 0$.) If we add now a P_2-component to T, it is necessary that $A_1/A_3 \leq 1/3$ if T is to decrease monotonically with θ. If we enforce the condition that the centroid of the body have no initial motion, this is equivalent to choosing C_1 as before, so that the coefficient of P_1 in the expression for $u_r(1,\theta)$ vanishes. Then, since the P_2-component is symmetric in $\cos\theta$, so is $u_r(1,\theta)$ and there is initial invagination at both poles.

Suppose, however, that we now add the P_3-component to T. Fixing C_1 as before we now have

$$u_r(1,\theta) = -\left[\frac{1}{19}A_2P_2(\cos\theta) + \frac{26}{145}A_3P_3(\cos\theta)\right].$$

We thus obtain in P_3 a *gastrulation mode* in the sense that invagination occurs *initially* at the vegetal pole; see Figure 5.5(b). Indeed, assuming that P_2-components are present, but that A_2 and A_3 are comparable in magnitude, we see that the radial motion in the gastrulation mode is about three times that in the P_2-component. One important question, which we shall consider in more detail below, concerns the manner in which the P_3-component could be created by nonlinear mechanisms even if only P_1 were present initially. Evidently if this could occur, the presence or absence of an initial gastrulation mode is academic, the relevant initial movement being the surface streaming toward the vegetal pole.

5.3.3. Modeling the Blastocoel. The solid sphere is the simplest but crudest representation of the amphibian blastula, the main deficiencies being the neglect of the blastocoel as well as the large endodermal mass that partially fills the interior.

We might attempt to treat the blastocoel in an axisymmetric problem by assuming that initially the surfaces of constant cell type lie on cones $\theta = \text{const}$ and the aggregate is a spherical shell $a < r < 1$. We also allow a volume potential $\phi(\theta)$ and assume that initially we have

$$T = \begin{cases} T_1(\theta), & r = 1, \\ T_0(\theta), & r = a. \end{cases}$$

The initial body force arising from this volume potential is then

$$f = -\nabla\phi\nabla \cdot \left(\frac{\nabla\phi}{|\nabla\phi|}\right) = -\frac{|\phi'|}{r}(\nabla \cdot \hat{\theta})\hat{\theta} = -\frac{|\phi'| \cot\theta}{r^2}\hat{\theta}.$$

This force expresses the fact that layers of constant cell type are actually ribbons encircling the axis of revolution and are under tension. The equations of motion, now involving f on the right-hand side, reduce after introduction of the stream function and elimination of the pressure to

$$L^2\psi = r^{-2}|\phi'| \cos\theta.$$

The general solution is found to have the form

$$\psi = r^2 F(\theta) + Q\cos\theta + \psi_h,$$

where ψ_h is the homogeneous solution having the general expansion in Legendre functions given above and F is determined from ϕ by quadratures. The term involving the constant Q is needed in addition to ψ_h because the latter term assumes that the radial velocity has zero spherical mean, whereas this will not be true if the blastocoel changes its volume during the movement. (The added component is the stream function of a spherically symmetric sink.) The pressure associated with the above solution has the form

$$p = \frac{1}{r}\int \cot\theta|\phi'(\theta)|d\theta + p_h$$

and does not involve Q.

The boundary conditions now include the effect of the surface stress from the internal interfaces and we have

$$\tilde{\sigma}_s \cdot \hat{r} = \sigma_s \cdot \hat{r} \mp \frac{|\phi'|}{r}\hat{r}, \quad r = 1, a,$$

where the upper sign is taken on $r = 1$.

The surface conditions then take the following form:

$$-p_h + 2\frac{\partial}{\partial r}\left(\frac{1}{r^2\sin\theta}\frac{\partial\psi_h}{\partial\theta}\right) + \frac{4Q}{r^3} =$$

$$\begin{cases} -2T_0(\theta) - |\phi'(\theta)| - \int \cot\theta|\phi'(\theta)|d\theta = S_0(\theta), & r = 1, \\ a^{-1}\left[2T_1 + |\phi'| - \int \cot\theta|\phi'(\theta)|d\theta - p_0\right] = a^{-1}S_1(\theta), & r = a. \end{cases}$$

$$-r\frac{\partial}{\partial r}\left(\frac{1}{r^2\sin\theta}\frac{\partial\psi_h}{\partial r}\right) + \frac{1}{r}\frac{\partial}{\partial\theta}\left(\frac{1}{r^2\sin\theta}\frac{\partial\psi_h}{\partial\theta}\right) =$$

$$\begin{cases} T_0' - \int \cot\theta |\phi'(\theta)| d\theta - \frac{F}{\sin\theta} = -S_2'(\theta), & r = 1, \\ a^{-1}\left[-T_1' - \int \cot\theta |\phi'(\theta)| d\theta - \frac{F}{\sin\theta}\right] = -a^{-1} S_3'(\theta), & r = a. \end{cases}$$

In terms of the coefficients (B_n, C_n, D_n, E_n), for $n \geq 1$ the conditions are

$$(n+1)(4n+6)B_n + 2n(2n-1)D_n \\ - 2n(n+1)[(n+1)B_n + (n-1)C_n - nD_n - (n+2)E_n] = A_n^{(0)},$$

$$(n+1)(4n+6)a^n B_n + 2n(2n-1)a^{-(n+1)} D_n - 2n(n+1)\big[(n+1)a^n B_n \\ + (n-1)a^{n-2} C_n - na^{-(n+1)} D_n - (n+2)a^{-(n+3)} E_n\big] = a^{-1} A_n^{(1)},$$

$$n(n+3)B_n + (n+1)(n-2)C_n + (n-2)(n+1)D_n + n(n+3)E_n \\ + n(n+1)(B_n + C_n + D_n + E_n) = A_n^{(2)},$$

$$n(n+3)a^n B_n + (n+1)(n-2)a^{n-2} C_n + (n-2)(n+1)a^{-(n+1)} D_n + n(n+3)a^{-(n+3)} E_n \\ + n(n+1)(a^n B_n + a^{n-2} C_n + a^{-(n+1)} D_n + a^{-(n+3)} E_n) = a^{-1} A_n^{(3)},$$

where we have set

$$S_i = \sum_{n=0} A_n^{(i)} P_n(\cos\theta), \quad i = 0, 1, 2, 3.$$

The constant terms give

$$4Q = A_0^{(0)}, \quad \frac{4Q}{a^3} = \frac{A_1^{(1)}}{a} - p_0.$$

Here p_0 is the blastocoel pressure. If this is taken as given (by cavity volume, rate of change of volume, etc.), then Q is thereby determined. Alternatively, we may fix Q by the requirement that blastocoel volume does not change initially, in which case these expressions determine the necessary blastocoel pressure.

We consider here only $n = 1$ and $\phi = 0$ in order to estimate the surface streaming. Then $-A_1^{(2)} = b$ is the coefficient of $\cos\theta$ in T_0, and $A_1^{(3)} = c$ is the coefficient in T_1. Then we find

$$B_1 = -\frac{1}{6}\frac{(b + ca^3)}{1 - a^5}, \quad E_1 = \frac{a^3(c + ba^2)}{6(1 - a^5)}, \quad D_1 = 0, \quad C_1 = -B_1 - E_1,$$

and, with $u_r(1, \theta)$ now due only to spherical expansion or contraction,

$$-u(1, \theta) = \frac{\sin\theta}{6(1 - a^5)}[(2 + 3a^5)b + 5a^3 c].$$

It follows that streaming is again toward the endoderm (vegetal pole) even for certain negative c. The last possibility is important because of the presence of the large mass of endoderm adjacent to the blastocoel. If we adopt the floating site model (Section 5.1.3) and confine ourselves to the special case where k is constant and N variable, then we see that if C engulfs B and B engulfs A, then $T_{AC} \geq T_{BC}$. If A represents endoderm, then the pattern of surface forces has the orientation shown

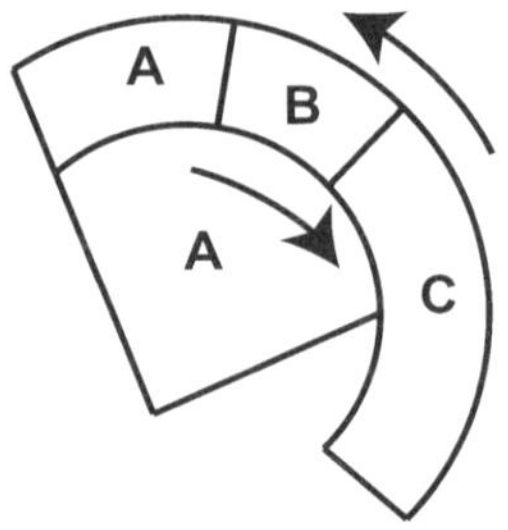

FIGURE 5.6. Blastopore formation.

in Figure 5.6. Negative c may therefore be relevant near the presumptive blastopore region, in which case the amount caused by this reversal of force is also of possible importance.

5.3.4. Gastrulation as an Instability. The problems considered above all start with a layering that is not static, and as such are not examples of instability of an initial equilibrium state. On the other hand, it is likely that one may enlarge the model so that the movement becomes part of an unstable system. One way that this can be done, in principle, is to invoke a surfactant that diffuses and is conserved. In this case the term involving Γ in the conservation equation for the surfactant (Section 5.1.1) contributes to the linearized equations and is responsible for a possible exponential instability. The linearized equations for surfactant and velocity fields are mutually coupled, so the instability has a dynamical component. For the case of a fluid sphere in equilibrium with a uniform surfactant, Greenspan [**84**] has shown that the system is unstable for sufficiently small surfactant diffusivity provided that surface tension increases with surfactant concentration. The second mode to be excited as diffusivity decreases is the P_2 mode, and its symmetry leads to invagination along an equatorial belt, and thereby to a mechanical model of cell cleavage. Of course in linear theory all perturbations may be reversed in sign and so another possibility is dimpling at the poles. Neither picture is applicable here since the P_2 mode does not have the symmetry of the gastrulation movement.

Another starting point may be based upon the late (unpublished) work of Alan Turing, who sought to establish the nature of the patterns induced by a diffusive instability in a spherical shell. As we have seen in Section 3.2, locally the reaction-diffusion equations describing a diffusive instability combine near the onset of instability to yield an equation for the level of a local morphogen m (actually a fixed linear combination of contributing chemicals). We may view this equation as having the following form:

$$\frac{\partial m}{\partial t} = \mathcal{F}(-\nabla^2)m.$$

Here $\mathcal{F}$ is assumed to have a single, slightly positive maximum at $-\nabla^2 = k_c^2$. As the radius R of the shell grows, the P_n^m modes are excited once $n(n+1) = Rk_c$ approximately. This happens first for the P_1 mode, and if surface tension is activated by the morphogen m, the instability can initiate gastrulation as interpreted in the present model.

But note that the two systems are actually not mutually coupled in the linear theory, the movements being only a response to the growing asymmetry in surface tension. Suppose for example that $u_\theta(1,\theta) = -\sin\theta$. Then the conservation equation

$$\frac{\partial T}{\partial t} + u_\theta \frac{\partial T}{\partial \theta} = 0$$

leads immediately to the conclusion that P_2, P_3, and higher-degree modes will grow by advection of the cells toward the vegetal pole. (It can be seen that the P_3 mode is excited with the appropriate sign.) Although this starts as algebraic growth, it is nevertheless possible that the nonlinear nature of the system will eventually cause sizeable and even "catastrophic" changes of form. We can see this in an approximate calculation by noting that if the P_1 mode in T creates a $\sin\theta$ streaming initially, we might suppose that this relation continues and we have

$$u_\theta = k\frac{\partial T}{\partial \theta},$$

where k is a constant. The equation for T then becomes

$$\frac{\partial T}{\partial t} + K\left(\frac{\partial T}{\partial \theta}\right)^2 = 0,$$

which has the trivial linearization $\frac{\partial T}{\partial t} = 0$. Nevertheless, the nonlinear response is nontrivial since whenever T has initially a smooth maximum, an infinite second derivative with respect to θ will appear in finite time. This follows from the manner in which shocks are formed for the equation

$$\frac{\partial g}{\partial t} + g\frac{\partial g}{\partial \theta} = 0, \quad g = 2k\frac{\partial T}{\partial \theta}.$$

5.3.5. The Stress Field. The movements of the gastrulation mode can be interpreted in terms of the local stress field, and this is helpful since the use of a fluid model obscures the fact that the actual displacement of cells may not be large. Consider the two-dimensional stress tensor

$$\sigma = \begin{pmatrix} 2\frac{\partial u}{\partial x} & \frac{\partial u}{\partial y} + \frac{\partial v}{\partial x} \\ \frac{\partial u}{\partial y} + \frac{\partial v}{\partial x} & 2\frac{\partial v}{\partial y} \end{pmatrix}.$$

We introduce two flows that are tangential at the lines $y = 0$ ($u = \frac{\partial \psi}{\partial y}$, $v = -\frac{\partial \psi}{\partial x}$):

(a) $$\psi = -\frac{1}{2}x(x^2 + y^2), \quad -\sigma = \begin{pmatrix} 1+2y & x \\ x & -(1+2y) \end{pmatrix},$$

(b) $$\psi = \frac{1}{2}y^2, \quad -\sigma = \begin{pmatrix} 0 & 1 \\ 1 & 0 \end{pmatrix}.$$

Here ψ is the two-dimensional stream function, and the streamline patterns have simple mapping properties (ψ is constant on trajectories). Thus Figure 5.7(a) corresponds to the immediate vicinity of the invagination while Figure 5.7(b) is typical of the initial region of surface streaming. These maps tell us how cell shapes must change locally, and therefore indicate how cell shape should change to account for

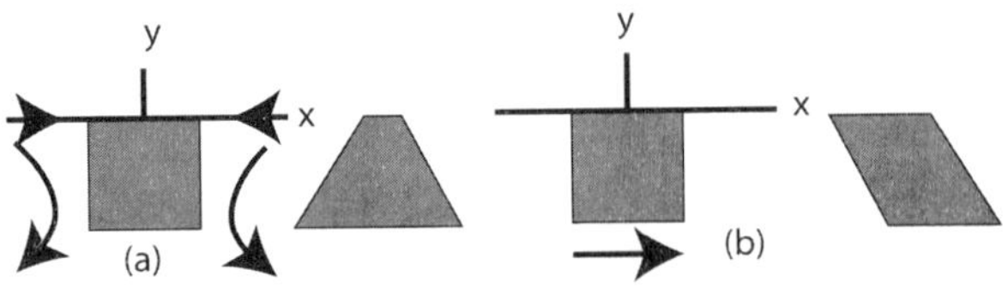

FIGURE 5.7. Deformation of cells under strain.

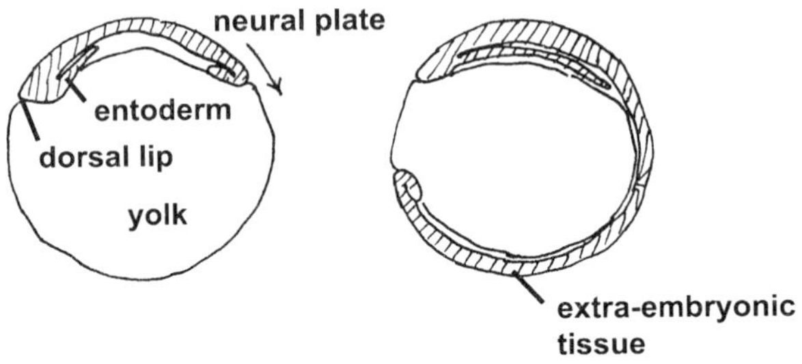

FIGURE 5.8. Teleost gastrulation.

effective surface tension. Case (a) gives one interpretation of Holtfreter's "bottle cells."

5.4. Cell Motion in Thin Layers

In the present section we consider several problems where cell movement takes place within a thin region. Although one reason for considering this approximation is purely mathematical—it offers situations where the nonlinear free-boundary problems of morphodynamics can be partially analyzed analytically—many events in early development do in fact involve tissue sheets or even monolayers of cells. Such problems are probably the simplest examples of movements of interacting cells.

5.4.1. Epiboly in Teleost Fishes. In embryos involving a large yolk the movements of gastrulation are quite different from those of the amphibians and may involve movements of individual cells, delamination of cell sheets, or epiboly. In the latter movement, the gut of the embryo is internalized not by invagination but rather by being coated by a spreading sheet of cells. A good example of this occurs in the teleost fishes (e.g., brook and rainbow trout). Here late gastrulation involves spreading around the yolk of extra-embryonic tissue; see Figure 5.8.

Although growth is probably contributing to the spreading, the resemblance to the coating of a substrate by a wetting agent suggests a differential adhesion model may be relevant.

To study this we make use of *lubrication theory* since that class of fluid problems deals for the most part with viscous flows in thin domains. Let us suppose that the domain has characteristic dimensions ℓ and L where $\ell/L = \epsilon \ll 1$, and L is typical of variations in the xy-plane. Assuming that L is the unit of length, it is clear that to resolve the z-dependence of the fields we should introduce a stretched

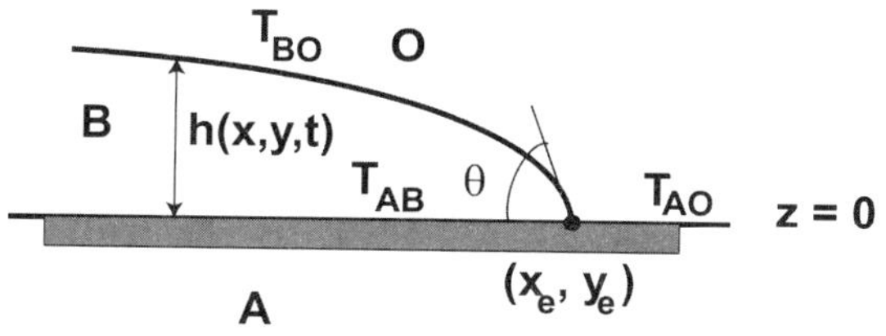

FIGURE 5.9. Model thin layer.

variable $\zeta = z/\epsilon$. The dimensionless Stokes' equations become, in these variables,

$$\left(\frac{\partial}{\partial x}, \frac{\partial}{\partial y}\right) p = \left(\frac{\partial^2}{\partial x^2} + \frac{\partial^2}{\partial y^2} + \frac{1}{\epsilon^2}\frac{\partial^2}{\partial \zeta^2}\right)(u, v),$$

$$\frac{1}{\epsilon}\frac{\partial p}{\partial \zeta} = \left(\frac{\partial^2}{\partial x^2} + \frac{\partial^2}{\partial y^2} + \frac{1}{\epsilon^2}\frac{\partial^2}{\partial \zeta^2}\right) w,$$

$$\frac{\partial u}{\partial x} + \frac{\partial v}{\partial y} + \frac{1}{\epsilon}\frac{\partial w}{\partial \zeta} = 0.$$

In order to retain the solenoidal condition in its entirety and to maintain a balance with pressure in the lateral momentum equation, we shall introduce $W = w/\epsilon$ as $O(1)$. We take $P = p\epsilon^2$ as $O(1)$ in order to retain a pressure balance in the xy-plane. In the limit $\epsilon \to 0$ the equations reduce in these variables to

$$\frac{\partial P}{\partial \zeta} = 0, \quad \nabla_2 P = \frac{\partial^2 \mathbf{u}}{\partial \zeta^2}, \quad \nabla_2 = \left(\frac{\partial}{\partial x}, \frac{\partial}{\partial y}\right),$$

$$\mathbf{u} = (u, v), \quad \nabla_2 \cdot \mathbf{u} + \frac{\partial W}{\partial \zeta} = 0.$$

Physically, these approximate equations tell us that in a thin layer the pressure penetrates and depends only on the lateral coordinates. As a result the lateral velocity has a simple structure through the layer and the transverse component W is only a response to in-plane divergence of $\mathbf{u}$. The flow can be described as laminar and quasi-two-dimensional.

Once these approximate equations are obtained, it is convenient to restore the original unstretched notation and write the general solution in the form

$$\mathbf{u} = \frac{1}{2}\nabla_2 p\, z^2 + \mathbf{F}\, z + \mathbf{G},$$

$$w = -\frac{1}{6}\left(\nabla_2^2 p\right) z^3 - \frac{1}{2}(\nabla_2 \cdot \mathbf{F})\, z^2 - (\nabla_2 \cdot \mathbf{G})\, x + C,$$

where $\mathbf{F}$, $\mathbf{G}$, and C (and of course p) are arbitrary functions of x, y, and t. The boundary conditions to be imposed depend upon the problem but will in principle determine all functions. In writing these we shall again make use of the thinness of the layer since this will be taken to imply that surface slope is small.

We therefore specialize to a layer of thickness $h(x, y, t)$, supported on a substrate $z = 0$; see Figure 5.9. We may regard the layer as composed of B-type cells, the substrate of A-type cells, and so introduce surface tensions T_{AO}, T_{BO}, and T_{AB}

defined as in Chapter 4. At the free surface the dominant term in the lateral stress is $\partial\mathbf{u}/\partial z$, and this must vanish since there is no lateral gradient of surface tension and surface slope is small:

$$\frac{\partial \mathbf{u}}{\partial z} = 0, \quad z = h(x, y, t).$$

The balance of normal stress at the free surface relates pressure and surface curvature. In terms of principal radii of curvature we have

$$\left(\frac{1}{R_1} + \frac{1}{R_2}\right) = \nabla \cdot \hat{\mathbf{n}} \equiv \nabla \cdot (\hat{\mathbf{z}} - \nabla_2 h) = -\nabla_2^2 h,$$

and therefore p is determined by h:

$$p = -T_{BO}\nabla_2^2 h.$$

In order to obtain relative movement of two cell types we must have slippage at the interface, and therefore it is necessary to depart from a strict fluid model. The precise conditions that correctly account for slippage in a continuum model are still a matter of controversy, but it is natural to assume slippage is proportional to stress on the substrate $z = 0$:

$$\mathbf{u} = K(h)\frac{\partial \mathbf{u}}{\partial z}, \quad z = 0.$$

Here $K(h)$ is presumably zero except when h is small. We follow Greenspan [**85**] in adopting

$$K(h) = \frac{k^2}{3h},$$

where k is a small number

At the moving edge of B-cells a similar approximation is made. Given that surface slope is small, and assuming that the equilibrium contact angle θ_e is also small, Greenspan takes

$$\mathbf{u}_e = k_e(\theta - \theta_e)\mathbf{n}_2,$$

where $\mathbf{n}_2$ is the outward (from B-cells) unit normal to the curve C of the edge, and k_e is a constant. Since

$$\theta \approx \tan\theta = |\nabla_2 h|, \quad \mathbf{n}_2 = -\frac{\nabla_2 h}{|\nabla_2 h|},$$

the edge condition becomes

$$\mathbf{u}_e = -K_e(\nabla_2 h)\left(1 - \frac{\theta_e}{|\nabla_2 h|}\right) \quad \text{on } C.$$

From the above conditions we have

$$\mathbf{u} = -T_{BO}\nabla_2(\nabla_2^2 h)\left(\frac{1}{2}z^2 - hz - \frac{k^2}{3}\right). \tag{5.2}$$

To obtain an evolution equation for h we use continuity of matter. Integrating the solenoidal condition with respect to z between 0 and h and using the fact that the free surface is a material, we have

$$\frac{\partial h}{\partial t} + \mathbf{u} \cdot \nabla_2 h = w, \quad z = h,$$

and there results

$$\frac{\partial h}{\partial t} + \nabla_2 \cdot (\mathbf{U}h) = 0, \quad \mathbf{U} = \frac{1}{h}\int_0^h \mathbf{u}\, dz.$$

From (5.2) we have

$$\mathbf{U} = \frac{1}{3} T_{BO}(h^2 + k^2)\nabla_2(\nabla_2^2 h),$$

and introducing this into the conservation equation we obtain Greenspan's conservation equation for h:

$$\frac{\partial h}{\partial t} + \frac{1}{3} T_{BO}\nabla_2 \cdot \left[(h^3 + k^2 h)\nabla_2 \nabla_2^2 h\right] \equiv \frac{\partial h}{\partial t} + N(h) = 0,$$

with edge condition

$$\mathbf{u}_e = (\dot{x}_e, \dot{y}_e) = \frac{k^2}{3} T_{BO}\nabla_2(\nabla_2^2 h) = -K_e(\nabla_2 h)\left(1 - \frac{\theta_0}{|\nabla_2 h|}\right).$$

We note that in terms of adhesive energy the force balance at edge equilibrium, $T_{AO} = T_{AB} + T_{BO}\cos\theta_0$, becomes

$$\frac{1}{2}(1 - \cos\theta_0) \approx \frac{\theta_0^2}{4} = 1 - \frac{E_{AB}}{E_{BB}}.$$

Consequently the derivation assumes near total engulfment of an A-aggregate by a B-aggregate.

Although the equation for h is still quite complicated, approximate solutions can be obtained by quasi-steady methods. Let h_0 satisfy

$$\nabla_2 \nabla_2^2 h_0 = 0,$$

and let the edge condition be in the form

$$(\dot{x}_e, \dot{y}_e) = -K_e(\nabla_2 h_0)\left(1 - \frac{\theta_0}{|\nabla_2 h_0|}\right).$$

This is the initial step of a recursion procedure that frequently gives good results even at zeroth order. If the equation is cast into a suitable dimensionless form, a small parameter may be defined so that the procedure becomes formally asymptotic. For one-dimensional spreading of a ribbon of cells, for example, the relevant solution (conserving cell number) is

$$h_0 = \frac{1}{2a}\left[1 - \left(\frac{x}{a}\right)^2\right],$$

and the edge condition gives an equation for $a(t)$:

$$\dot{a} = K_e\left(\frac{1}{a^2} - \theta_0\right).$$

For $\theta_0 = 0$ the speed of the edge varies as $t^{-2/3}$, while for $\theta_0 > 0$, a approaches $\pm\theta_0^{-1/2}$ according to

$$-\bar{a} + \frac{1}{2}\ln\left|\frac{1+\bar{a}}{1-\bar{a}}\right| = \theta_0^{\frac{2}{3}} K_e t + \text{const}, \quad \bar{a} = a\theta_0^{\frac{1}{2}}.$$

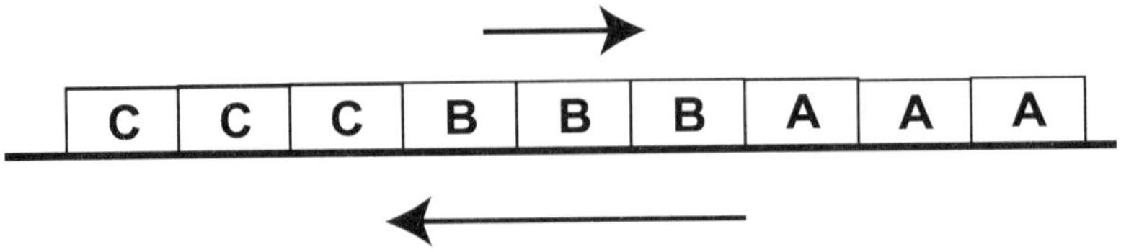

FIGURE 5.10. Adhesion in a monolayer.

5.4.2. Sorting in Monolayers. The tendency of embryonic cells to spread and form a monolayer on flat substrate allows a study of an essentially two-dimensional sorting out process. Under these conditions one may, moreover, observe the paths of individual cells. It is found that membrane ruffling and pseudopodal activity occurs when cells have access to empty substrate, but that there is a certain degree of contact inhibition; i.e., cells bordered by other cells tend not to move.

The fact that monolayering occurs at all, rather than rounding up into balls, for example, indicates that an interpretation using the differential adhesion hypothesis must assume a high adhesive energy between cells and substrate. As we have noted in the last section in connection with the amphibian dorsal lip, this can lead to a reversal of direction of the surface stress. Assuming C engulfs B and B engulfs A, but that strong bonds are formed with substrate (in appropriate site models we may take substrate to be a fourth cell type with adhesion exceeding that of A), then tension at the substrate will be in the opposite sense (and presumably larger) than tension at the "free surface"; see Figure 5.10. In this case A engulfs B and B engulfs C in a monolayer.

This reversal has in fact been observed by Steinberg and Garrod in cultures of chick liver and limb bud cells [**143**]. In three-dimensional experiments liver engulfs limb bud, but, in a monolayer, islands of liver are surrounded by limb bud. After sorting out, the limb bud cells also tend to be flatter than liver cells.

Although it is not difficult to formulate continuum models of the monolayer, to a large extent the analysis is conditioned by the velocity-force relationship one chooses. The end result is necessarily that the islands shrink as much as possible. The corresponding "rounding up" depends upon the relatively feeble cell-cell contacts around the boundary of an island, and indeed the experiments indicate that this is not a prominent feature of monolayer sorting.

5.4.3. Neurulation. The elongation of the amphibian neural plate is another example where essentially a monolayer of cells is involved. However, the movements cannot be explained purely on the basis of plate cells alone, according to a model due to Jacobson and Gordon [**98**]. In this model the formation of a "keyhole" shape from a circular disc is affected by a combination of deepening of the cell layer (causing shrinkage of lateral area) and elongation of the cells of the underlying notochord; see Figure 5.11(a). Such a flow can be represented in two dimensions by a harmonic stream function ψ. We indicate in Figure 5.11(b) the velocity field when ψ consists of a fluid source at the end of the notochord and a balancing line of sinks along it, the resulting motion being confined to a disc. In the numerical simulation

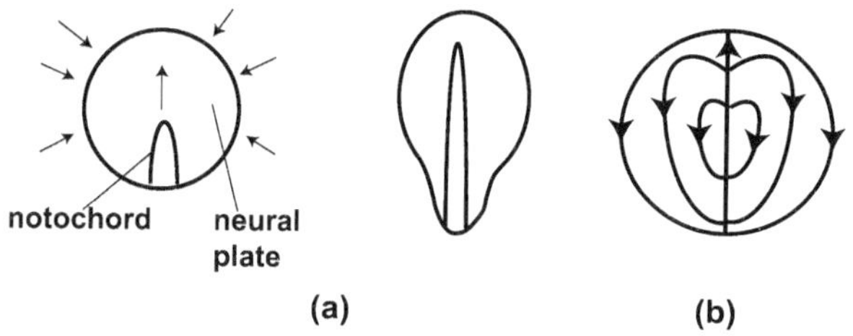

FIGURE 5.11. Keyhole formation.

this motion was incorporated into the algorithm for movement due to cell shrinkage. However, it is of interest to see to what extent this picture is consistent with the most obvious differential adhesion model, given that cell shrinkage, suitably chosen, can mimic the action of an effective surface tension.

To do this we divide up the model into two parts (essentially paralleling the construction of Jacobson and Gordon). The upper layer consists of neural plate cells. These cells will be thought of as riding on a lower layer of mesoderm and notochord. The upper layer can either be taken to be completely passive, or else can increase or decrease the area of contact with the lower layer through the process of epiboly as described above. The only difference is in the boundary condition imposed at the bottom of the upper layer, given that the lower layer may be in motion.

We propose that the lower layer moves under the action of stresses set up at the interface separating the two layers. To describe the process analytically, we shall assume that the lower layer rests on a rigid substrate. Having in mind that the lower layer is thick enough to justify a fluid description, we simply assume adherence to the substrate ($z = 0$).

We now apply lubrication theory to the lower layer. If T is the interfacial tension,

$$\frac{\partial \mathbf{u}}{\partial z} \sim \nabla_2 T, \quad \nabla_2^2 T \sim p,$$

where we are only comparing orders of magnitude. Given that $\frac{\partial \mathbf{u}}{\partial z}$ and p are comparable in the lower layer, p must be negligible in lubrication theory since to be significant it must be comparable with $\partial^2 \mathbf{u}/\partial z^2$, i.e., larger by a factor of ϵ^{-1} times our original ordering. Thus we have

$$\mathbf{u}(x, y, z, t) = \mathbf{F}(x, y, t)z + \mathbf{G}(x, y, t),$$

subject to the conditions

$$\mathbf{u}(x, y, 0, t) = 0, \qquad \frac{\partial \mathbf{u}}{\partial z}(x, y, h(x, y, t), t) = \nabla_2 T(x, y, t).$$

The solution gives

$$\mathbf{u}(x, y, h, t) = h\nabla_2 T(x, y, t).$$

To make surface tension a cell property we have

$$\frac{\partial T}{\partial t} + \mathbf{u} \cdot \nabla_2 T = \frac{\partial T}{\partial t} + h(\nabla_2 T)^2 = 0,$$

while the conservation of matter now gives us

$$\frac{\partial h}{\partial t} + \nabla_2 \cdot \left(\frac{h^2}{2}\nabla_2 T\right) = 0.$$

The last two equations comprise a coupled pair of evolution equations. If we regard the upper layer as a finite patch riding on an infinite lower layer, there are no further conditions, and the motion of the interface is completely determined from the initial distribution of T.

This is still a complicated system but some slight progress can be made if we assume h is close to a constant h_0. In that case T satisfies, approximately,

$$\frac{\partial T}{\partial t} + h_0(\nabla_2 T)^2 = 0,$$

with h then determined by

$$\frac{\partial h}{\partial t} \approx -\frac{1}{2}h_0^2 \nabla_2^2 T.$$

This in fact realizes the irrotational (curl-free) motion envisaged by Jacobson and Gordon, since now $\mathbf{u} = h_0 \nabla_2 T$.

To indicate a method for solving this equation for T, it is convenient to consider a two-dimensional vector field $\mathbf{q}(x, y, t) = (q_1, q_2)$, which solves the equation of two-dimensional pressureless hydrodynamics,

$$\frac{\partial \mathbf{q}}{\partial t} + \mathbf{q} \cdot \nabla_2 \mathbf{q} = 0.$$

If indeed $\mathbf{q}$ is interpreted as a velocity field, this simply states that the acceleration of particles is zero. Let us suppose that initially $\mathbf{q}$ is irrotational, $\mathbf{q}_0 = \mathbf{q}(x, y, 0) = \nabla\phi$. Then from the vector identity

$$\mathbf{q} \cdot \nabla_2 \mathbf{q} = \frac{1}{2}\nabla_2 q^2 - \mathbf{q} \times \nabla_2 \times \mathbf{q} = -\frac{\partial \mathbf{q}}{\partial t}$$

we see that the irrotational property is preserved under evolution of the field. Given such a field we may then define $\mathbf{q}$ by $\mathbf{q} = 2\mathbf{u} = 2h_0 \nabla_2 T$.

Now the $\mathbf{q}$-equation can be solved along characteristics, given the initial values $\mathbf{q}_0$. If the latter is determined implicitly by the system

$$(F_1, F_2) = \mathbf{F}(x, y, \mathbf{q}_0) = 0,$$

then the solution is determined implicitly by

$$\mathbf{F}(x - q_1 t, y - q_2 t, \mathbf{q}) = 0.$$

(Try this with $\mathbf{q}_0 = (y, x)$.) In general, $\mathbf{q}$ does not remain regular with respect to x, y for more than a finite time, even starting from analytic initial data. (A related one-dimensional problem was encountered in Section 5.3.4.)

The initial gradient of T is imposed by the interfacial surface tension, and the arrangement of tissue beneath the neural plate suggests that a reversal of tractions (relative to the upper surface of the upper layer) takes place. We thus regard notochord (A) to be very adhesive relative to mesoderm (B) but comparable to neural plate (C), the only condition being that $T_{BC} > T_{AC}$. Replacing the discontinuous arrangement by a continuous one with interfacial tension T, lines T = const should decrease rapidly into the notochord, but also outward along its length, the latter property being contrary to the model of Jacobson and Gordon. However, their numerical results indicate that the thrusting movement seems to be dominated by the field near the tip of the notochord. In any event the characteristic keyhole shape is easily obtained with suitable initial T. There is a general expansion of the layer so that the auxiliary shrinkage of neural plate cells is again required. It is interesting that along the length of notochord any reasonable assignment of T will have $\nabla_2^2 T$ positive near the midline and negative on either side. From the equation for h this predicts the formation of a groove bordered by two folds. Such a feature is suggestive of the early h neural groove, although it is not clear what corresponding changes would occur in the upper layer.

It is perhaps worth noting that the movements considered here are similar to those involved in the formation of the primitive streak in the blastula of birds. This movement would appear to involve two layers of cells as above, but with possible contributions from stresses acting toward the presumptive streak along the outer surface. Coupled with a reversal of stress along the interface, such a system suggests a "push-pull" model of groove formation and invagination in layered tissue.

5.5. Dynamics of Adhesion-Driven Structure

The reaggregation of disaggregated tissue was mentioned as one of the dramatic—if artificial—effects of cell-cell adhesion. Perhaps the simplest example of this is sponge reaggregation, in which one can watch single cells condense to pairs, k-cell clusters and l-cell clusters to $(k+l)$-cell clusters, etc., until a complete organism reforms. We suppose—a vast oversimplification—that all cells are the same and that the combination rate is fixed by experimental conditions. Then if $p_k = N_k / \sum k' N_{k'}$ is the ratio of k-cell clusters to total cells, we model the dynamics by

$$\dot{p}_k = \frac{1}{2} \sum_{l=1}^{k-1} f(k-l, l) p_{k-l} p_l - \sum_{l=1}^{\infty} f(k, l) p_k p_l .$$

(This is the chemical kinetics, Smoluchowski, master, or Boltzmann equation, depending on the context.) Note that $\sum k \dot{p}_k = 0$, consistent with $\sum k p_k = 1$. In the sponge case, there are a number of possible models. If there is simply a biomolecule-mediated attraction between each cell of one cluster and each of the other, then $f(k,l) \propto kl$. If this is only true of surface cells, then something like $f(k,l) \propto k^{2/3} l^{2/3}$ holds. Or if the rate for cluster pairs is fixed, e.g., by filopod (observed) long-range contact and shortening, then $f(k,l)$ is constant.

In the third model, scaling time suitably,

$$\dot{p}_k = \frac{1}{2} \sum_{l=1}^{k-1} p_{k-l} p_l - \sum_{l=1}^{\infty} p_k p_l .$$

On defining the generating function

$$G(x,t) = \sum_{k=1}^{\infty} p_k(t) x^k ,$$

where we assume that $p_1 = 1$ at time zero, so that $G(x,0) = x$ together with $G'(1,t) = 1$, we have

$$\dot{G}(x,t) = \frac{1}{2}[G(x,t)]^2 - G(x,t)G(1,t), \quad G(x,0) = x.$$

Of the many ways to solve this, the most direct is to first take $x = 1$, so that

$$\dot{G}(1,t) = -\frac{1}{2}[G(1,t)]^2, \quad G(1,0) = 1,$$

or $G(1,t) = \frac{2}{2+t}$, and then

$$\dot{G}(x,t) = \frac{1}{2}[G(x,t)]^2 - \frac{2}{2+t}G(x,t).$$

Hence

$$\frac{\dot{G}(x,t)}{[G(x,t)]^2} + \frac{2/(2+t)}{G(x,t)} = \frac{1}{2}$$

or

$$\frac{2}{2+t}\left[\frac{1}{G(x,t)}\right] = \frac{1}{2} + \frac{d}{dt}\left[\frac{1}{G(x,t)}\right],$$

with the obvious solution

$$\frac{1}{G(x,t)} = \frac{2+t}{2} + a(x)(2+t)^2.$$

Inserting $G(x,0) = x$, then $a(x) = \frac{1}{4x} - \frac{1}{4}$, and

$$G(x,t) = \frac{2x/(2+t)}{x + \frac{(1-x)(2+t)}{2}} = \frac{4x/(2+t)^2}{1 + \frac{2x}{2+t} - x}.$$

Now p_k is the coefficient of x^k in $G(x,t)$, so

$$p_k = \frac{4}{(2+t)^2}\left(1 - \frac{2}{2+t}\right)^{k-1} = \frac{4}{(2+t)^2}\left(\frac{t}{2+t}\right)^{k-1}.$$

Then $p_1 = 4/(2+t)^2$, $p_2 = 4t/(2+t)^3$, etc., which is pretty close to observation, rather than that with k- and l-dependent $f(k,l)$.

There was no particular spatial context in the above. Now let us proceed to the matter of dynamical patterns in which cell interactions and diffusion play major roles—but inertia does not. We deal with a single-species system at the level at which a continuum picture makes sense and in the regime in which cell production and destruction are not of consequence. In the resulting conservation law for cell density,

$$\dot{n}(\mathbf{r},t) + \nabla \cdot j(\mathbf{r},t) = 0,$$

it is only necessary to add to the Fick law diffusion,

$$j_{\text{diff}}(\mathbf{r},t) = -D\nabla n(\mathbf{r},t),$$

the flow due to any force field $f(\mathbf{r},t)$ that may be present. In the high damping region of cellular motion, the kinetics

$$m\dot{v} + \gamma v = f$$

is of course replaced by $v = \frac{f}{\gamma}$, or

$$j_f(\mathbf{r},t) = \frac{1}{\gamma} n(\mathbf{r},t) f(\mathbf{r},t),$$

so that the dynamics takes the form

$$\dot{n}(\mathbf{r},t) + \nabla \cdot \left[\frac{1}{\gamma} n(\mathbf{r},t) f(\mathbf{r},t) - D\nabla n(\mathbf{r},t)\right] = 0.$$

We will now consider several models for the force field $f(\mathbf{r},t)$:

(1) A leading estimate would say that for intercell potential of interaction $\varphi(\mathbf{r}-\mathbf{r}')$,

$$f(\mathbf{r},t) = -\int n(\mathbf{r},t)\nabla\varphi\big(\mathbf{r}-\mathbf{r}'\big)d\mathbf{r}',$$

so that

$$\dot{n}(\mathbf{r},t) = \nabla \cdot \left[D\nabla n(\mathbf{r},t) + \frac{1}{\gamma} n(\mathbf{r},t) \int n\big(\mathbf{r}+\mathbf{r}',t\big)\nabla\varphi\big(\mathbf{r}'\big)d^3\mathbf{r}'\right].$$

For a typical adhesive trap

$$\varphi(\mathbf{r}) = \varphi_0\theta(a - |\mathbf{r}|)(a - |\mathbf{r}|), \quad \nabla\varphi(\mathbf{r}) = -\varphi_0\hat{r}\theta(a - |\mathbf{r}|),$$

this becomes

$$\dot{n}(\mathbf{r},t) = \nabla \cdot \left[D\nabla n(\mathbf{r},t) - \frac{\varphi_0}{\gamma} n(\mathbf{r},t) \int_{|\mathbf{r}'|\le a} n\big(\mathbf{r}+\mathbf{r}',t\big)\hat{r}'\, d^3\mathbf{r}'\right].$$

Numerical solution in one dimension indeed produces an increasing set of sharp condensations.

(2) An empirical modification keeps the density from becoming singular by imposing an upper bound n_μ on its appearance:

$$n(\mathbf{r},t) \mapsto n(\mathbf{r},t)\left[1 - \frac{1}{n_\mu} n(\mathbf{r},t)\right].$$

This results in a much more reasonable pattern of condensation. Extension to mixtures and to two and three dimensions is direct, and reproduces cell sorting experiments well.

(3) A nonuniform equilibrium or quasi-equilibrium configuration for a given form of free energy requires a (self-consistent) potential field $u(\mathbf{r})$. Then the basic system free energy $F[n]$ increases to

$$\begin{aligned} F_{\mu-u}[n] &= F[n] - \mu \int n(\mathbf{r})d^3\mathbf{r} + \int u(\mathbf{r})n(\mathbf{r})d^3\mathbf{r} \\ &= F_\mu[n] + \int u(\mathbf{r})n(\mathbf{r})d^3\mathbf{r}. \end{aligned}$$

The equilibrium equation for $n(\mathbf{r})$ is then obtained by minimizing $F_{\mu-u}$, so that

$$\delta F_{\mu-u}[n] = F_{\mu-u}[n+\delta n] - F_{\mu-u}[n] = 0.$$

Now in general the variational difference must be linear in δn, so that we can write

$$\begin{aligned}\delta F[n] = F[n+\delta n] - F[n] &= \int C[n,\mathbf{r}]\delta n(\mathbf{r})d^3\mathbf{r} \\ &\equiv \int \frac{\delta F[n]}{\delta n(\mathbf{r})}\delta n(\mathbf{r})d^3\mathbf{r},\end{aligned}$$

where $\frac{\delta F}{\delta n}$ is the functional (Fréchet) derivative. In the above, we see by applying $\frac{\delta}{\delta n(\mathbf{r})}$ that

$$0 = \frac{\delta F_\mu[n]}{\delta n(\mathbf{r})} + u(\mathbf{r}), \quad \text{so} \quad u(\mathbf{r}) = -\frac{\delta F_\mu[n]}{\delta n(\mathbf{r})}.$$

Consequently,

$$f(\mathbf{r}) = -\nabla u(\mathbf{r}) = \nabla \frac{\delta F_\mu[n]}{\delta n(\mathbf{r})},$$

where, assuming continued temporal equilibrium (local), t is simply an implicit parameter.

The result is then that

$$\dot{n}(\mathbf{r},t) + \nabla \cdot \left[\frac{1}{\gamma} n(\mathbf{r},t)\nabla \frac{\delta F_\mu[n]}{\delta n(\mathbf{r},t)} - D\nabla n(\mathbf{r},t)\right] = 0,$$

and in the (Van der Waals) approximation that

$$F_\mu[n] = \int \left(\phi_\mu[n] + \frac{b}{2}|\nabla n|^2\right) d^3\mathbf{r},$$

we end up with

$$\begin{aligned}&\dot{n}(\mathbf{r},t) - D\nabla^2 n(\mathbf{r},t) \\ &\qquad + \frac{1}{\gamma}\nabla \cdot \left[n(\mathbf{r},t)\nabla\phi'_\mu(n(\mathbf{r},t)) - bn(\mathbf{r},t)\nabla\nabla^2 n(\mathbf{r},t)\right] = 0,\end{aligned}$$

the fourth differential order Cahn-Hilliard equation, used for understanding the dynamics of droplet formation.

(4) Going in the opposite direction, and quite generally, if $n(\mathbf{r},t)$ equilibrates by minimizing $F_\mu[n]$, we seek a path in time that carries out the minimization—and of course use an empirical scale for time. Suppose we hypothesize

$$\dot{n}(\mathbf{r},t) = -\gamma \frac{\delta F_\mu[n]}{\delta n(\mathbf{r},t)}$$

(see, e.g., [**36**]); then by the chain rule,

$$\frac{d}{dt}F_\mu[n(\mathbf{r},t)] = \int \frac{\delta F_\mu[n]}{\delta n(\mathbf{r},t)}\dot{n}(\mathbf{r},t)d^3\mathbf{r} = -\gamma \int \left(\frac{\delta F_\mu[n]}{\delta n(\mathbf{r},t)}\right)^2 d^3n.$$

Hence, $F_\mu[n(\mathbf{r},t)]$ decreases until $\delta F_\mu[n]/\delta n(\mathbf{r},t) = 0$, as desired. In particular, for our primitive form of $F_\mu[n]$, this becomes

$$\dot{n}(\mathbf{r},t) + \gamma\phi'_\mu(n(\mathbf{r},t)) = b\nabla^2 n(\mathbf{r},t),$$

masquerading as a reaction-diffusion equation, with properties quite similar to those of Cahn-Hilliard.

5.6. Cell-Substrate Adhesion

Mesenchymal cells (connective tissue, bone, cartilage, hair papillae, etc.) constitute a diffuse tissue. They exert strong forces via cell processes on the extracellular matrix (ECM), distorting it over long distances. If the mutual adhesion is nonuniform, cells will move up the associated gradient (haptotaxis). Finally, the ECM is elastically bound to the substratum. When the traction forces of the cells on the ECM are strong enough, an instability can develop, leading to periodic projections, e.g., feathers, scales, teeth. Let us see how this situation can be modeled, starting by including most everything in sight, and then reducing the enormous complexity to something more manageable.

Denote the cell density per unit volume by $n(\mathbf{r},t)$, as usual, the cell current density by $j(\mathbf{r},t)$, and also the mitotic cell production rate per unit volume by $M(\mathbf{r},t)$. Cell conservation is then given by

$$\dot{n} = -\nabla\cdot j + M.$$

Similarly, let the ECM density be $\rho(\mathbf{r},t)$, and its displacement from an initial unstressed position by $u(\mathbf{r},t)$. The contributions to j consist of the following:

(1) $j_1 = -D\nabla\bar{n}$, the diffusion current due to cell-cell interaction. Since this can be long range, we use the locally averaged density

$$\bar{n}(\mathbf{r},t) = \int w(\mathbf{r}')n(\mathbf{r}+\mathbf{r}')d^3\mathbf{r}'.$$

For slow variation of n within the range of w, we proceed in familiar fashion. Assuming

$$\int w(\mathbf{r}')d^3\mathbf{r}' = 1, \quad \int \mathbf{r}'w(\mathbf{r}')d^3\mathbf{r}' = 0,$$

$$\int \mathbf{r}'\mathbf{r}'w(\mathbf{r}')d^3\mathbf{r}' = -\gamma I,$$

we have, on second-order Taylor expansion,

$$j_1 = -D\nabla n + D'\nabla\nabla^2 n.$$

(2) $j_2 = \alpha n\nabla\rho$ for motion up the adhesive density gradient.

(3) $j_3 = n\dot{u}$ because the moving ECM carries resting cells along with it.

Finally, the mitotic production term is given by the standard saturating form terminating at a given N:

$$M = rn(N - n).$$

Hence

$$\dot{n} = \nabla \cdot [D\nabla n - D'\nabla\nabla^2 n - \alpha n \nabla \rho - n\dot{u}] + rn(N - n).$$

Next, we proceed to the dynamics of the ECM, assumed source-free on the time scale of interest, so by conservation ($v = \dot{u}$),

$$\dot{\rho} + \nabla \cdot (\rho \dot{u}) = 0.$$

Force balance is more complex. It says (neglecting inertial forces and imagining static equilibrium)

$$\nabla \cdot \sigma + \rho F = 0,$$

where σ is the ECM stress tensor and F the force per unit volume exerted by the substrate, here simply the elastic

$$F = -su.$$

The stress tensor has several contributions due to the viscoelastic nature of the ECM and of course the cell interactions. We will just quote the result of traditional analysis:

$$\sigma \equiv \mu_1 \dot{\epsilon} + \mu_2 \dot{\theta} I + E'(\epsilon + \nu' \theta I) + \frac{\tau n}{1 + \lambda \rho^2}(\rho + \beta \nabla^2 \rho I).$$

Here,

$$\epsilon = \frac{1}{2}[\nabla u + (\nabla u)^{\mathrm{T}}]$$

is the strain tensor, $\theta = \nabla \cdot u$ is the dilation, μ_1 is the shear viscosity, μ_2 is the bulk viscosity, $E' = E/(1 + \nu)$, where E is the Young's modulus, is the stress/strain ratio for a filament, ν is the Poisson ratio, $\nu' = \nu/(1 - 2\nu)$, and τ is the coefficient of traction.

An example of vertebrate skin organs is an epidermal layer underlain by dermis of the ECM and motile dermal cells, which can aggregate periodically to form papillae. The question is that of the stability of the quiescent system where (scaled) $n = \rho = 1$, $u = 0$. For this purpose, we linearize: $n \mapsto 1 + n$, $\rho \mapsto 1 + \rho$, where ρ, n, and u are now infinitesimal, and simplify by setting $\lambda = 0$ (no contact inhibition). This yields

$$\begin{aligned} \dot{n} - D_1 \nabla^2 n + D_2 \nabla^4 n + \alpha \nabla^2 \rho + \dot{\theta} + rn &= 0, \\ \nabla \cdot [\mu_1 \dot{\epsilon} + \mu_2 \dot{\theta} I + (\epsilon + \nu' \theta I) + \tau(n + \rho + \beta \nabla^2 \rho) I] - su &= 0, \\ \dot{\rho} + \dot{\theta} &= 0. \end{aligned}$$

But even this is unnecessarily complicated; fortunately, the phenomenology is maintained if we now choose $D_2 = \beta = 0$ (no long-range effects), $\alpha = 0$ (no cell-ECM adhesion), $r = 0$ (no mitosis), and $\mu_1 = \mu_2 = \mu$. Hence

$$\dot{n} = D_1 \nabla^2 n - \nabla \cdot \dot{u}, \quad \nabla \cdot \left[\mu\left(\dot{\epsilon} + \dot{\theta} I\right) + \left(\epsilon + \nu' \theta I\right) + \tau(n + \rho) I\right] = su,$$
$$\dot{\rho} + \dot{\theta} = 0.$$

To check stability, insert $e^{\lambda t + \imath \mathbf{k}\cdot\mathbf{r}}$ for the common $(\mathbf{r}, t)$ dependence of amplitudes, obtaining after a bit of algebra

$$\lambda[\mu k^2 \lambda^2 + b(k^2)\lambda + c(k^2)] = 0,$$

where

$$b(k^2) = \mu D_1 k^4 + k^2(1 - 2\tau) + s, \quad c(k^2) = k^2[k^2(1 - \tau) + s]D_1$$

(factors of $1 + \nu'$ have been absorbed into parameter definitions). The stable λ at $b(k^2) > 0$ bifurcates into complex conjugate unstable roots when $b(k^2)$ passes down through zero. But $\min_k b(k^2) < 0$ when $\tau > \tau_c = \frac{1}{2} + (\mu s D_1)^{1/2}$, with a corresponding

$$k_c = \left(\frac{s}{\mu D_1}\right)^{\frac{1}{4}},$$

claimed to correspond quite well to observable mesenchymal projections.

5.7. Imaginal Disc Evagination

Under adhesion, distinct cell populations will have sorted out, leaving a population homogeneous in patches, which can behave as units of still lower resolution tissue.

For example, the insect imaginal leg disc starts as a set of rings of cells and goes unstable to become a set of cylindrical leg elements. Why, how, and to what cylinders? How, from observation, is by rearrangement, changing the shape of the patches to minimize the free energy of the three-dimensional distorted membrane minus the energy due to adhesion? Let us assume vertical and horizontal patch junctions with mean excess adhesive surface energy proportional to boundary length. Suppose there are N_L bands and N_c patches per band, both fixed, and cylinder radius $y(x)$. Then the vertical boundary per unit area is the reciprocal of the mean horizontal height, i.e., $1/(L/N_L)$, while the horizontal boundary per unit area is $1/(2\pi y/N_c)$, with associated energies per unit area

$$E_L = \alpha_L \frac{N_L}{L} = \frac{\alpha}{L}, \quad E_c = \alpha_c \frac{N_c}{2\pi y} = \frac{\beta}{y}.$$

As for the elastic energy of deformation, this is traditionally taken as (per unit area)

$$E_l = \frac{D}{2}(K_1 + K_2)^2 - 2(1 - \nu)K_1 K_2,$$

where ν is the Poisson ratio, and K_1 and K_2 are the principal radii of curvature. Here,

$$K_1 = \frac{y''}{\left[1 + (y')^2\right]^{3/2}}$$

and, since the radius is extended by $[1+(y')^2]^{1/2}$,

$$K_2 = -\frac{1/y}{\left[1+(y')^2\right]^{1/2}}.$$

Imposing the condition of fixed area

$$A = 2\pi \int_0^L y[1+(y')^2]^{\frac{1}{2}}\,dx$$

by a Lagrange parameter λ, we then have to minimize

$$E = 2\pi \int_0^L y\left[1+(y')^2\right]^{\frac{1}{2}} E(x)dx,$$

where, setting $D = 2$ and combining ν and D,

$$E(x) = \frac{(y'')^2}{[1+(y')^2]^3} + \frac{2\gamma y''}{y[1+(y')^2]^2} + \frac{1}{y[1+(y')^2]} + \lambda + \frac{\alpha}{L} + \frac{\beta}{y}.$$

The quantity to be minimized has the form

$$I = \int_0^L f(y, y', y'')dx,$$

where, assuming that the leg segment leads to another segment, we would want $y'(0) = y'(L) = 0$. Thus, the variational condition becomes

$$\begin{aligned}
0 &= \delta I \\
&= \int_0^L (f_y \delta y + f_{y'} \delta y' + f_{y''} \delta y'')dx \\
&= \int_0^L f_y \delta y\,dx + f'_y \delta y\Big|_0^L - \int_0^L \frac{d}{dx} f_{y'} \delta y\,dx + f_{y''} \delta y'\Big|_0^L - \int_0^L \frac{d}{dx} f_{y''} \delta y'\,dx \\
&= \int_0^L \left(f_y - \frac{d}{dx} f_{y'} + \frac{d^2}{dx^2} f_{y''}\right) \delta y\,dx + \left(f_{y'} - \frac{d}{dx} f_{y''}\right) \delta y\Big|_0^L
\end{aligned}$$

so that

$$f_y - \frac{d}{dx} f_{y'} + \frac{d^2}{dx^2} f_{y''} = 0$$

and $f_{y'} - (d/dx) f_{y''} = 0$ at $x = 0, L$. This is a bit complex, but it is easily seen that $y = R$ constant does satisfy it, provided that

$$\frac{1}{R^2} = 2\alpha_L \frac{N_L}{L} + \lambda,$$

and that, inserting the minimum E condition,

$$\frac{1}{R} = N_L \alpha_L \frac{R}{L} - \frac{N_L \alpha_L}{2\pi},$$

a relation between L and R for different leg elements, which works quite well.

Of course, this is all at a very gross level; for the cellular signaling events that are crucial to the structure, see, e.g., [**56**].

CHAPTER 6

Chemotaxis

6.1. Initiation of Slime Mold Aggregation

Up to now our study of morphogenetic movement has been based exclusively on the differential adhesion hypothesis. In essence such models describe a pattern of forces (derivable from cell type) that are viewed as responsible for the movement. As we have indicated from the outset, this is one of many possible descriptions and in the present section we consider how the motion of cells might be directly mediated by an auxiliary field. In the principal case where the auxiliary field is chemical, movement is said to be directed by *chemotaxis*.

At the cellular level it is known that the aggregation process in the slime mold *Dictyostelium discoideum* (see the discussion in Section 3.1 of the life cycle of this organism) involves chemical cues. It is conceivable that during vertebrate gastrulation a similar mechanism directs the movements of invaginating cells. In the case of slime mold the Keller-Segel model outlined below (and see [**105, 106**]) views the initial phase of aggregation as a chemical-dynamical instability of a system at equilibrium, and as such is similar to the instability that can occur when surface tension is controlled by a surfactant chemical [**105**].

The chemical attractant has been called *acrasin*, and is now known to be 3′,5′-cyclic AMP. The cells also produce a second chemical, which acts as an *acrasinase*, rendering the acrasin chemically inactive. In the Keller-Segel model it is assumed that acrasin and acrasinase react as an enzyme-substrate pair according to

$$\rho + \eta \underset{k_{-1}}{\overset{k_1}{\rightleftharpoons}} c \overset{k_2}{\rightarrow} \eta + \text{product},$$

where ρ, η, and c are concentrations of acrasin, acrasinase, and complex, respectively. If $a(x, y, t)$ denotes the number density of amoebae on the feeding surface, and production of chemicals is in proportion to the density, then a reasonable system of diffusion-reaction equations has the form

$$\frac{\partial \rho}{\partial t} = -k_1\rho\eta + k_{-1}c + af(\rho) + v_0\nabla^2\rho,$$

$$\frac{\partial \eta}{\partial t} = -k_1\rho\eta = (k_{-1} + k_2)c + aF(\rho, \eta) + \delta\nabla^2\eta,$$

$$\frac{\partial c}{\partial t} = k_1\rho\eta - (k_{-1} + k_2)c + \lambda\nabla^2 c.$$

where v_0, δ, and λ are constant coefficients of diffusion, and $f > 0$ and F are production functions for the two chemicals. If Michaelis-Menten kinetics is imposed on the enzymatic reaction, with η_0 the initial concentration of acrasinase, then

$$\eta = \frac{\eta_0}{1+K\rho}, \qquad K = \frac{k_1}{k_{-1}+k_2}, \qquad \eta + c = \eta_0,$$

approximately, and the system reduces to an equation for ρ:

$$\frac{\partial \rho}{\partial t} = af(\rho) - g(\rho) + \nu_0 \nabla^2 \rho,$$

where

$$g(\rho) = \frac{\rho \eta_0 k_2 K}{1+K\rho}.$$

An essential feature of the model is the conservation equation for cells. We may write this as

$$\frac{\partial a}{\partial t} + \nabla \cdot \mathbf{J} = Q,$$

where $\mathbf{J}$ is a flux vector and Q a source density. The flux consists of a part due to chemotaxis in response to gradients of ρ and a diffusion with coefficient $\mu(a,\rho)$:

$$\mathbf{J} = \kappa(a,\rho)\nabla\rho - \mu(a,\rho)\nabla a.$$

Thus we have two equations for ρ and a. Letting subscript 0 denote an equilibrium homogeneous state in which all quantities are independent of x, y, and t, and writing $a = a_0 + a'$, $\rho = \rho_0 + \rho'$, etc., the linearized system becomes (with Q identically zero)

$$\frac{\partial \rho'}{\partial t} = \alpha_0 \rho' + f_0 a' + \nu_0 \nabla^2 \rho', \quad \alpha_0 = a_0 \frac{df}{d\rho}(\rho_0) - \frac{dg}{d\rho}(\rho_0),$$

$$\frac{\partial a'}{\partial t} = -\kappa_0 \nabla^2 \rho' + \mu_0 \nabla^2 a'.$$

To test for exponential instability, assume all quantities vary as $e^{\sigma t + ik_1 x + ik_2 y}$; the determinantal condition is then $\sigma^2 + A\sigma + B = 0$, where

$$A = (\mu_0 + \nu_0)k^2 - \alpha_0, \quad k^2 = k_1^2 + k_2^2,$$
$$B = \mu_0 \nu_0 k^4 - (\mu_0 \alpha_0 + \kappa_0 f_0)k^2.$$

Since $A^2 - 4B = (\mu_0 k^2 - \nu_0 k^2 + \alpha_0)^2 + 4\kappa_0 f_0 k^2 \geq 0$, there are no complex roots, indicating no oscillatory response to chemotaxis in this model. Also we note that at $k = 0$ the roots are $\sigma = 0, \alpha_0$. We therefore require that $\alpha_0 < 0$ to avoid this nonphysical instability.

The necessary and sufficient condition for a growing solution for *given* k is then that $B < 0$; that is,

$$\mu_0 \alpha_0 + \kappa_0 f_0 > \mu_0 \nu_0 k^2.$$

A sufficient condition for instability at *some* k (however small) is simply that $\mu_0 \alpha_0 + \kappa_0 f_0 > 0$ or

$$\mu_0 a_0 f_\rho(\rho_0) + \kappa_0 f_0 > \mu_0 g_\rho(\rho_0).$$

This states that the equilibrium is unstable whenever either acrasin production or chemotactic response is sufficiently large compared to acrasinase production. The progress of the instability can be interpreted as chemotactic streaming toward any center of acrasin production, and hence to a further increase of acrasin production.

The model also yields the result that "marginal" instability occurs in the $k = 0$ mode, corresponding to infinite wavelength. That is, if $\alpha_0 < 0$ and $\Delta = \mu_0\alpha_0 + \kappa_0 f_0$ is small and negative, and increases through zero to a small positive value, then large wavelengths are excited. From the quadratic form in σ, we may locally maximize growth rate by taking $k^2 = O(\Delta), \sigma = O(\Delta^2)$ and using the approximate expression $\sigma \approx (1/\alpha_0)(\mu_0\nu_0 k^4 - \Delta k^2)$. Thus,

$$\sigma_{\max} = \frac{\Delta^2}{4|\alpha_0|\mu_0\nu_0}, \quad k^2_{\max} = \frac{\Delta}{2\mu_0\nu_0}.$$

It is difficult to reconcile this behavior with the finite territory size that is observed to emerge during the aggregation process. Segel and Stoeckley discuss the matter in some detail and propose a number of modifications [**139**]. One might regard the spacing as identical to the wavelength of maximum growth rate as obtained above, but this makes territory size dependent upon initial conditions (initial Δ). If a threshold is introduced, the linear stability problem acquires a nonlinear character and a maximal spacing is predicted. Perhaps more relevant is the possibility of obtaining a finite critical wavelength from an expanded system of reaction-diffusion equations, but in that case a chemical wavelength may only reflect diffusive instability, the cell pattern arising simultaneously by chemotaxis toward the chemical pattern. Therefore a "true" chemotactic instability may well be associated with zero critical k even in fairly large systems (cf. the following section). It should be noted that Segel and Stoeckley do provide an example of an enlarged system with finite territory size in linear theory.

6.1.1. Pulsatile Chemotaxis. The Keller-Segel model attempts to describe global cell movement in terms of the chemical fields, and in doing, this one effectively averages over space and time scales that are short compared to the parameters of the chemotactic pattern. Cohen and Robertson have taken an interesting contrasting viewpoint and formulated a model of the early stages of aggregation based upon the observation that amoebae locomotion actually consists of a series of sudden movements suggestive of pulsatile chemotactic signaling [**49, 50**]. It appears that certain amoebae, the future centers of the aggregation, begin to secrete acrasin in pulses with period T. Nearby amoebae respond to these pulses by pulsing themselves as well as by directed movement toward the source of the pulse.

Putting aside for the moment the chemotactic response, the propagation of pulsatile waves (which has general implications in the organization of patterns) is assumed to involve a threshold level of ambient acrasin. If the mean separation R of amoebae is defined by $a^{-1} = \pi R^2/4$, where A is the local number density, then the diffusion (in the plane of the substrate) of an acrasin pulse of strength s will cause concentration $\rho(R, t)$ to be observed by a nearby amoeba, where (with $\tau = 4\nu t/R^2$)

$$\rho(R, t) = \frac{s}{4\pi\nu t} e^{-\frac{R^2}{4\nu t}} = \frac{sa}{4}\frac{e^{-\frac{1}{\tau}}}{\tau} = \frac{sa}{4}g(\tau) \le \frac{sa}{4}.$$

Thus if there is a threshold value ρ_c that triggers an acrasin pulse, a pulsatile wave will propagate only if a rises above a threshold $a_c = 4e\rho_c/s$. Once this threshold is reached, the next pulse is triggered at time $T_p = (R^2/4\nu)\tau_p + T_d$ where

$$\frac{4\rho_c}{sa} = g(\tau_p),$$

and T_d is the delay between attainment of threshold and pulsing. The speed of propagation of the wave is

$$v = \frac{R}{T_p} = \frac{(4/\pi a)^{\frac{1}{2}}}{\frac{\tau_p}{\pi \nu a} + T_d} = v(a).$$

The initial pulse is assumed to orient the direction of chemotactic response for succeeding pulses, so that we may assume that the response is movement a distance $\Delta(a)$ toward toward the center of aggregation. Early in aggregation it is likely that Δ is a constant, but it will become zero for dense packing. It is not known exactly when, following passage of a pulse, the movement begins, but it is reasonable to assume that there will be some delay followed by a response lasting a time T_c. For *D. discoideum* we have the estimates $T, T_d, T_c = 300, 15, 100$ sec and $\Delta = 2\times 10^{-3}$ cm.

This model is especially attractive from the standpoint of direct simulation, since there is little difficulty in following a large number of interacting organisms. In order to study the equations in the continuum limit, consider one-dimensional aggregation at $x = 0$. Since there is no random component of the movement (diffusion of organisms) the solution actually contains at $x = 0$ a delta function of steadily increasing amplitude. Let us assume that the duration T_c of movement is actually small compared to the period T, so we may think of movement as an instantaneous response to a passing wave. If the nth pulse initiates a wave of responses at position $x_n(t)$, then we take

$$\frac{dx_n}{dt} = v(a_n) = \frac{\Delta(a_n)}{T_c},$$

where $a_n(x)$ is the acrasin distribution between the fronts of the waves for the nth and $(n-1)^{\text{st}}$ pulses. Also

$$\frac{a_n(x) - a_{n-1}(x)}{T} = \frac{a_{n-1}(x + \Delta(a_{n-1})) - a_{n-1}(x)}{T}.$$

In order to pass to the limit, let $a(x, nT) = a_n(x)$ and take Δ as small. Then, with $\Delta(a)/T = -Q'(a)$,

$$\frac{\partial a}{\partial t} + \frac{\partial Q(a)}{\partial x} = 0,$$

which is an equation for $a(x, t)$. With a solution given, the paths of pulsatile waves may be determined by $\dot{x} = ka^{-1/2}$. Conversely, the pattern of these organizing waves can be regarded as responsible for the chemotactic pattern $a(x, t)$.

6.2. Other Aspects of Chemotaxis

The instability leading to initiation of aggregation in the Keller-Segel model for slime mold is unusual in the infinite critical wavelength, and it is of interest to understand whether or not this phenomenon should be regarded as typical of pattern formation involving active transport, chemotaxis, or other mechanisms leading to directed movement. Also, it is of interest to extend the linear instability analysis to a nonlinear regime where the course of such an instability can be followed. We take up these questions in the present section.

6.2.1. An Augmented Turing System. Let us consider a system of n reaction-diffusion equations incorporating source terms proportional to a cell density a. The cell density will obey a conservation equation describing chemotaxis in response to a combined chemical gradient. The system is therefore a generalization of both the single-attractant slime mold model as well as the reaction-diffusion model of pattern formation.

The *linearized* version of this system is (now dropping the primes used in Section 6.1)

$$\frac{\partial \rho_i}{\partial t} = f_i a + \sum_{j=1}^{n} \alpha_{ij}\rho_j + \nu_i \nabla^2 \rho_i, \quad i = 1, 2, \ldots, n,$$

$$\frac{\partial a}{\partial t} = -\sum_{i=1}^{n} \kappa_i \nabla^2 \rho_i + \mu \nabla^2 a,$$

where certain of the κ_i may be zero if not all chemicals affect movement. If $(a, \rho) = (\hat{a}, \hat{\rho})e^{i(\mathbf{k}\cdot\mathbf{r}+\sigma t)}$ and $\hat{a}$ is eliminated, the vector $\hat{\rho}$ satisfies $L\hat{\rho} = 0$, where $L = \sigma^2 + A\sigma + B$ and

$$\begin{aligned} A &= P + k^2 Q, & B &= \mu k^2(k^2 S - R), \\ P_{ij} &= -\alpha_{ij}, & R_{ij} &= \mu^{-1} f_i \kappa_j + \alpha_{ij}, \\ Q_{ij} &= (\mu + \nu_i)\delta_{ij}, & S_{ij} &= \nu_i \delta_{ij} \quad \text{no summation.} \end{aligned}$$

To examine the onset of nonoscillatory instability, we consider the case of infinitesimal σ, which reduces the problem to the properties of B. The problem is then completely analogous to a problem in pattern formation involving the augmented matrix $\tilde{\alpha}_{ij} = R_{ij}$.

This suggests how to approach the questions raised earlier regarding the prevalence of infinite critical wavelength. If we can show that the augmented matrix $\tilde{\alpha}$ can generate a diffusive instability with finite chemical wavelength even though α cannot, then the critical wavelength will be finite even though the mechanism of instability involves chemotaxis. We will show that while this situation cannot arise with one chemical (as in the reduced Keller-Segel model) it can appear with two chemicals. In that case we recall that the conditions for a diffusive instability are

$$\alpha_{11} + \alpha_{22} < 0, \quad D = \alpha_{11}\alpha_{22} - \alpha_{12}\alpha_{21} > 0,$$

$$\nu_1 \alpha_{22} + \nu_2 \alpha_{11} > 2(\nu_1 \nu_2 D)^{\frac{1}{2}}.$$

If the last inequality is reversed the instability does not occur. If we choose $\nu_1 = 1$, $\nu_2 = 4$, $\mu^{-1} f_1 = \mu^{-1} f_2 = 1$, $\kappa_1 = 1$, $\kappa_2 = 0$, and

$$\alpha = \begin{pmatrix} 2 & -3 \\ 5 & -5 \end{pmatrix}, \qquad \nu_1 \alpha_{22} + \nu_2 \alpha_{11} = 3 < 2\sqrt{5},$$

then

$$\tilde{\alpha} = \begin{pmatrix} 3 & -3 \\ 6 & -5 \end{pmatrix}, \qquad \nu_1 \tilde{\alpha}_{22} + \nu_2 \tilde{\alpha}_{11} = 7 > 2\sqrt{3}.$$

Thus instability occurs in the second case but not the first.

It therefore appears that the chemotaxis terms can lead to instability at finite wavelength provided there is already in the diffusion-reaction kinetics a destabilization with decreasing (large) wavelength, even though instability is not realized at *any* wavelength without chemotaxis terms. In a sense the chemotaxis realizes the capacity of the system for diffusive instability.

A more extreme case would be the "absolutely stable" one where α is symmetric negative definite. Let us suppose that the null space associated with the numerically largest eigenvalue of α is one-dimensional, spanned by the eigenvector v. If we choose both f and κ to be parallel to v, and if the product of their lengths is chosen to make the maximum eigenvalue of R equal to zero, it then follows that

$$\rho \cdot B \cdot \rho > 0, \quad k \neq 0,$$

and so the instability sets in with infinite wavelength. In practice, of course, domain size and shape will determine the critical conditions that must be attained for instability at finite wavelength.

6.2.2. Nonlinear Stability Theory. The equations of the Keller-Segel model will now be examined from the standpoint of nonlinear stability theory. We start with (see Section 6.1)

$$\frac{\partial \rho}{\partial t} = a f(\rho) - g(\rho) + \nu \nabla^2 \rho,$$
$$\frac{\partial a}{\partial t} = \nabla \cdot [\mu(a, \rho) \nabla a - \kappa(a, \rho) \nabla \rho].$$

From the determinantal condition $\sigma^2 + A\sigma + B = 0$ for the growth rate, we recall that near the onset of instability we have the ordering $\sigma \sim k^4 \sim \Delta^2$, where $\Delta = \mu_0 \alpha_0 + \kappa_0 f_0 = \epsilon r$ will now play the role of a small expansion parameter. It will turn out that the nonlinear effect will be significant when the amplitude of perturbations becomes comparable to ϵ, so we assume expansions of the form

$$a(\mathbf{x}, t; \epsilon) = a_0 + \epsilon a_1(\tilde{\mathbf{x}}, \tilde{t}) + \epsilon^2 a_2(\tilde{\mathbf{x}}, \tilde{t}) + \cdots,$$
$$\rho(\mathbf{x}, t; \epsilon) = \rho_0 + \epsilon \rho_1(\tilde{\mathbf{x}}, \tilde{t}) + \epsilon^2 \rho_2(\tilde{\mathbf{x}}, \tilde{t}) + \cdots,$$

where a_0 and ρ_0 are equilibrium values, and we have introduced $\tilde{\mathbf{x}} = (\tilde{x}, \tilde{y}) = \epsilon^{1/2} \mathbf{x}, \tilde{t} = \epsilon^2 t$. Introducing these expansions and collecting terms of like order we have

$$\epsilon(f_0 a_1 + \alpha_0 \rho_1) + \epsilon^2 \left[f_0 a_2 + \alpha_0 \rho_2 + \nu \tilde{\nabla}^2 \rho_1 + \rho_1 a_1 f'(\rho_0) - \frac{1}{2} g''(\rho_0) \rho_1^2 \right] + O(\epsilon^3) = 0,$$

$$\epsilon^2 \left[\mu_0 \tilde{\nabla}^2 a_1 - \kappa_0 \tilde{\nabla}^2 \rho_1 \right] + \epsilon^3 \left\{ \mu_0 \tilde{\nabla}^2 a_2 - \kappa_0 \tilde{\nabla}^2 \rho_2 - \frac{\partial a_1}{\partial \tilde{t}} + \tilde{\nabla} \cdot \left[\left(\frac{\partial \mu}{\partial a} \right)_0 a_1 \tilde{\nabla} a_1 + \left(\frac{\partial \mu}{\partial \rho} \right)_0 \rho_1 \tilde{\nabla} a_1 - \left(\frac{\partial \kappa}{\partial a} \right)_0 a_1 \tilde{\nabla} \rho_1 - \left(\frac{\partial \kappa}{\partial \rho} \right)_0 \rho_1 \tilde{\nabla} \rho_1 \right] \right\} + O(\epsilon^4) = 0.$$

Adding $\kappa_0 \tilde{\nabla}^2$ times the first equation to α_0 times the second, and using $\mu_0 \alpha_0 + f_0 \kappa_0 = \epsilon r$, we obtain

$$r \tilde{\nabla}^2 a_1 - \alpha_0 \frac{\partial a_1}{\partial \tilde{t}} + \nu \kappa_0 \tilde{\nabla}^4 \rho_1 + Q(a_1, \rho_1) = O(\epsilon),$$

where the quadratic terms are collected in Q. In this expression we may now substitute $\mu_0 a_1 / \kappa_0$ for ρ_1, to obtain

$$-\alpha_0 \frac{\partial a_1}{\partial \tilde{t}} + r \tilde{\nabla}^2 a_1 + \nu \mu_0 \tilde{\nabla}^4 a_1 + q \tilde{\nabla}^2 a_1^2 = 0,$$

where q is a constant. Now $\alpha_0 < 0$ to ensure stability of the chemical system in the absence of amoebae, and $\nu \mu_0 > 0$, so under the scaling

$$\tilde{\nabla}^2 = \frac{1}{\nu \mu_0} \nabla^{*2}, \quad \frac{\partial}{\partial \tilde{t}} = -\frac{1}{\alpha_0 \nu \mu_0} \frac{\partial}{\partial t^*}, \quad a = q^{-1} a^*,$$

we have

$$\frac{\partial a_1^*}{\partial t^*} + r \nabla^{*2} a_1^* + \nabla^{*4} a_1^* + \nabla^{*2} (a_1^*)^2 = 0.$$

It can be shown that this nonlinear equation is also obtained locally in a general augmented nonlinear Turing system.

Now dropping stars and specializing to one dimension, we seek the *steady* solutions, assumed to be periodic with wavelength unity. By requiring $a_x = a_{xxx} = 0$ at $x = 0, 1$, we obtain after two integrations

$$a_{xx} + ra + a^2 = C = \text{const}.$$

Multiplying by a_x and integrating again yields

$$a_x^2 = -\frac{2}{3} a^3 - r a^2 + 2 C a + D.$$

Under the shift $a = A + h$, we move the middle root of the cubic to the origin to obtain

$$A_x^2 = \frac{2}{3} (A + A_1) A (A_0 - A), \quad 0 \le A \le A_0,$$

where A_0 and A_1 are positive numbers. Since amoebae are conserved, the mean of the perturbation a must vanish, giving $h = -\bar{A}$. The relation of r to the sum of the roots then yields

$$r = 2\bar{A} + \frac{2}{3}(A_1 - A_0).$$

The integral

$$\pm\sqrt{\frac{3}{2}}x = \int_0^A [A(A_0 - A)(A + A_1)]^{-\frac{1}{2}}\, dA$$

may be inverted in terms of the elliptic function sn (analogous to sine). We then have

$$A = A_0 \frac{(1-k^2)sn^2(\xi, k)}{1 - k^2 sn^2(\xi, k)}, \quad k^2 = \frac{A_0}{A_0 + A_1}, \quad \xi = \frac{x}{k}\sqrt{\frac{A_0}{6}},$$

where now k is a parameter that varies between 0 and 1. The half-period $K(k)$ may be related to the amplitude A_0 if the x-period is fixed as above at unity. We may then also obtain r in terms of k:

$$r = 48K\left(E + \frac{1}{3}k^2 K - \frac{2}{3}K\right), \quad A_0 = 24K^2k^2.$$

Here

$$E(k) = \int_0^{\pi/2} (1 - k^2 \sin^2\theta)^{\frac{1}{2}}\, d\theta,$$

$$K(k) = \int_0^{\pi/2} (1 - k^2 \sin^2\theta)^{-\frac{1}{2}}\, d\theta,$$

Thus for k small the value of r is positive, corresponding to the enhanced chemotaxis that is needed to sustain a pattern of finite (but large) wavelength. But as k increases, increasing the amplitude, r actually *decreases*. Thus in this model the chemotactic instability is *subcritical*, this being in part a consequence of bifurcation from infinite wavelength.

We expect to find, however, that these new equilibria are unstable to small perturbations, this being a direct consequence of the subcritical nature of the instability. If the parameters are such that the homogeneous equilibrium is linearly stable, but only just so, it is possible nevertheless for instability to occur as a result of a perturbation of finite amplitude. These results do not tell us what the ultimate fate of the pattern will then be, only that it will differ substantially from homogeneity.

6.2.3. A Special Case. We have examined the beginning of a branch of nonlinear equilibria, and in order to see how it might be continued, we specialize to the simplest nonlinear system (with all parameters being constant and α normalized to be -1):

$$fa - \rho + \nu\rho_{xx} = 0, \quad \mu a_{xx} - \kappa(a\rho_x)_x = 0.$$

For stationary patterns on a surface, with zero cell flux around the boundary, we must have zero flux everywhere,

$$a_x - \lambda a\rho_x = 0, \quad \lambda = \frac{\kappa}{\mu},$$

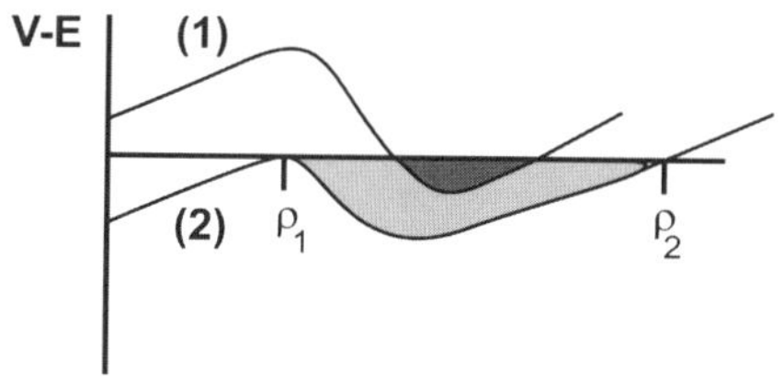

FIGURE 6.1. Subcritical bifurcation.

and therefore $a = (A\lambda/2f)e^{\lambda\rho}$, where A is arbitrary. The integral of the system then has the form

$$\nu(\rho_x)^2 + V(\rho) = E, \quad V(\rho) = ae^{\lambda\rho} - \rho^2,$$

where E is "total energy." The oscillatory pattern near the subcritical bifurcation corresponds to the shaded potential well on curve (1); see Figure 6.1. A highly nonlinear case is the "solitary bump" represented by curve (2).

Let us study somewhat further the family of solutions having the latter form. To obtain a double root at ρ_1 we must have

$$E = -\rho_1^2 + Ae^{\lambda\rho_1}, \quad 2\rho_1 = \lambda Ae^{\lambda\rho_1}.$$

At $x = \pm\infty$ we must satisfy $\rho = \rho_1, a = a_1 = \frac{\rho_1}{f} = (A\lambda/2f)e^{\lambda\rho_1}$, and therefore

$$A = \frac{2\rho_1}{\lambda}e^{-\lambda\rho_1}, \quad E = -\rho_1^2 + 2\frac{\rho_1}{\lambda}.$$

Thus ρ_2 satisfies

$$\rho_1^2 - \frac{2\rho_1}{\lambda} = \rho_2^2 - Ae^{\lambda\rho_2} = \rho_2^2 - \frac{2\rho_1}{\lambda}e^{\lambda(\rho_2-\rho_1)}.$$

Without loss of generality we now take $\rho_1 = 1$, in which case the solution $\rho_2(\lambda)$ of the last equation is monotone decreasing to zero at $\lambda = 1$. As $\lambda \to 0$,

$$\rho_2^2 \sim \frac{2}{\lambda}e^{\lambda\rho_2} \sim \left(\frac{\ln\lambda}{\lambda}\right)^2,$$

and also

$$a_2 = \frac{1}{f}e^{\lambda(\rho_2-1)} \sim \frac{(\ln\lambda)^2}{2\lambda f}.$$

If indeed these solutions are unstable, as seems likely, the indication is that an isolated local concentration of cells will continue to grow indefinitely supplied by the ambient cells, and to contract to an ever-smaller region. This would realize the time-dependent singular solutions envisaged by Nanjundiah [**124**].[1]

For the corresponding radially symmetric solutions Nanjundiah shows that an additional instability of the time-dependent chemotaxis can occur, in which the angular distribution of cell density becomes uneven. This may be associated with

[1] It would be of interest to examine the possibility of self-similar solutions of the nonlinear equations representing locally such a buildup of cell density. Added note: This problem was subsequently tackled by the authors; see the Supplemental Notes below.

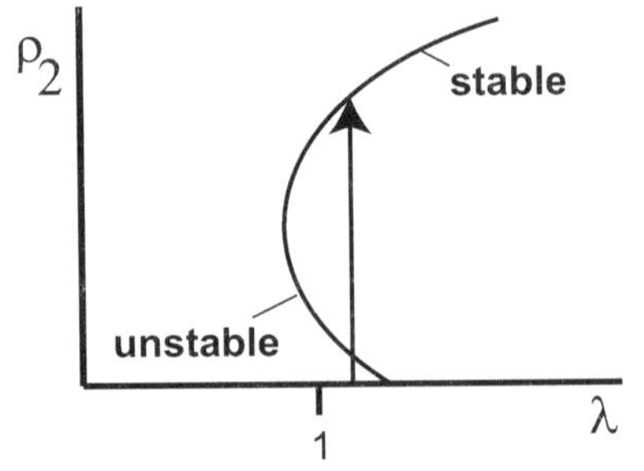

FIGURE 6.2. Aggregation by a subcritical bifurcation.

the *streaming* that occurs during the final stages of slime mold aggregation, the flux of cells being in thin rivulets that coalesce as the center is approached; see the aggregation phase in Figure 3.1.

The situation is quite different if solutions are (fully) spatially periodic, since the number of cells then available to a single center of aggregation is finite. If $\rho_0 = 1$ the one-dimensional periodic solutions of infinitesimal amplitude appear once $E + V(1) = V'(1) = 0$ and $-V''(1) = 2(1 - \lambda) < 0$. The periodic solutions thus branch out from the λ-axis in the region $\lambda > 1$, the initial wavelength being

$$\int_{\rho_1}^{\rho_2} \frac{d\rho}{\sqrt{E - (\lambda - 1)(\rho - 1)^2}} = \frac{\pi}{\sqrt{\lambda - 1}} = L_0.$$

A given branch is now fixed by this wavelength and the total cell number in the territory, $L_0 a_0 = L_0/f$. These two conditions imply a relation between $a_2 - a_0$ and λ, but it is difficult to make this explicit. It is possible that in this case the bifurcation diagram contains a second branch of stable steady solutions, although this has yet to be demonstrated for this model. If this were true, it would be possible to view aggregation with finite territory size as a finite amplitude instability through the unstable branch, culminating in a stable concentration of cells; see Figure 6.2.

Some evidence for these conclusions is provided by an asymptotic time-independent solution to the equations when λ is *large*. We set

$$\frac{A\lambda}{2f} = \lambda B e^{-\lambda \rho_m},$$

where ρ_m is the maximum value of ρ. Assuming that $x = \pm L_0/2$ are maxima, we neglect chemotaxis in the interval $0 \le |x| < L_0/2$ and therefore have

$$\rho = \frac{\cosh(x/\sqrt{\nu})}{2f\sqrt{\nu}\sinh(L_0/2\sqrt{\nu})} = \rho^*(x).$$

The periodic extension has a discontinuity of derivative at the maxima. To smooth this out, we need a boundary-layer approximation near the maxima. Near $x = L_0/2$ let

$$\xi = \frac{\left(x - \frac{L_0}{2}\right)\lambda}{\sqrt{\nu}}, \qquad w = (\rho - \rho_m)\lambda.$$

Then as $\lambda \to \infty$ with B fixed the approximate equation for w is

$$Be^{w} + w_{\xi\xi} = 0.$$

Integrating once and requiring that w_ξ vanish at $\xi = 0$, we have

$$Be^{w} + \frac{1}{2}w_\xi^2 = B.$$

The solution is

$$w = \ln \operatorname{sech}^2\left(\sqrt{\frac{B}{2}}\xi\right), \qquad B = \frac{1}{8f^2\nu}.$$

$$\rho = \frac{1}{\lambda}\left[w\left(\frac{\lambda}{\sqrt{\nu}}\left(x - \frac{L_0}{2}\right)\right) + w\left(\frac{\lambda}{\sqrt{\nu}}\left(x + \frac{L_0}{2}\right)\right)\right] + \rho^*(x) = \frac{L_0}{2f\nu}.$$

The corresponding cell density near $\frac{L_0}{2}$ is

$$a = \lambda B \operatorname{sech}^2\left(\sqrt{\frac{B}{2}}\xi\right).$$

Thus the amoeba density is exponentially small except near the points of aggregation.

CHAPTER 7

Cell Proliferation

7.1. Homogeneous Population

The development of structure is of course the epitome of embryogenesis of multicelled organisms. Perhaps more primitive but even more blatant is the sheer increase in size of the organism. The two, however, are clearly related as soon as we focus upon the growth of substructures. We now turn to quantitative aspects of the cell division process that is responsible for growth, focusing, however, upon macroscopic rather than biochemical aspects of the process. A major objective will be to assess the control of cell division and, in particular, the relative roles of genetic, external, autonomous, and stochastic factors.

There are a number of different types of cell division, ranging from simple *cleavage*—binary division without intermediate growth—through *fragmentation* into many daughter cells after parasitic cytoplasmic growth in a host cell (viruses, malaria, etc.). The division can be strictly nuclear (*noncellular*) with no cytoplasmic walls, as in early insect cleavage or the growth of a cellular slime mold, or with full cytoplasmic dividing walls. The division can be very *unequal*, as in *budding*, with a small daughter forming as a surface bleb and pinching off from its mother after reaching her size, or as in *filamentous growth* of fungi and some algae, which is like budding in a sharply defined geometry. However, our main interest will be in *fission*, the dominant mode in bacterial, animal, and plant cells, in which the parent cells grow to about double their original size, then split into similar daughters by constriction (protozoan, animal) or construction of a cross-wall (bacterium, plant).

7.1.1. Stem Cell Proliferation. Cell division may or may not be associated with differentiation. The simplest case is that of *exponential growth* in the absence of an inhibitory environment and its substantial modification when cell death, mitotic inhibition, and varying cell cycle time are taken into account. A qualitative change occurs when *cell differentiation* can take place along the way, performing a discontinuous change in the growth parameters. The special case that we will consider in this section is that of *stem cell* proliferation, in which only the undifferentiated stem cells continue to divide, the differentiated ones being at the end of their developmental line, or differentiating without division a few more times (or even dividing thereafter a *fixed* number of times). Proliferation via stem cells is the norm in *cell renewal systems*, such as that of red blood cell production, requiring continued supply of a depleted differentiated cell population, or in systems requiring a reliable switching on of a quiescent differentiated cell type, as in *regeneration* of organs of animals, or indeed in the *immunological* system.

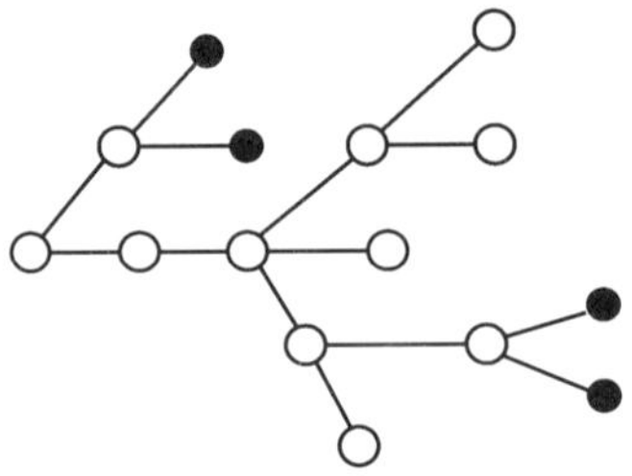

FIGURE 7.1. Tree model of birth-death process.

The first quantitative work on stem cell systems seems to have been done on the mouse red blood cell forming system [**153**]. A cell mixture containing stem cells was injected into sterilized spleen as the culture medium, and the stem cell number distribution of the colonies formed after a definite time (number computed by reinjecting into spleen and counting colonies) was tallied. This distribution was compared with those resulting from numerical simulations employing simple differentiation probabilities as a control mechanism.

More recently the regeneration of *hydra* has been examined from a similar viewpoint to decide between simple probabilistic mechanisms and those involving more explicit control by changing microscopic and macroscopic environment [**67**]. During regeneration in whole or in part (or with normal asexual budding) the nonproliferating *nerve* and *nematoblast* cells are produced (at 1:3 ratio) by I (*interstitial*) cells as stem cell populations. The remaining *epithelial, gland*, and *mucous* cells form a separate proliferating population. The proportion of nerve, nematoblast, and interstitials varies along the body column in a way that is consistent with a proliferation probability ranging only from .6 to .7. Here the quantitative regeneration experiments seed stem cells into N-mustard sterilized aggregates with I-cell colonies directly tallied after a given time.

The experimental information in practice is restricted to the stem cell population developing from a single stem cell, and not to differentiated end products. We can take advantage of this by regarding a stem cell either as producing two stem cells (one stem cell is "born") or two differentiated cells (one stem cell "dies"), and it is the time development of this "*birth-death*" process (see Figure 7.1) that we want to investigate. We shall examine two extreme models, depending upon the assumed dynamics of cell birth and death. In the first model, we assume a spatially independent constant rate of cell production per cell of p per unit time and of death $q < p$ per unit time. $P_N(t)$, the probability of N cells at time t, can increase by birth from any of $N-1$ cells in a population of $N-1$ cells, or by death from a population of $N+1$ cells, or decrease by birth or death from an N-cell population. Hence we have the *master equation*

$$\dot{P}_N(t) = p(N-1)P_{N-1}(t) + q(N+1)P_{N+1}(t) - pN P_N(t) - qN P_N(t),$$

and for an initial single stem cell

$$P_N(0) = \delta_{N,1}.$$

These are most expeditiously solved by constructing the *generating function*

$$Q(x,t) = \sum_0^\infty x^N P_N(t),$$

x^N serving as the "tag" of an N-cell population. Now on multiplying by N and summing over N,

$$\begin{aligned}\frac{\partial}{\partial t} Q(x,t) &= \left(px + \frac{q}{x}\right) x \frac{\partial}{\partial x} Q(x,t) - (p+q)x \frac{\partial}{\partial x} Q(x,t) \\ &= (px-q)(x-1)\frac{\partial}{\partial x} Q(x,t),\end{aligned}$$

with $Q(x,0) = x$. To solve, we have

$$\left(\frac{\partial t}{\partial x}\right)_Q = -\frac{\partial Q/\partial x}{\partial Q/\partial t} = -\frac{1}{(px-q)(x-1)},$$

with $t = 0$ at $Q = x$, or integrating,

$$t = \frac{1}{p-q}\left(\ln\frac{1-x}{q-px} - \ln\frac{1-Q}{q-pQ}\right),$$

from which

$$Q(x,t) = \frac{q(e^{(p-q)t}-1) - (qe^{(p-q)t}-p)x}{(pe^{(p-q)t}-q) - p(e^{(p-q)t}-1)x}.$$

This model can now be assessed in various ways. To start with, $Q(x,\infty) = q/p$, so that $P_0(\infty) = q/p$, the *extinction probability*, while $P_N(\infty) \to 0$ as $N \to \infty$ (coefficient of x^N in q/p equals 0), but in such a way that $\sum_1^\infty P_N(\infty) = 1 - q/p$ in order to maintain the normalization of probability, $\sum_0^\infty P_N(\infty) = 1$. Next, the population mean is

$$\bar{N}(t) = \sum N P_N(t) = \left.\frac{\partial Q(x,t)}{\partial x}\right|_{x=1},$$

so that

$$\bar{N}(t) = -\frac{qe^{(p-q)t}-p}{p-q} + \frac{p(e^{(p-q)t}-1)}{(p-q)^2}(p-q)$$

or

$$\bar{N}(t) = e^{(p-q)t},$$

yielding steady exponential growth. Similarly, for the variance

$$\sigma^2(t) = \sum N^2 P_N(t) - \bar{N}^2(t) = Q''(1,t) + Q'(1,t) - [Q'(1,t)]^2,$$

a brief computation yields

$$\sigma^2(t) = e^{(p-q)t}\,\frac{e^{(p-q)t}-1}{p-q}.$$

But the mean and variance as functions of time are not experimentally available, only the distribution at a fixed time t, the conclusion of the experiment. Taking the coefficient of x^N we now find

$$P_0(t) = \frac{q(1-e^{-(p-q)t})}{p-qe^{-(p-q)t}},$$

$$P_N(t) = \frac{(p-q)^2 e^{(p-q)t}}{p(e^{(p-q)t}-1)(pe^{(p-q)t}-q)} \left(\frac{1-e^{-(p-q)t}}{1-\frac{q}{p}e^{-(p-q)t}} \right)^N$$

for $N > 0$. The asymptotic distribution is most easily visualized in terms of the normalized variable

$$w = \frac{N}{\bar{N}(t)},$$

with corresponding probability distribution

$$P(w,t) = \bar{N}(t) P_{\bar{N}(t)w}(t),$$

normalized so that $\sum P(w,t)\Delta w = 1$, $\Delta w = \frac{1}{\bar{N}(t)}$, or $\int P(w,t)dw = 1$ as $t \to \infty$. Explicitly, we find

$$P(w,\infty) = \frac{q}{p}\delta(w) + \left(1-\frac{q}{p}\right)^2 e^{(1-q/p)w},$$

giving extinction grafted onto exponential falloff. Experimental comparison is facilitated by rewriting in terms of the measurable parameters $\bar{N}(t)$ and $\sigma^2(t)$ at the final time; this yields

$$P(w,\infty) = \frac{1-r^2}{1+r^2}\delta(w) + \left(\frac{2}{1+r^2}\right)^2 e^{-\frac{2w}{1+r^2}} \quad \text{where } r^2 \equiv \frac{\sigma^2(t)}{\bar{N}(t)(\bar{N}(t)-1)}$$

and a somewhat messier expression for finite t.

This first model does not make much biological sense: the probability of a stem cell dividing for the first time between T and $T+dt$ would be

$$\rho(T)dt = (1-p\,dt)^{\frac{T}{dt}} p\,dt,$$

or

$$\rho(T) = pe^{-pT},$$

i.e., a *Poisson process*, which is far from consistent with the well-known minimum time for a cell to divide. We are then led to a second extreme model, in which the cell lifetime is fixed—say one time unit after which it either gives birth or dies, now with respective probabilities p and $q = 1-p < p$. This is a substantially harder problem, although setting it up is easy enough. We again define the generating function

$$Q_t(x) = \sum x^N P_N(t), \quad Q_0(x) = x,$$

for integer t. Since each cell at time t becomes 0 cells at time $t+1$ with probability p or 2 cells with probability q, the time step is readily seen to be equivalent to the replacement

$$x \to f(x) = q + px^2.$$

Thus

$$Q_{t+1}(x) = Q_t(f(x)), \quad Q_0 = x,$$

or iterating,

$$Q_t = f(f(f \cdots (x) \cdots))) \equiv f^t(x).$$

Let us assess this model as we did the previous one. First, since

$$Q_{t+1}(x) = f(Q_t(x)),$$

$P_0(t) = Q_t(0)$ satisfies

$$P_0(t+1) = f(P_0(t)),$$

and the extinction probability $P_0(\infty)$ must be a *fixed point* of the transformation $x \to f(x)$:

$$P_0(\infty) = f(P_0(\infty)).$$

Here then $q + pP_0^2 = P_0 = (q+p)P_0$, so that

$$P_0(\infty) = \frac{q}{p},$$

as well as the possibility $P_0 = 1$.

The latter, however, is an unstable fixed point and will not be attained. That is, suppose we are close to a fixed point x_0:

$$P_0(t) = x_0 + \Delta_t$$

with Δ_t small. Will we get closer or further away? Now

$$\Delta_{t+1} = P_0(t+1) - x_0 = f(x_0 + \Delta_t) - x_0 = f(x_0) + \Delta_t f'(x_0) + \cdots - x_0,$$

or

$$\Delta_{t+1} = f'(x_0)\Delta_t,$$

so that

$$x_0 \begin{cases} \text{is stable} & \text{if } |f'(x_0)| < 1, \\ \text{is unstable} & \text{if } |f'(x_0)| > 1. \end{cases}$$

In the present case, $f'(1) = 2p > 1$ is unstable, while $f'(q/p) = 2q < 1$ is stable. It is of course also necessary to show that the sequence $P_0(t)$ converges to the stable fixed point, but here indeed

$$\left(\frac{q}{p} - P(t+1)\right) = p\left(\frac{q}{p} + P(t)\right)\left(\frac{q}{p} - P(t)\right) < 2q\left(\frac{q}{p} - P(t)\right)$$

decreases monotonically to 0.

Next, how does the population grow? We again need $\bar{N}(t) = Q_t'(1)$, and this is immediately obtainable, using the fact that 1 is a fixed point, stable or not: $f(1) = 1$, and that $Q_t(1) = 1$. From $Q_{t+1}(x) = f(Q_t(x))$, we have directly by differentiation $Q_{t+1}'(1) = Q_t'(1)f'(Q_t(1)) = Q_t'(1)f'(1)$. But $Q_0'(1) = 1$, and so

$$\bar{N}(t) = m^t,$$

where $m = f'(1) = 2p$ is the multiplication factor, obvious from the fact that $q+px^2$ takes one cell into $1 \cdot 0 + 2 \cdot p = 2p$ on the average. In fact, all population moments are found the same way, e.g.,

$$\overline{N^2}(t) = \sum N^2 P_N(t) = \left(x\frac{\partial}{\partial x}\right)^2 \sum x^N P_N(t)\Big|_{x=1} = Q_y''(1) + Q_t'(1).$$

But

$$Q''_{t+1}(x) = Q''_t(x)f'(Q_t(x)) + \left[Q'_t(x)\right]^2 f''(Q_t(x))$$

or

$$Q''_{t+1}(1) = mQ''_t(1) + m \cdot m^{2t}.$$

Since $Q''_0(1) = 0$, it follows from $\frac{Q''_{t+1}(1)}{m^{t+1}} = \frac{Q''_t(1)}{m^t} + m^t$ that

$$Q''_t(1) = m^t\left(\frac{m^t - 1}{m - 1}\right),$$

yielding a relative standard deviation

$$\frac{\sigma(t)}{\bar{N}(t)} = \left(\frac{\bar{N}^2(t) - \bar{N}(t)^2}{\bar{N}(t)^2}\right)^{\frac{1}{2}} = \left(\frac{2-m}{m-1}(1 - m^{-t})\right)^{\frac{1}{2}} \to \left(\frac{2-m}{m-1}\right)^{\frac{1}{2}}$$

as $t \to \infty$.

Finally, what is the population distribution, even asymptotically? In fact, to solve the finite-t case, it is enough to solve the asymptotic problem. Suppose

$$\phi_t(s) = \left\langle e^{-sN/m^t}\right\rangle = \sum \left(e^{-s/m^t}\right)^N P_N(t) = Q_t\left(e^{-s/m^t}\right)$$

is the moment-generating function scaled by the population mean m^t, and

$$\phi(s) = \phi_\infty(s)$$

its asymptotic limit, which can be shown to exist. Then from $Q_{t+1}(x) = f(Q_t(x))$, $Q_t(1) = 1$, and $Q'_t(1) = m$, it follows that

$$\phi(ms) = f(\phi(s)), \quad \phi(0) = 1, \ \phi'(1) = -1,$$

allowing ϕ to be determined in principle. If this could also be done in practice, then from

$$f(x) = \phi(m\phi^{-1}(x))$$

we would have $Q_t(x) = f^t(x) = (\phi m \phi^{-1})^t(x)$, or

$$Q_t(x) = \phi(m^t\phi^{-1}(x)),$$

completely explicit. For example, the Poisson case is generated by

$$\text{Poisson:} \quad \phi(s) = \frac{p - q + qs}{p - q + ps}.$$

The asymptotic ϕ is, however, not known exactly in any other significant case, although it can be expanded about the Poisson form [**129**]. Approximations are required, and the simplest by far is an intelligent fixed point expansion. We therefore return to the form

$$Q_{t+1}(x) = f(Q_t(x)),$$

which itself can be used to sequentially improve any approximation to Q_t. The idea is to approximate the nonlinear recursion by a linear fractional recursion that is trivial to evaluate.

Suppose we are interested primarily in moments. Then the fixed point $x_0 = 1$ is appropriate. The present case $f(x) = px^2 + q$ illustrates the general case very well. We write $y = px^2 + q$ as

$$\begin{aligned} y - 1 = px^2 + q - 1 &= m(x-1) + \frac{m}{2}(x-1)^2 \\ &= m(x-1)\left(1 + \frac{1}{2}(x-1)\right) \cong \frac{m(x-1)}{1 - \frac{1}{2}(x-1)}. \end{aligned}$$

But if

$$Y = F(X) = \frac{\alpha X}{1 - \beta X}, \qquad Z = G(Y) = \frac{\alpha' Y}{1 - \beta' Y},$$

then

$$Z = GF(X) = \frac{\alpha\alpha' X}{1 - (\beta + \alpha\beta')X}.$$

It readily follows that

$$F^t(X) = \frac{\alpha^t X}{1 - \beta\frac{\alpha^t - 1}{\alpha - 1}(x-1)},$$

and so in the present case

$$Q_t(x) \cong 1 + \frac{m^t(x-1)}{1 - \frac{1}{2}\frac{m^t - 1}{m-1}(x-1)}.$$

Indeed, for this approximation, we have that $Q_t(1) = 1$, $Q_t'(1) = m^t$, and $Q_t''(1) = m^t(m^t - 1)/(m-1)$ are now exact, and even $P_0(\infty) = Q_\infty(0) = 3 - 2m = 3 - 4p = 1 - 2\epsilon$ for $p = \frac{1}{2} + \epsilon$ compares tolerably to $q/p = (1-2\epsilon)/(1+2\epsilon)$ for small ϵ. Explicitly, taking coefficients of x^N,

$$\left.\begin{aligned} P_0(t) &= 1 - \frac{m^t}{1 + \frac{1}{2}\frac{m^t-1}{m-1}} \sim 3 - 2m \\ P_N(t) &\sim \frac{4(m-1)^2}{m^t} e^{-\left(2(m-1)/m^t\right)N} \end{aligned}\right\} \quad \text{for large } t.$$

This is not merely similar to the Poisson case; in terms of $r^2 = \sigma^2/\bar{N}(\bar{N}-1) = (2-m)/(m-1)$ and scaled cell number, it becomes

$$P(w) = \frac{1 - r^2}{1 + r^2}\delta(w) + \left(\frac{2}{1 + r^2}\right)^2 e^{-\frac{2w}{1+r^2}},$$

identical with the Poisson case. Although there are certainly differences at small N, Till et al. find that experimental, Poisson, and fixed generation time models done by Monte Carlo simulation give very close to the same curve for cumulative probability [**153**]. They conclude that the red blood cell system at least has its stem cell population under pure stochastic control; however, it is not clear that the extinguished populations were properly controlled in their comparison. David and MacWilliams, on the other hand, find in their work on hydra that the resetting of stem cell proliferation rate cannot be attributed to some global organismic feedback, but must have quite local control [**55**]. Thus local inhomogeneities must be attended to as well, and we will do so in due course.

7.1.2. Simple Population Growth. The distribution of cell generation times is neither Poisson nor of a fixed period. As a cell goes through its normal life cycle: G_1 (variable, absent in cleavage), S (synthesis), G_2 (mitotic apparatus prepared), M (mitosis), there is a fair amount of fluctuation in the duration of each period, even if the means are genetically determined. There is general agreement that *mitosis* (division) starts when some biochemical reaches a critical value, but the nature and control of this biochemical is not known. Models, however, are not in short supply. Older models tended to regard the *mitogen* as a mitotic inhibitor produced when daughters separate, whose concentration decreases as the cell enlarges until it no longer inhibits. Newer models focus more on the possibility of autonomous periodicity, i.e., by a *biochemical oscillation*. Let us briefly consider the genesis and consequences of one of these, denoting an activator mitogen concentration by y and its oscillating partner by x, both measured with respect to static equilibrium concentrations.

For an ideal linear oscillator, one would have equations of motion,

$$\dot{x} = -wy, \quad \dot{y} = wx,$$

giving rise to $(x, y) = r(\cos\theta, \sin\theta)$ where $\dot{\theta} = w$ and $\dot{r} = 0$. Growing or shrinking oscillation amplitude would be generated instead by

$$\dot{x} = \alpha x - wy, \quad \dot{y} = wx + \alpha y,$$

and we may again pass to polar coordinates, now with $\dot{\theta} = w$ and $\dot{r} = \alpha r$. Indeed, nothing changes if w and α are functions of r: the sets

$$\dot{x} = \frac{v(r)}{r}x - w(r)y, \quad \dot{y} = w(r)x + \frac{v(r)}{r}y,$$

and

$$\dot{r} = v(r), \quad \dot{\theta} = w(r),$$

are equivalent. A self-generated stable oscillation can now be produced, e.g., by $v(r) = a(1 - r^2)$, since $\alpha = a\frac{1-r^2}{r} > 0$ for $r < 1$ will expand the orbit spatially to $r = 1$; $\alpha < 0$ will similarly contract. This, together with $w(r) = \frac{w}{1+r}$, constitutes Goodwin's model [**79**].

Suppose now that the origin is shifted so that the true concentrations $X = 1 + x$ and $Y = 1 + y$ are nonnegative on the limit cycle $r = 1$. Does this possible biochemical system have any bearing on reality? For this purpose, Kauffman investigated the acellular slime mold *physarum*, in which cell walls do not form but nuclei divide, resulting in a *synctitium* [**103**]. The common cytoplasm means that all nuclei have the same biochemical milieu, and so the mitotic cycle is completely synchronous, creating a spatially homogeneous biochemical system. Under normal conditions, mitosis is supposed to occur when Y crosses the threshold Y_0 periodically; see Figure 7.2. To test this, one applies heat shock before mitosis. This degrades the enzymes producing X and Y, dropping the system point to 1 and so increasing

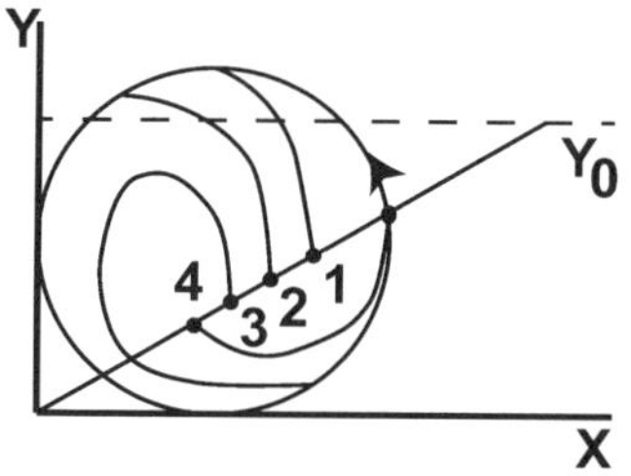

FIGURE 7.2. Oscillation of mitogen concentration.

the phase required to reach Y_0—a delay in mitotic time for one division. Further degradation produces further but still small delay, as with curve 2. Still further degradation causes the spiral to miss Y_0 completely, resulting in the loss of one cycle time (curve 3), increasing with further degradation, but then decreasing as one crosses to the other side of the spiral, becoming only a half cycle delay at curve 4. The observations are totally in accord with prediction.

In other than synctitial aggregates, biochemical concentrations and mechanical form can be quite variable, and it is to the resulting distribution of cycle times and desynchronization that we now look, but in the context of a homogeneous nondifferentiating population, corresponding to the few relevant (bacterial and protozoal) experiments available. As we know, however, birth and death translate at once to stem cell proliferation as well.

Suppose that a cell population divides synchronously at $t = 0^-$; we tally cell number normalized by cell number at $t = 0$, so that $N(0) = 1$. If there is a fixed cell cycle time γ until the next binary division, then clearly

$$N(t) = 2^{[t/\gamma]},$$

where $[\cdot]$ denotes the integer part, with growth of $\ln N$ as a staircase.

But there will in general be a distribution of cycle times $P(\gamma)$. We first consider the extreme case in which the precise value of γ is *inherited* by each daughter, the population of γ-values now yielding (in contrast to the above)

$$N(t) = \int_0^\infty P(\gamma) 2^{[t/\gamma]}\, d\gamma = \sum_{\mu=0}^{\infty} 2^\mu \int_{t/(\mu+1)}^{t/\mu} P(\gamma) d\gamma,$$

and the steps are increasingly smeared as generations continue. If $P(\gamma) = 0$ for $\gamma <$ some γ_0 (certainly the case), then in fact $\ln N(t) \to \frac{t}{\gamma_0} \ln 2$ for large t. In this case, however, synchrony is never entirely lost: expanding for large t,

$$
\begin{aligned}
N(t) &= \int_{\gamma_0}^{\infty} P(\gamma) e^{[t/\gamma]\ln 2}\, d\gamma = \int_0^{1/\gamma_0} \frac{1}{\lambda^2} P\Big(\frac{1}{\lambda}\Big) e^{[\lambda t]\ln 2}\, d\lambda \\
&= \int_0^{1/\gamma_0} \big(\gamma_0^2 P(\gamma_0) + \cdots\big) e^{[\lambda t]\ln 2}\, d\lambda \\
&= \frac{1}{t}\gamma_0^2 P(\gamma_0)\bigg(\bigg[1 + \bigg\{\frac{t}{\gamma_0}\bigg\}\bigg] 2^{[t/\gamma_0]} - 1 + \cdots\bigg),
\end{aligned}
$$

where $\{x\} \equiv x - [x]$ (an exact result if $P(\gamma) \propto \frac{1}{\gamma^2}$ for $\gamma > \gamma_0$).

The residual synchrony is due of course to the sharpness of the division at $t = 0$. Suppose we start instead with a steady state cell culture, achieved, e.g., by a constant rate of cell death: if γ is fixed, λ denotes death rate, and $0 < \theta < 1$ refers to position in the cell cycle, the conservation law and mitotic amplification conditions

$$
\frac{\partial}{\partial t} N(\theta, t) + \frac{\partial}{\partial \theta}\Big(\frac{1}{\gamma} N(\theta, t)\Big) = -\lambda N(\theta, t), \quad N(0, t) = 2N(1^-, t),
$$

have a time-independent solution when $\lambda = \frac{\ln 2}{\gamma}$ that is given by

$$
N(\theta) = 2^{-\theta} \ln 2
$$

(normalized so that $\int_0^1 N(\theta) d\theta = 1$). If this is "released" at time 0 by turning off λ, we easily find $N(\theta, t) = 2^{(t/\gamma - \theta)} \ln 2$ or $N(t) = 2^{t/\gamma}$. Now for a γ-distribution with $P(\gamma) = 0$ for $\gamma < \gamma_0$,

$$
\begin{aligned}
N(t) &= \int_{\gamma_0}^{\infty} P(\gamma) 2^{t/\gamma}\, d\gamma = \int_0^{1/\gamma_0} \frac{1}{\lambda^2} P\Big(\frac{1}{\lambda}\Big) e^{\lambda t \ln 2}\, d\lambda \\
&= \int_0^{1/\gamma_0} \big(\gamma_0^2 P(\gamma_0) + \cdots\big) e^{\lambda t \ln 2}\, d\lambda \\
&= \frac{1}{t}\gamma_0^2 P(\gamma_0)\bigg[\frac{2^{t/\gamma_0} - 1}{\ln 2} + \cdots\bigg],
\end{aligned}
$$

with no oscillation at all.

Let us switch to the opposite extreme case of no hereditary influence: whatever the mother's generation time, each daughter will have a fixed generation time distribution $g(\gamma)$. In this case, not only is the function $N(t)$ not stepwise, but no vestige of synchrony can appear asymptotically. It is simplest to find the density $N_M(t)$ of cells maturing per unit time at time t. If a cell has gone through j divisions from a cell born at time 0, and these divisions occurred at times $t_1, t_2, \ldots, t_j$, such a path has probability $g(t - t_j) g(t_j - t_{j-1}) \cdots g(t_2 - t_1) g(t_1)$. Thus

$$
N_M(t) = \sum_{j=0}^{\infty} 2^j \int_0^{\infty} \cdots \int_0^{\infty} g(t - t_j) \cdots g(t_1) dt_1 \cdots dt_j.
$$

Laplace transforming, with $\tilde{g}(p) = \int_0^\infty e^{-tp} g(t)dt$, we have

$$\tilde{N}_M(p) = \sum_{j=0}^{\infty} 2^j (\tilde{g}(p))^{j+1} = \frac{\tilde{g}(p)}{1 - 2\tilde{g}(p)}.$$

The asymptotic time dependence is hence governed by the root of largest real part of

$$1 - 2\tilde{g}(p) = 0.$$

But if $g(\gamma) = 0$ when $\gamma \leq \gamma_0$, it follows from $g(\gamma) \geq 0$ and $\int g(\gamma)d\gamma = 1$ that

$$\int_{\gamma_0}^{\infty} e^{-\frac{\ln 2}{\gamma_0}} g(\gamma)d\gamma < \frac{1}{2} < \int_{\gamma_0}^{\infty} e^{0\gamma} g(\gamma)d\gamma.$$

Further, if p_0 is the largest real root and $\mathrm{Re}(p) > p_0$, we have

$$\mathrm{Re} \int_{\gamma_0}^{\infty} g(\gamma)e^{-p\gamma}\, d\gamma < \int_{\gamma_0}^{\infty} g(\gamma)e^{-p_0\gamma}\, d\gamma = \frac{1}{2},$$

which cannot be. Hence the root of largest real part of $\tilde{g}(p) = \frac{1}{2}$ is real, between 0 and $\ln(2/\gamma_0)$, and a pure exponential growth law ultimately dominates.

The population growth rate has been studied in a number of biological contexts, but extensive quantitative data is restricted to specialized systems such as the protozoal organism *tetrahymena*, with the presumption that the qualitative characteristics are typical of isolated growth in higher organisms as well. Observations indicate that neither extreme analyzed above is valid, but that the rate of decay of synchrony requires something like a generation time distribution of the two daughters depending parametrically on the mother's cycle time γ, i.e., $g(\gamma', \gamma''|\gamma)$.

To allow for the study of the steady state as well, we also assume a probability $h(\gamma'|\gamma)$ that a cell destined to divide after cycle time γ dies instead after cycle time γ'. If the generation time is set at birth, a very direct representation of the branching process can be carried out as follows. We group the cells in the μth generation, and characterize each dividing cell by generation time γ and *laboratory time of division t*. Such a cell is then "tagged" with a multiplicative weight $f(\gamma, t)$—an indeterminate *function*. Thus the mitotic event described above, with probability $g(\gamma', \gamma''|\gamma)$ for daughters to have generation times γ' and γ'', replaces a (γ, t)-cell by two that divide at $t + \gamma'$ and $t + \gamma''$. Hence the transition to the next generation is specified by the replacement

$$T_g: \quad f(\gamma, t) \to \iint g(\gamma', \gamma''|\gamma) f(\gamma', t + \gamma') f(\gamma'', t + \gamma'') d\gamma'\, d\gamma''.$$

Similarly, a moribund cell will be given a multiplicative weight $\bar{f}(\gamma, t)$ if it has a generation time γ but *dies* at t. Cell death clearly is accounted for by

$$\overline{T}_h: \quad f(\gamma, t) \to \left(1 - \int h(\gamma'|\gamma)d\gamma'\right) f(\gamma, t) + \int h(\gamma'|\gamma)\bar{f}(\gamma', t - \gamma + \gamma')d\gamma'.$$

If we now weight any n^{th} generation cellular configuration $(\gamma_1, t_1, \gamma_2, t_2, \dots)$ by both its probability and the product of its associated tags f and $\bar{f}$, and then integrate over all configurations, the result is the generating functional $W_\mu[f, \bar{f}]$ of degree

$\leq 2^\mu$ in the weights $\{f(\gamma,t), \bar{f}(\gamma,t)\}$. The $\mu \to \mu+1$ generation transformations must be completed by observing that

$$T_g\colon\ \bar{f}(\gamma,t) \to 1, \qquad \overline{T}_h\colon\ \bar{f}(\gamma,t) \to \bar{f}(\gamma,t),$$

and then we have simply

$$W_{\mu+1} = \overline{T}_h T_g W_\mu.$$

For an initial progenitor distribution $P(\gamma,t)$ $(= g(\gamma)\delta(t-\gamma)$ for division at $t=0)$, the sequence must start at

$$W_0 = \int P(\gamma,t)\overline{T}_h f(\gamma,t) d\gamma\, dt.$$

It is now convenient to construct a full generating function by summing over μ,

$$W = \sum_{\mu=0}^{\infty} W_\mu,$$

which thereby satisfies $W = W_0 + \overline{T}_h T_g W$, or equivalently

$$\overline{T}_h^{-1} W = \overline{T}_h^{-1} W_0 + T_g W,$$

our basic relation.

Since

$$\overline{T}_h^{-1}\colon\ f(\gamma,t) \to \frac{f(\gamma,t) - \int h(\gamma'|\gamma)\bar{f}(\gamma',t-\gamma+\gamma')d\gamma'}{1-\int h(\gamma'|\gamma)d\gamma'},$$

we have in full and gory detail

$$\begin{aligned} W&\left[\frac{f(\gamma,t)}{1-\int h(\gamma'|\gamma)d\gamma'} - \int \frac{h(\gamma'|\gamma)}{1-\int h(\gamma'|\gamma)d\gamma'}\bar{f}(\gamma',t-\gamma+\gamma')d\gamma', \bar{f}(\gamma,t)\right] \\ &= \int P(\gamma t) f(\gamma,t) d\gamma\, dt \\ &\quad + W\left[\iint g(\gamma',\gamma''|\gamma) f(\gamma',t+\gamma') f(\gamma'',t+\gamma'') d\gamma'\, d\gamma'', 1\right], \end{aligned}$$

from which all properties of the growth process can be obtained by observing the correspondence

$$\hat{n}(\gamma,t) = f(\gamma,t)\frac{\delta}{\delta f(\gamma,t)}, \qquad \hat{\bar{n}}(\gamma,t) = \bar{f}(\gamma,t)\frac{\delta}{\delta \bar{f}(\gamma,t)},$$

in the sense that the operator multiplies each term in the generating functional by the associated number. Here, we shall be content with a determination of the time development of mean population size. Since $W_\mu[1,1] = 1$, the desired mean values are given by

$$E(\gamma,t) = \sum_\mu \langle \hat{n}(\gamma,t)\rangle_\mu = \left.\frac{\delta W}{\delta f(\gamma,t)}\right|_{f=\bar{f}=1},$$

$$\bar{E}(\gamma,t) = \sum_\mu \langle \hat{\bar{n}}(\gamma,t)\rangle_\mu = \left.\frac{\delta W}{\delta \bar{f}(\gamma,t)}\right|_{f=\bar{f}=1}.$$

Let us apply these operations directly to the generating functional equation. Using

$$\frac{\delta T_g f(\gamma,t)}{\delta f(\gamma',t')} = 2\delta(t+\gamma'-t')\int g(\gamma',\gamma''|\gamma)f(\gamma'',t+\gamma'')d\gamma'',$$

and the corresponding $\frac{\delta}{\delta \bar{f}}$, we find at once

$$\frac{E(\gamma,t)}{1-\int h(\gamma'|\gamma)d\gamma'} = P(\gamma,t) + 2\int g(\gamma,\gamma''|\gamma')E(\gamma',t-\gamma)d\gamma'\,d\gamma'',$$

and

$$\bar{E}(\gamma,t) - \int \frac{h(\gamma,\gamma')}{1-\int h(\gamma''|\gamma')d\gamma''}E(\gamma',t+\gamma'-\gamma)d\gamma' = 0.$$

Let us now concentrate on the population of dividing cells. Only the quantities

$$h(\gamma) \equiv \int h(\gamma'|\gamma)d\gamma', \quad g(\gamma|\gamma') = \int g(\gamma,\gamma''|\gamma')d\gamma''$$

appear, the total death rate of an inherent γ-cell and the generation time γ distribution of a single daughter when her sister's parameter is not known. In terms of these, then

$$E(\gamma,t) = (1-h(\gamma))P(\gamma,t) + 2(1-h(\gamma))\int g(\gamma|\gamma')E(\gamma',t-\gamma)d\gamma'. \tag{7.1}$$

But $E(\gamma,t)$, the number of cells of generation time that divide at t, is not normally observed. $N(\gamma,t)$, the number of γ-cycle cells present at time t (or better, its integral over γ) is rather the quantity to focus upon. Since the latter is just the (relative) number of γ-cells that divide between t and $t+\gamma$, we must append the relation

$$N(\gamma,t) = \int_t^{t+\gamma} E(\gamma,t')dt'. \tag{7.2}$$

But equations (7.1) and (7.2) cannot be solved in closed form. They can be simplified materially by working with the Laplace transforms

$$\Gamma(\gamma,p) \equiv \int_0^\infty e^{-pt}E(\gamma,t)dt,$$

$$\tilde{N}(\gamma,p) \equiv \int_0^\infty e^{-pt}N(\gamma,t)dt.$$

Applying the Laplace transform,

$$\tilde{N}(\gamma,p) = \int_0^\infty \int_t^{t+\gamma} e^{-pt}E(\gamma,t')dt'\,dt = \int_0^\infty \int_{t'-\gamma}^{t'} e^{-pt}\,dt E(\gamma,t')dt',$$

yielding the new pair of relations

$$\Gamma(\gamma,p) = (1-h(\gamma))\tilde{P}(\gamma,p) + 2\big(1-h(\gamma)\big)e^{-\gamma p}\int_0^\infty g(\gamma|\gamma')\Gamma(\gamma',p)d\gamma',$$

$$\tilde{N}(\gamma,p) = \frac{e^{\gamma p}-1}{p}\Gamma(\gamma,p).$$

Consider the asymptotic form of the growth. All observables will have the behavior e^{pt} where p is the pole of $\Gamma(\gamma, p)$ of largest real part, but the inhomogeneous equation for $\Gamma(\gamma, p)$ will have a pole precisely when (except for very special nonphysical $\Gamma(\gamma, p)$) the homogeneous equation is solvable. Hence we must investigate the solution of

$$\Gamma(\gamma, p) = 2(1 - h(\gamma))e^{-\gamma p} \int_0^\infty g(\gamma|\gamma')\Gamma(\gamma', p)d\gamma'.$$

For example, when can the process have a steady state? Then $p = 0$ must yield a solution:

$$K(\gamma, \gamma') = \delta(\gamma - \gamma') - 2(1 - h(\gamma))g(\gamma|\gamma')$$

is singular but otherwise stable, an eigenvalue problem for the strength of $h(\gamma)$, contrasting strongly with the realistic case in which the effective death rate depends nonlinearly on population size.

Let us drop the death term $h(\gamma)$ and look at the extreme cases previously analyzed. For direct inheritance with minimum generation time γ_0, we have

$$g(\gamma|\gamma') = \delta(\gamma - \gamma') \quad \text{for } \gamma' \geq \gamma_0,$$

and the pole condition reduces to

$$(1 - 2e^{-\gamma p})\Gamma(\gamma, p) = 0 \quad \text{for } \gamma \geq \gamma_0.$$

The roots of largest real part are given by $\gamma = \gamma_0$ and

$$p = \frac{1}{\gamma_0}(\ln 2 + 2\pi n i),$$

so that the dominant exponential increase has the form

$$E(\gamma, t) \sim e^{\frac{t}{\gamma_0}\ln 2} \sum A_n e^{2\pi i n t/\gamma_0},$$

showing the nondecaying residual synchrony previously obtained. In the opposite extreme of innate variability with no hereditary factor,

$$g(\gamma|\gamma') = g(\gamma) \quad \text{with } g(\gamma) = 0 \text{ for } \gamma < \gamma_0,$$

we have

$$\Gamma(\gamma, p) = 2e^{-\gamma p}g(\gamma) \int \Gamma(\gamma', p)d\gamma',$$

or

$$\int \Gamma(\gamma, p)d\gamma = \int \Gamma(\gamma', p)d\gamma' \int 2e^{-\gamma p}g(\gamma)d\gamma,$$

yielding the requirement

$$\int_{\gamma_0}^\infty e^{-\gamma p}g(\gamma)d\gamma = \frac{1}{2}$$

previously derived and the resulting absence of oscillations.

Now what of the intermediate cases? Simplest is a linear combination of the extremes:

$$g(\gamma|\gamma') = \beta\delta(\gamma - \gamma') + (1 - \beta)g(\gamma), \quad \gamma, \gamma' \geq \gamma_0,$$

and indeed this is completely solvable. Set $P(\gamma, t) = g(\gamma)\delta(t-\gamma)$ so that $\tilde{P}(\gamma, p) = g(\gamma)e^{-\gamma p}$. Then we have

$$\begin{aligned}\Gamma(\gamma, p) &= 2e^{-\gamma p}\int_0^\infty g(\gamma|\gamma')\Gamma(\gamma', p)d\gamma' + \tilde{P}(\gamma, p)\\ &= 2e^{-\gamma p}\left[\beta\Gamma(\gamma, p) + (1-\beta)g(\gamma)\int_{\gamma_0}^\infty \Gamma(\gamma', p)d\gamma'\right] + g(\gamma)e^{-\gamma p}\end{aligned}$$

or

$$\Gamma(\gamma, p) = \frac{g(\gamma)e^{-\gamma p}}{1-2\beta e^{-\gamma p}}\left(1 + 2(1-\beta)\int_{\gamma_0}^\infty \Gamma(\gamma', p)d\gamma'\right).$$

Integrating, we recover the unknown integral:

$$\int_{\gamma_0}^\infty \Gamma(\gamma', p)d\gamma' = \frac{\int_{\gamma_0}^\infty g(\gamma)\frac{e^{-\gamma p}}{1-2\beta e^{-\gamma p}}d\gamma}{1-2(1-\beta)\int_{\gamma_0}^\infty g(\gamma)\frac{e^{-\gamma p}}{1-2\beta e^{-\gamma p}}d\gamma},$$

so that

$$\Gamma(\gamma, p) = \frac{\frac{g(\gamma)}{e^{\gamma p}-2\beta}}{1-2(1-\beta)\int_{\gamma_0}^\infty \frac{g(\gamma')}{e^{\gamma' p}-2\beta}d\gamma'}.$$

The synchronized poles from $e^{\gamma p} - 2\beta$, $p = (1/\gamma_0)(\ln 2\beta + 2n\pi i)$, remain for $\beta > \frac{1}{2}$, but there is always a real pole with larger real part: $\int_{\gamma_0}^\infty g(\gamma')/(e^{\gamma' p} - 2\beta)d\gamma'$ goes from ∞ to 0 as p goes from $(1/\gamma_0)\ln 2\beta$ to ∞. Thus there is necessarily an asymptotic loss of synchrony. The resulting $\overline{N}(t)$ can be fitted easily to experimental data. While there is thus no compelling reason to generalize further, it is not hard to show that

$$g(\gamma|\gamma') = \beta\delta(\gamma-\gamma') + (1-\beta)g(\gamma - c\gamma'), \quad \gamma, \gamma' \geq 0,$$

for arbitrary c is solvable as well. The real problem is that of the distribution of population size at a given time, but this remains generally analytically intractable.

7.2. Environmental Control of Cell Division

A dilute suspension of cells well supplied by nutrient maintains an exponential growth pattern for cell number. This can be broken in cultures of single cell organisms by metabolic products that inhibit division or are even lethal. But in a developing multicellular organism, it appears to be principally the cellular environment that controls the division of a given cell. We will now look at some aspects of this control, first empirical, then mechanistic, and finally as part of a well-studied developmental system.

7.2.1. Macroscopic Inhomogeneity. There are well authenticated instances of structures arising from orientation of division, e.g., blastula formation and insect imaginal disc development. *Unoriented division* can also serve as an effective structure regulating device if it is responsive to local or global aspects of structure. The nature of this response has been studied in detail in a nearly isolated initially homogeneous developing system, that of chick limb (wing or hind limb) bud. In its earliest stage, this is a loosely packed but uniform *mesenchymal* bump enclosed by an *ectodermal* jacket, and with a quite high uniform mitotic index: 12/100 mesenchymal cells in mitosis at a given time. Later on (Hamburger-Hamilton stage 25), the indices at the *distal* (far) end and the edges along the *dorso-ventral* (back-front) axis decrease, a decrease associated to some extent with an enlargement of nuclear volume (most of the cell). The important observation is that the mitotic index seems to go inversely as the cell density. Since we perforce use a sample of one (the bud has to be fixed and sectioned) and there are relatively few mitoses, and we would like to assess a local phenomenon, it is clear that the statistics of the data must be considered with some care. Let us do so.

The problem is simply that of establishing a reliable estimate of the presumed underlying field $m(\mathbf{x})$ of *mitotic probability per cell* (the net cell population density is comparatively no problem), and the experimental data must be regarded as a two-dimensional slice divided into grid squares of centers $\{\mathbf{x}\}$, with *observed mitotic number* $M(\mathbf{x})$ in each grid square tallied separately. Similarly, *total cell number* $\rho(\mathbf{x})$ in each square is tallied. Assuming constant time of mitosis τ throughout, the *mitotic index* $M(\mathbf{x})/\rho(\mathbf{x})$ also represents division rate τ/γ as provoked by environmental causes (presumably acting at least as early as the S-phase, a time lag which should in principle be taken into account). The grid division is of course chosen fine enough to have a chance at getting at the underlying $m(\mathbf{x})$, but coarse enough that there are at least a few mitoses per square.

For a large homogeneous population, the observed mitotic index would equal the probability m of a cell being in mitosis. To estimate the underlying $m(\mathbf{x})$ for an inhomogeneous system, suppose that we have determined in some fashion

$$P(\{M(\mathbf{x})\}|\{m(\mathbf{x})\}),$$

the probability that the set $\{M(\mathbf{x})\}$ ($\mathbf{x}$ the center of a grid square) will be observed if $\{m(\mathbf{x})\}$ (continuous $\mathbf{x}$) is known. If the mitoses in different squares are *uncorrelated*, i.e., few boundary mitoses, easy to achieve, and no clustering or anticlustering—there is, however, evidence of mitotic chains—then we may write instead

$$P(\{M(\mathbf{x})\}|\{m(\mathbf{x})\} = e^{\sum_{\mathbf{x}} \phi(M(\mathbf{x})|m(\mathbf{x}))},$$

and we shall do so. Finally, $m(\mathbf{x})$ must be chosen from some class of possibilities, conveniently specified by parameters $\{c_\lambda\}$. Examples would be

$$m(\mathbf{x}) = \sum c_\lambda u_\lambda(\mathbf{x}), \quad \left(\sum c_\lambda u_\lambda(\mathbf{x})\right)^2, \quad e^{\sum c_\lambda u_\lambda(\mathbf{x})}.$$

The $\{c_\lambda\}$ can then be determined by imposing the condition of *maximum likelihood*: $P(\{M(\mathbf{x})\}|m(\mathbf{x})\}) = \max$ over $\{c_\lambda\}$, so that now

$$\sum_{\mathbf{x}} \frac{\partial\phi(M(\mathbf{x})|m(\mathbf{x}))}{\partial m(\mathbf{x})} \frac{\partial m(\mathbf{x})}{\partial c_\lambda} = 0.$$

If the mitotic events are truly independent, the observed $M(\mathbf{x})$ should be Poisson:

$$e^{\phi(M|m)} = \frac{(m\rho)^M}{M!} e^{-m\rho}.$$

Choosing, e.g., $m(\mathbf{x}) = e^{\sum c_\lambda u_\lambda(\mathbf{x})}$, we then have

$$\sum_{\mathbf{x}} (M(\mathbf{x}) - m(\mathbf{x})\rho(\mathbf{x}))u_\lambda(\mathbf{x}) = 0,$$

which can only be solved numerically.

Instead we may use a Gaussian approximation

$$e^{\phi(M|m)} = \frac{1}{\sigma\sqrt{2\pi}} e^{-\frac{(M-m\rho)^2}{2\sigma^2}},$$

together with the parametric form $m(\mathbf{x}) = \sum c_\lambda u_\lambda(\mathbf{x})$. This yields at once the same expression

$$\sum_{\mathbf{x}} (M(\mathbf{x}) - m(\mathbf{x})\rho(\mathbf{x}))u_\lambda(\mathbf{x}) = 0,$$

but now equivalent to the least squares estimation

$$\frac{\partial}{\partial c_\lambda} \sum_{\mathbf{x}} (M(\mathbf{x}) - m(\mathbf{x})\rho(\mathbf{x}))^2 = 0.$$

The Gaussian approximation corresponding to Poisson should, however, have a Poisson variance $\sigma^2 = m\rho \sim M$. It is just as easy to include an intrinsic fluctuation: $\sigma^2 = \sigma_0^2 + M$, now modifying the foregoing to

$$\sum_{\mathbf{x}} \frac{M(\mathbf{x}) - m(\mathbf{x})\rho(\mathbf{x})}{\sigma_0^2 + M(\mathbf{x})} u_\lambda(\mathbf{x}) = 0,$$

which at least represents a linear system of equations for the $\{c_\lambda\}$.

Simplest by far, and not incompatible with experimental accuracy, the Gaussian approximation with fixed σ can be used, coupled with the cell density weighted parametric expression

$$\overline{M}(\mathbf{x}) \equiv m(\mathbf{x})\rho(\mathbf{x}) = \sum_\lambda c_\lambda u_\lambda(\mathbf{x}).$$

Now $\rho(\mathbf{x})$ drops out, leaving us with

$$\sum_{\mathbf{x}} \Big(M(\mathbf{x}) - \sum_{v} c_v u_v(\mathbf{x}) \Big) u_\lambda(\mathbf{x}) = 0.$$

In particular, suppose that the $u_\lambda(\mathbf{x})$, which are p in number and complete under complex conjugation, are orthonormal on the N grid points $\mathbf{x}$:

$$\sum_{\mathbf{x}} u_\lambda^*(\mathbf{x}) u_v(\mathbf{x}) = \delta_{\lambda v}.$$

Then we have the immediate solution

$$c_\lambda = \sum_{\mathbf{x}} u_\lambda^*(\mathbf{x}) M(\mathbf{x})$$

or

$$\overline{M}(\mathbf{x}) = \sum_{y} \Big(\sum_\lambda u_\lambda(\mathbf{x}) u_\lambda^*(\mathbf{y}) \Big) M(\mathbf{y}).$$

There are now two modifications we shall consider. First, we may want to drop the assumption that the space $(u_1(\mathbf{x}), \ldots, u_p(\mathbf{x}))$ into which $\overline{M}(\mathbf{x})$ must fall is accurately known and imagine that only the probability of it being in such a space is known. This extension can be incorporated in many ways, the simplest being to replace the averaging function connecting $\overline{M}$ and M by $\sum_1^\infty w_\lambda u_\lambda(\mathbf{x})u_\lambda^*(\mathbf{y})$ where $w_1 = 1$, $w_\lambda \to 0$ as $\lambda \to \infty$, and $\sum_1^\infty w_\lambda = p$. Now

$$\overline{M}(\mathbf{x}) = \sum_{\mathbf{y}} W(\mathbf{x}, \mathbf{y}) M(\mathbf{y}),$$

where

$$W(\mathbf{x}, \mathbf{y}) = \sum w_\lambda u_\lambda(\mathbf{x}) u_\lambda^*(\mathbf{y}).$$

In the "pure case" above, $\overline{M}$ is precisely the projection of M onto the space $\{u_\lambda(\mathbf{x}), \lambda = 1, \ldots, p\}$, but now we have instead just a general average, diagonal in the $\{u_\lambda(\mathbf{x})\}$.

For example, if $\{u_{\lambda_1\lambda_2}(x_1, x_2)\}$ is the Fourier set on a square of $N = (2L+1)^2$ unit squares, truncated so that only $p = (k+1)^2$ frequencies are retained,

$$u_{\lambda_1\lambda_2}(x_1, x_2) = \frac{1}{2L+1} e^{\frac{2\pi i}{2L+1}(\lambda_1 x_1 + \lambda_2 x_2)},$$

where $|\lambda_1|$ and $|\lambda_2|$ are integers $\leq K$, then the pure projection weight is the running average

$$W(x_1 - y_1, x_2 - y_2) = \frac{1}{N} \prod_{\alpha=1}^{2} \frac{\sin \frac{2\pi}{2L+1}\left(K + \frac{1}{2}\right)(x_\alpha - y_\alpha)}{\sin \frac{2\pi}{2L+1}(x_\alpha - y_\alpha)},$$

with coefficients of both positive and negative sign. But if the discontinuity in $w_{\lambda_1\lambda_2}$ is smoothed to at least a Gaussian tail, a true nonnegative averaging function can be created. A simple form is

$$w_{\lambda_1\lambda_2} = \left(\cos \frac{\pi \lambda_1}{2L+1}\right)^{2r} \left(\cos \frac{\pi \lambda_2}{2L+1}\right)^{2r},$$

with an effective $p = \frac{N}{\pi r}$, yielding a weight function

$$W(x_1 - y_1, x_2 - y_2) = N \left(\frac{1}{2}\right)^{4r} \binom{2r}{r - x_1 + y_1} \binom{2r}{r - x_2 + y_2}.$$

This represents an iterated nearest-neighbor average, a standard data-smoothing procedure, and is in fact what Summerbell and Lewis used in their mitotic field estimations [**146**].

7.2.2. Boundary Corrections. The second modification is one of convenience. For comparison purposes, one would like to use a standard set of expansion functions, e.g., a set orthonormal on a large square circumscribing a limb bud. There will then be a sharp discontinuity of mitotic index on leaving the tissue that is not of biological significance but tends to mess up the estimation. Clearly, the function set should instead be orthogonalized over the area of the physical limb $\mathcal{L}$. This causes no difficulty in principle. We still have

$$\sum_{\mathbf{x}\in\mathcal{L}} \Big(M(\mathbf{x}) - \sum_{\nu} c_\nu u_\nu(\mathbf{x})\Big) u_\lambda^*(\mathbf{x}) = 0,$$

but the *overlap matrix*

$$T_{\lambda\nu} = \sum_{\mathbf{x}\in\mathcal{L}} u_\lambda^*(\mathbf{x}) u_\nu(\mathbf{x})$$

is no longer the identity. If the $u_\nu(\mathbf{x})$ were Fourier, the $T_{\lambda\nu}$ would depend only on $\lambda-\nu$, but unlike an infinite matrix of this form, a finite section is in general invertible only numerically. Thus the solution, in the form

$$\bar{M}(\mathbf{x}) = \sum_{\mathbf{y}\in\mathcal{L}} \Big(\sum_{\lambda,\nu} u_\nu(\mathbf{x}) T^{-1}_{\nu\lambda} u_\lambda^*(\mathbf{y}) \Big) M(\mathbf{y}),$$

is not available in simple analytic form. However, the study of approximations to properties of finite sections of regular, diagonalizable, infinite matrices is an old one, under the topic of Toeplitz matrices. Let us consider this briefly.

Classical Toeplitz matrix theory examines matrices of the form

$$T_{\lambda\nu} = f(\lambda - \nu).$$

The doubly infinite matrices, $-\infty \le \lambda, \nu \le \infty$, are trivial to deal with, since they are diagonalized by a unitary Fourier transform:

$$\begin{aligned} \widetilde{T}_{\theta u} = \frac{1}{2\pi} \sum_{\lambda\nu} e^{i\lambda\theta} T_{\lambda\nu} e^{-i\nu\phi} &= \frac{1}{2\pi} \sum_{\lambda\nu} e^{i(\lambda-\nu)\theta} f(\lambda-\nu) e^{i\nu(\theta-\phi)} \\ &= \tilde{f}(\theta)\delta(\theta-\phi), \end{aligned}$$

where θ and ϕ are angles, and $\tilde{f}$ is the Fourier transform. The interest derives from finite sections; i.e., if Q is a projection

$$Q_{\lambda\nu} = \begin{cases} \delta_{\lambda\nu} & \text{if } 1 \le \lambda, \nu \le p, \\ 0 & \text{otherwise,} \end{cases}$$

then the properties of

$$T^{(p)} \equiv QTQ$$

are desired.

The determinant is typical. Let us not specialize T as above, but instead assume that it is positive definite, as indeed it is for an overlap matrix. Now clearly

$$D = \det_{p\times p} QTQ = \det(\overline{Q} + QTQ) \quad \text{where } \overline{Q} = I - Q.$$

The simplest formal way of evaluating a determinant is as

$$\det A = e^{\operatorname{tr} \ln A},$$

where $\ln A$ refers to any branch of the logarithm. We can now "turn on" the matrix from I to $\bar{Q} + QTQ$ by writing

$$\ln D(t) = \operatorname{tr} \ln(\overline{Q} + Qe^{t \ln T} Q),$$

so that $\ln D(0) = 0$, $\ln D(1) = \ln D$. Since $\frac{\partial}{\partial t} \operatorname{tr} g(A(t)) = \operatorname{tr}(A'(t)g'(A(t)))$, and

$$\frac{\partial}{\partial t} A(t)^{-1} = -A(t)^{-1} A'(t) A(t)^{-1},$$

we have

$$\begin{aligned}
\ln D(t) &= \operatorname{tr} \ln A(t), \\
\frac{\partial}{\partial t} \ln D(t) &= \operatorname{tr} Q \ln T e^{t \ln T} Q A(t)^{-1}, \\
\frac{\partial^2}{\partial t^2} \ln D(t) &= \operatorname{tr} Q (\ln T)^2 e^{t \ln T} Q A(t)^{-1} \\
&\quad + 2 \operatorname{tr} Q \ln T e^{t \ln T} Q A(t)^{-1} Q \ln T e^{t \ln T} Q A(t)^{-1}.
\end{aligned}$$

Thus a Maclaurin series in t at $t = 1$ yields

$$\ln D = 0 + \operatorname{tr} Q \ln T Q + \frac{1}{2} \operatorname{tr} (\ln T)^2 Q + \operatorname{tr} Q \ln T Q \ln T Q + \cdots$$

or

$$\ln D = \operatorname{tr} Q \ln T Q + \frac{1}{2} \operatorname{tr} Q \ln T \overline{Q} \ln T Q + \cdots .$$

This in fact is the classical Szegö result for the classical Toeplitz matrices: if $T_{\lambda\nu} = f(\lambda - \nu)$, then

$$\begin{aligned}
(\ln T)_{\lambda\nu} &= \frac{1}{2\pi} \iint e^{-i\lambda\theta} \ln \tilde{f}(\theta) f(\theta - \phi) e^{i\nu\phi} \, d\theta \, d\phi \\
&= \frac{1}{2\pi} \int \ln \tilde{f}(\theta) e^{i(\nu-\lambda)\theta} \, d\theta \equiv k(\nu - \lambda),
\end{aligned}$$

so that

$$\ln D = pk(0) + \frac{1}{2} \sum_{\lambda > p \text{ or } \lambda < 1}^{1 \le \nu \le p} k(\nu - \lambda) k(\lambda - \nu) + \cdots$$

or

$$\ln D = pk(0) + \sum_{1}^{\infty} \lambda k(\lambda)^2 + \cdots ,$$

and the remainder is $o(p)$. The Maclaurin development is not, however, itself an asymptotic expansion, a multiple scattering transformation being required to convert it to one.

We need the inverse, not the determinant. But

$$A^{-1}_{\lambda\nu} = \frac{\partial}{\partial A_{\nu\lambda}} \ln \det A,$$

and in particular $(\overline{Q} + QTQ)^{-1} = \overline{Q}_{\lambda\nu} + (\frac{\partial}{\partial T_{\nu\lambda}}) \ln D$, so that

$$(T^{(p)})^{-1}_{\lambda\nu} = \frac{\partial}{\partial T_{\nu\lambda}} \ln D.$$

To carry this out, we require

$$\frac{\partial}{\partial T_{\nu\lambda}}(\ln T)_{\nu'\lambda'} = \frac{\partial}{\partial T_{\nu\lambda}} \int_0^{X\to\infty} (x+T)^{-1}_{\nu'\lambda'}\,dx$$
$$= -\int_0^{\infty} (x+T)^{-1}_{\nu'\nu}(x+T)^{-1}_{\lambda\lambda'}\,dx.$$

Hence

$$\frac{\partial}{\partial T_{\nu\lambda}}\operatorname{tr}(B\ln T) = -\int_0^{\infty} ((x+T)^{-1}B(x+T)^{-1})_{\lambda\nu}\,dx,$$

yielding at once

$$(T^{(p)})^{-1} = -\int_0^{\infty} (x+T)^{-1} \times \left[Q + \frac{1}{2}\overline{Q}\ln TQ + \frac{1}{2}Q\ln T\overline{Q} + \cdots\right](x+T)^{-1}\,dx,$$

our desired result. Here Q projects onto the desired p-state space, and the sensitivity of the resulting $\overline{M}$ to input data is very easy to assess.

7.2.3. Local Control of Mitosis. What now are the results of mitotic index field estimations? The primary system, as we have indicated, is the chick limb bud, which under normal circumstances shows a substantial drop in overall mesenchyme mitotic index over time, accompanied by a large spatial inhomogeneity in index, higher on both sides along a dorsal-ventral axis—i.e., high at the boundaries. Associated with this is an inverse change in overall mesenchymal density ρ, in turn associated with a change in nuclear diameter and cell size. If all of the data is pooled, a least-squares fit of m as a function of $1/\rho$ yields an approximate relationship

$$m = -4 + \frac{0.13}{\rho}, \quad \rho \text{ in (micron)}^{-2}.$$

The fluctuations are high but the trend is evident.

If the above relation is causal, and if the cell density controls the mitotic rate, how does it do so? Experiments indicate that the control is quite local, eliminating the possibility of highly diffusible inhibitory *chalones*. The most obvious local model would be that of cell-cell contact inhibition of division, using a cell-cell "collision" as the basic event. Suppose that while the cell cycle is γ_0 for collison rate $R < R_0$ at low density, normal development takes place in the region

$$\gamma = kR, \quad R > R_0,$$

each collision causing a mitotic delay. For cells regarded as spheres of diameter a at low density ρ, cells of relative velocity v would collide at a rate $\pi a^2\rho v$ per unit time. If the mean relative velocity is c, and the density is high so that we must take into account the restriction of each cell to a mean volume of $1/\rho - v_0$, $v_0 = \pi a^3/6$, we have instead

$$\gamma = \frac{k\pi a^2 c\rho}{1-\rho v_0}.$$

Hence for the mitotic index $m = \frac{\tau}{\gamma}$,

$$m = \frac{\tau}{\pi k a^2 c}\left(-v_0 + \frac{1}{\rho}\right).$$

This expression is certainly qualitatively correct. The quantitative similarity with experiment is more tenuous. For one thing, a and v_0 have a dependence on ρ, as we have seen. For another, the measured density $\rho_{\rm obs}$ per unit surface at fixed section thickness d is not ρ, but is related to it: for thickness d and acceptance angle θ for recognizing a nucleus, the cell center can cover a distance of $d + 2 \times \frac{a}{2}\cos\theta$; hence

$$\rho_{\rm obs} = (d + a\cos\theta)\rho.$$

And of course the variation of c is not known. These matters of interpretation have not been considered in great detail.

Is there any other evidence for a simple contact inhibition model? One possibility is the nature of the fluctuations in mitotic index. From a thermodynamic viewpoint, collision rate is pressure in suitable units, so that

$$\beta P = \frac{\rho}{1 - \rho v_0}.$$

Choose volume Ω to examine; the cell number N in Ω is variable in nominally identical environments, determined by an effective cell potential μ in the external region. Write in standard thermodynamic fashion

$$e^{\beta P \Omega} = \sum \frac{e^{\beta\mu N}}{N!} Q_N(\Omega)$$

(the $N!$ taking care of the indistinguishability of different cells). Then indeed if an N-cell configuration is weighted as indicated, we will have

$$\begin{aligned}\overline{N} &= \frac{\sum N e^{\beta\mu N} Q_N/N!}{\sum e^{\beta\mu N} Q_N/N!}\\ &= \frac{\partial}{\partial \beta\mu}\ln e^{\beta P\Omega} = \frac{\partial P\Omega}{\partial \mu}.\end{aligned}$$

Similarly,

$$\begin{aligned}\overline{N^2} - (\overline{N})^2 &= \left(\frac{\partial}{\partial\beta\mu}\right)^2 \ln e^{\beta P\Omega} = \frac{\partial^2}{\partial(\beta\mu)^2}\beta P\Omega\\ &= \frac{\partial\overline{N}}{\partial\beta\mu} = \frac{\partial\overline{N}}{\partial\beta P\Omega}\frac{\partial\beta P\Omega}{\partial\beta\mu} = \overline{N}\frac{\partial\overline{N}}{\beta P\Omega} = \left(\rho \Big/ \frac{\partial\beta P}{\partial\rho}\right)\Omega.\end{aligned}$$

Of course we are interested only in those cells that are at the division phase of this cycle. Assuming that the interaction properties of all cells are substantially the same, the only difference is that ρ must be replaced by $m\rho$ and P by the "partial pressure" mP. Hence $(\delta m)^2 = (\delta N/\rho\Omega)^2 = (\delta N)^2/\rho^2\Omega^2 = m\rho\Omega(1-\rho v_0)^2/\rho^2\Omega^2$:

$$\delta m^2 = \frac{m}{\rho\Omega}(1 - \rho v_0)^2.$$

This agrees quite well with the experimental data spread, but not well enough to pin down the crucial $1 - \rho v_0$ factor.

Mesenchymal division is then consistent with contact inhibition. Let us proceed to ectoderm-esenchyme interaction. It has been known for some time that the *apical ectodermal ridge* (AER) is crucial to the development of the limb bud. One performs various operations and examines the fully developed limb after all detailed modeling has occurred to see what types of cells have been affected. Then successively late AER *removal* only annuls structures distal to successively more distal locations, but prior proximal structures are complete; an extra AER doubles up the same distal structures; replacement by *earlier or later* AER yields no defects. But the AER control is exerted on the mesoderm directly next to it: leg mesoderm inserted makes a distal leg; even thigh mesoderm does the same, but nonlimb mesoderm stops growth. Further, young AER-mes on old mesenchyme makes a complete new detailed distal portion, and old or young ablates segments.

The general conclusion has been that the wing mesoderm in the initial limb bump has already been *restricted* genetically via differentiation to form wing. The succeeding differentiations fill in the finer detail, the first level being type of limb structure (humerus, phalanges, etc.), which peel off in precisely this time sequence as long as the cells are near the AER, but stop when nearby AER activation is lost. After being so restricted (with implied poor regulation), their further division is under control of local environment. If this is so, it might be possible, knowing the mitotic index variation $m(x, t)$ in space and time, to take a later limb with cells identifiable as to segment type and trace their progenitors (without cell marking!) back to the initial bump to show the correlation of segment type with time under the influence of the AER.

This fate map has been constructed by Lewis, working back from a stage early enough that one can assume simple elongation along the proximal-distal axis accompanied by a scale change due to the varying transverse area $A(x, t)$ of the bud [**110**]. Thus the motion of a progenitor can be traced by finding the rate at which the volume between it and $x = 0$, the AER location, increases:

$$\frac{dx}{dt} = \frac{1}{A(x,t)} \int_0^x A(x', t)\rho(x', t)m(x', t)dx'/\tau.$$

Then one can ask for the number of doublings of a progenitor cell while within an influence distance (300 μm works nicely) of the AER. The result is that each of the five segment types spends one more division time in the AER-mes than its proximal neighbor: the dorsal-ventral specification clock runs only in the AER-mes. Thereafter, shaping is done by contact.

But there is also obvious evidence for a third level of pattern modification—that in the anterior-posterior direction: radius-ulna, digits, etc. In part, this seems "standard" repetitive pattern formation—the number of digits is determined by the width of the AER, polydactyl mutant mesoderm alone being ineffective. However, the anterior-posterior organization appears primarily controlled by a diffusible activator from a posterior *zone of polarizing activity* (ZPA), explants of which can produce multiple structures, posterior always meaning towards a ZPA. Distributing the ZPA by dissociation and mixing yields a symmetric structure. And finally, the ZPA and AER are not at all independent: the pressure of the former is required for the maintenance of the latter. It appears then that the basic ingredients of a limb bud analysis during its period of "independent functioning"—i.e., before vascularization, etc., are known in outline and in some detail, but an overall analysis has yet to be carried out.

7.3. Stem Cell Dynamics

We have thus far omitted explicit reference to the cell population growth that is crucial to embryonic development. But this is controlled growth in that as it occurs, the cells that are produced themselves undergo differentiation to other cell types. A typical tactic goes back at least to hydra, most of whose cell types are self-proliferating, but one pair, the neurons and the nematocytes (or poison dart vehicles), is produced from the interstitial *stem cells*, a close relation between the nervous and defense systems, which is maintained as one goes up the phyla. Very fine control can be exerted by adjusting the ratio of differentiation events to the cell types involved. Meanwhile, however, and strictly analogously, the stem cell has to decide whether to differentiate to an "end" cell and hence be "lost" to the stem cell population, or just propagate. There is substantial evidence that control is stochastic, only the probability of propagation being adjusted. This *branching-process*, going back to Galton and Watson, is what we now examine, elaborating on our discussion in 7.1.1.

Assume constant unit generation time. We only count stem (S) cells, so at each division, a cell either splits in two or "dies," producing two end (E) cells. An obvious solution strategy to analyze the dynamics is to regard this process as chemical kinetics, and so we have the "reactions" per generation

$$S \xrightarrow{p} 2S, \quad S \xrightarrow{q} 2E,$$

in which case

$$\dot{S} = pS - qS \quad \text{or} \quad S = S_0 e^{(p-q)t}.$$

In other words, the population expands exponentially if $p > q$, dies exponentially if $p < q$, and is stable if $p = q$. But at the beginning of the process, only a few cells are involved, and speaking of "chemical concentrations" makes little sense. In fact, there is a nonzero probability of S fluctuating to zero by chance along the way, and then it can never recover—it has become extinct. Actually the Galton-Watson process was originally applied to the extinction of royal family names, carried only by the father. It is a *stochastic* process, typically unsolvable analytically, but with many systematic iterative solution methods. We will focus on one of these that is particularly visualizable, appearing in many other contexts, and really attends to the detailed structure of the population.

Give each stem cell a tag or multiplicative weight x. Starting with one cell x, the next generation has weight x^2 with probability p, or $x^0 = 1$ with probability q, represented therefore by the transformation

$$x \mapsto f(x) = q + px^2.$$

At the next division, we would have

$$f(x) \mapsto f_2(x) = q + p(q + px^2)^2 = q(1 + pq) + 2p^2qx^2 + p^3x^4,$$

and so forth (for an exponential distribution of division times, we would need instead $q\,dt$, $[1 - (p + q)]dt$, and $p\,dt$ for the next generation weights, so that if $p + q + r = 1$, then $x \mapsto x + (q - rx + px^2)dt$).

In general, for k similar cells produced with probability p_k (clearly not at a single division, but perhaps in a standard time period), we would write

$$x \mapsto f(x) = \sum_{k=0}^{\infty} p_k x^k$$

with the normalization $f(1) = 1$. If Z_N is the number of cells at generation N, starting with one at generation zero, this is equivalent to

$$p_k = \Pr(Z_1 = k \mid Z_0 = 1), \quad f(x) = \mathrm{E}(x^{Z_1} \mid Z_0 = 1),$$

and, for the Nth generation, correspondingly to

$$f_N(x) \equiv \sum p_{N,k} x^k = \mathrm{E}(x^{Z_N} \mid Z_0 = 1).$$

Clearly, $f_N(x) = f_{N-1}(f(x))$, so together with $f_0(x) = x$,

$$f_N(x) = \underbrace{f \cdot f \cdot f \cdots f(x)}_{N},$$

leading to the general

$$f_{a+b}(x) = f_a(f_b(x)).$$

Now how does one find the explicit structure of the population for arbitrary "time" N? This is not easy, but it is very much constrained by the behavior of the moments. Define the *multiplicative factor*

$$m = \mathrm{E}(Z_1 \mid Z_0 = 1) = \sum k p_k = f'(1)$$

($m = p - q = 2p - 1$ in our elementary example) and also the *second factorial moment*

$$Q = \sum k(k-1)p_k = f''(1) = \sigma^2(Z_1 \mid Z_0 = 1) + m^2 - m.$$

Then

$$m_N = \frac{d}{dx} f_N(x)\Big|_{x=1} = \frac{d}{dx} f_{N-1}(f(x))\Big|_{x=1} = f'_{N-1}(1) f'(1) = m_{N-1} m,$$

so $m_N = m^N$, of course. Similarly,

$$Q_N = m^{N-1}\left(\frac{m^N - 1}{m - 1}\right) Q.$$

The ideal asymptotic behavior as $N \to \infty$ is of obvious importance. One can prove (but these are not too helpful):

(1) $m < 1$ (subcritical—the population goes extinct): $\lim_{N\to\infty} \Pr(Z_N = k \mid Z_N > 0)$ exists.
(2) $m = 1$ (critical): $\lim_{N\to\infty} \Pr(Z_N/N > z \mid Z_N > 0) = e^{-2z/\sigma^2}$ or

$$\frac{1}{N}\left[\frac{1}{1-f_N(x)} - \frac{1}{1-f(x)}\right] \to \frac{1}{2}\sigma^2.$$

(3) $m > 1$ (supercritical—the population diverges): if $f(x_0) = x_0 \neq 1$, and $\gamma = f'(x_0)$, then $\lim_{N\to\infty}[f_N(x) - x_0]$ converges.

Are there cases that one can solve exactly? One is

$$f(x) = -1 + 2x^2 = \cos(2\cos^{-1} x),$$

so that

$$f_N(x) = \cos(2^N \cos^{-1} x),$$

but the "probabilities" here are not nonnegative. Another is the exponential probability distribution

$$f(x) = \frac{2m}{Q+2m}\sum_{k=0}^{\infty}\left(\frac{Q}{Q+2m}\right)^k x^k = 1 + \frac{\frac{Q}{2m}(x-1)}{1-\frac{Q}{2m}(x-1)},$$

parametrized so that $f(1) = 1$, $f'(1) = m$, and $f''(1) = Q$. Here matters are different; being bilinear rational, it does iterate to

$$f_N(x) = 1 + \frac{\frac{m^N-1}{m-1}\frac{Q}{2m}(x-1)}{1-\frac{m^N-1}{m-1}\frac{Q}{2m}(x-1)},$$

but this doesn't do too well as an approximation, e.g., for $f(x) = px^2 + q$.

A small $m - 1$ expansion is a viable approach, since $m = 1$ (i.e., $p = \frac{1}{2}$) is often closely approximated in real systems, it being the threshold between nonproliferation and proliferation of the differentiated types—the former held in check by ordinary population saturation effects *if* we know what the population pressure is. So assume $m > 1$; then m^N establishes the scale for Z_N, suggesting that we consider

$$\varphi_N(S) = \mathrm{E}(e^{-SZ_N/m^N}) = f_N(e^{-S/m^N}).$$

Clearly $\varphi_{N+1}(mS) = f(\varphi_N(S))$, and one can show that φ_N converges to some φ, which thereby satisfies

$$\varphi(mS) = f(\varphi(S)), \quad \varphi(0) = 1, \ \varphi'(0) = 1,$$

uniquely determining φ. If φ is known (it has to be monotonically decreasing), then so is φ^{-1}, and we see that

$$f(x) = \varphi(m\varphi^{-1}(x)),$$

a nonlinear similarity transformation with the consequence that

$$f_N(x) = \varphi(m^N\varphi^{-1}(x)).$$

The problem then is to find φ. Suppose $r \equiv m - 1$ is small, and imagine that we are in the vicinity of $x = 1$. Set

$$f(x) = x + rU(x)$$

so that $U(1) = 0$, $U'(1) = 1$, and define $S(x) = \varphi^{-1}(x)$. Using $\varphi^{-1}(f(x)) = m\varphi^{-1}(x)$, we have

$$(1+r)S(\varphi) = S(\varphi + rU(\varphi)) = S(\varphi) + rU(\varphi)S'(\varphi) + \frac{1}{2}r^2[U(\varphi)]^2 S''(\varphi) + \cdots,$$

which we write as

$$S'(\varphi) = \left[1 + \frac{1}{2}rU(\varphi)\frac{d}{d\varphi}\cdots\right]^{-1}\frac{S(\varphi)}{U(\varphi)} = \left[1 - \frac{1}{2}rU(\varphi)\frac{d}{d\varphi}\cdots\right]\frac{S(\varphi)}{U(\varphi)},$$

a first-order ODE, which, since $S/U \to S'/U' \to -1$ as $\varphi \to 1$, integrates to

$$S(\varphi) = -U(\varphi)\exp\left\{\frac{1}{1+\frac{r}{2}}\int_1^{\varphi}\left[\frac{1-U'(\varphi)}{U(\varphi)} + \cdots\right]d\varphi\right\}.$$

Another example: $f(x) = q + px^2$, so that $m = 2p$ and $r = p - q$. We readily find that

$$S(\varphi) = r\left(\frac{1-q}{p\varphi - q}\right)\left(1 + \log\frac{p\varphi - q}{r} + \cdots\right),$$

or

$$\varphi(S) = \frac{q}{p} + \frac{r}{p}\frac{r}{r + pS} + \frac{r^2}{(r+pS)^2}S\log\frac{r}{r+pS} + \cdots,$$

and this works quite well even in extreme cases such as $p = 1$, $q = 0$, $r = 1$, for which the exact answer is trivial.

7.4. Sol-Gel Transformation

A branched structure with no loops is termed a tree, and if a starting vertex is identified, a rooted tree. Basic stem cell proliferation is an example of a tree that develops in time (or generation number). The ordering parameter can have a spatial meaning as well, but it is analyzed similarly. An example is that of the molecular network of the extracellular matrix, involving two or more molecular species, each of which has a fixed number of available binding sites for another species. Far into the development of the network, it remains a tree because geometric considerations keep the unbound interior sites from being filled. A typical model assumes an unlimited source of bivalent A at concentration A and trivalent B at concentration B. Possible reactions are as follows: an addition of an A-bond to a complex X,

$$A + X \underset{1}{\overset{k}{\rightleftharpoons}} X',$$

with an addition probability given by $p/q = X'/X = kAX/X = kA$, or $p = kA/(1+kA)$; and conversely, addition of a B to an A-bond of Y,

$$B + Y \underset{1}{\overset{k'}{\rightleftharpoons}} Y',$$

with $p' = k'B/(1+k'B)$.

Now identify each resulting tree by its composition, e.g., A^6B^3, and weight A^iB^j by $P(i,j)$, its probability of occurrence, producing the generating function $G_A(A,B)$ if one starts with A, and $G_B(A,B)$ if one starts with B. Since A adds to a tree at a given B with probability p, and does not add with probability q, each A addition carries out the transformation

$$B \mapsto B(q+pA).$$

An internal B can accept two A's, so

$$B \mapsto B(q+pA)^2,$$

while for an initial B,

$$B \mapsto B(q+pA)^3.$$

Similarly, for B addition, we have, for internal A, $A \mapsto A(q'+p'B)$, and for initial A, $A \mapsto A(q'+p'B)^2$.

We want to find the asymptotic or equilibrium distribution of trees. Hence, the addition process must go on forever. To carry this out, denote the weighted sum of internal s-level A-initiated trees by $Q_s(A,B)$, and s-level B-initiated trees by $R_s(A,B)$. Clearly,

$$Q_s(A,B) = A(q'+p'R_{s-1}(A,B)), \quad R_s(A,B) = B(q+pQ_{s-1}(A,B))^2,$$

and, in particular,

$$R_s(A,B) = B[q+pA(q'+p'R_{s-2}(A,B))]^2.$$

Hence, if the process converges, we have

$$R_\infty(A,B) = B[q+pA(q'+p'R_\infty)]^2,$$

and then, e.g., $G_A(A,B) = A[R_\infty(A,B)]^2$.

For small p or p', the addition process becomes rapidly less likely as the size of the tree increases, and so the tree ensemble represents a solution of clusters, a *sol*. But for large p and p', a tree is more likely to percolate out to infinity—a *gel*. To see if there is indeed such a transition, we must find the mean number of A's in the network, $\langle n_A \rangle$. Since $G_A(A,B) = \sum P_A(i,j)A^iB^j$, then $G_A(1,1) = 1$ and $Q_\infty(1,1) = 1$, but also

$$\langle n_A \rangle = \sum iP_A(i,j) = \left.\frac{\partial G_A}{\partial A}\right|_{A=B=1}.$$

Directly from the equation for R_∞, we have, using $R_\infty = A = B = 1$,

$$\left.\frac{\partial R_\infty}{\partial A}\right|_{A=B=1} = 2\left(p + pp'\left.\frac{\partial R_\infty}{\partial A}\right|_{A=B=1}\right),$$

so

$$\left.\frac{\partial R_\infty}{\partial A}\right|_{A=B=1} = \frac{2p}{1-2pp'},$$

and

$$\langle n_A \rangle = \left.\frac{\partial G_A}{\partial A}\right|_{A=B=1} = 1 + \frac{4p}{1-2pp'},$$

reaching gelation at $pp' = \frac{1}{2}$. More generally, if η_A and η_B are the mean addition rates per generation ($2p$ and p' in the present case), it can be shown that the gelation point occurs at $\eta_A \eta_B = 1$.

7.5. Mesoscopic Viewpoint

7.5.1. Lubrication Theory and Epiboly. An early massive tissue movement in embryonic development is that of gastrulation, in which a hollow ball of cells (blastula)—or its equivalent in the almost two-dimensional embryo of heavily yolky eggs—create an invagination into which a layer of cells streams, guided by an adhesive gradient of the underlying tissue and corresponding gradient of tension. This movement, termed *epiboly*, can be modeled by standard *lubrication theory*, the dynamics of viscous flow in thin domains.

If the cell density is a constant ρ and the local velocity is $\mathbf{u}$, then conservation,

$$\dot{\rho} + \nabla \cdot \rho \mathbf{u} = 0,$$

reduces to $\nabla \cdot \mathbf{u} = 0$. If the fluid is inertialess Stokes flow ($\dot{\mathbf{u}} + \mathbf{u} \cdot \nabla \mathbf{u} \to 0$), viscous and pressure forces per unit volume must cancel:

$$\nabla P + \eta \nabla^2 \mathbf{u} = 0,$$

where η is the viscosity. Hence, separating z from the transverse components $\mathbf{x}$ with $\mathbf{r} = (\mathbf{x}, z)$ and $\mathbf{u} = (\mathbf{v}, w)$, we have

$$\nabla_T \cdot \mathbf{v} + \frac{\partial w}{\partial z} = 0, \quad \nabla_T P = \eta\left(\nabla_T^2 \mathbf{v} + \frac{\partial^2 \mathbf{v}}{\partial z^2}\right), \quad \frac{\partial P}{\partial z} = \eta\left(\nabla_T^2 w + \frac{\partial^2 w}{\partial z^2}\right).$$

Now, for h the scale of the film height, we scale w and z by h and P by $\frac{1}{h^2}$ so that to lowest order in h,

$$\nabla_T \cdot \mathbf{v} + \frac{\partial w}{\partial z} = 0, \quad \nabla_T P = \eta \frac{\partial^2 \mathbf{v}}{\partial z^2}, \quad \frac{\partial P}{\partial z} = 0.$$

In terms of $P(\mathbf{x})$ these integrate at once to

$$\mathbf{v} = \frac{1}{2} z^2 \nabla_T \frac{P}{\eta} + z\mathbf{A} + \mathbf{B}, \quad w = -\frac{1}{6} z^3 \nabla_T^2 \frac{P}{\eta} - \frac{1}{2} z^2 \nabla_T \cdot \mathbf{A} - z \nabla_T \cdot \mathbf{B} + C,$$

where $\mathbf{A}$, $\mathbf{B}$, C, and P are unknown functions of $\mathbf{x}$.

Now we need boundary conditions. Since there is no flow in the z-direction at $z = 0$, $w(\mathbf{x}, 0) = 0$; at $z = h(\mathbf{x})$, the flow must follow the boundary: $w(\mathbf{x}, h(\mathbf{x})) = \dot{h}(x)$. Substituting into $w(\mathbf{x}, z)$, then, $C = 0$ and

$$\dot{h} + \frac{1}{6} h^3 \nabla^2 P + \frac{1}{2} h^2 \nabla \cdot \mathbf{A} + h \nabla \cdot \mathbf{B} = 0.$$

If we neglect pressure and assume no slip against the substrate that supplies the tension force $-\nabla T(\mathbf{x})$, then

$$\mathbf{v}(\mathbf{x},0)=0, \quad \eta\frac{\partial \mathbf{v}}{\partial z}(\mathbf{x},z)\bigg|_{z=0}=-\nabla T(\mathbf{x}),$$

and our expression for $\mathbf{v}(\mathbf{x},z)$ then tells us that

$$\mathbf{A}=-\frac{1}{\eta}\nabla T, \quad B=0.$$

We conclude that

$$\dot{h}=\frac{1}{2\eta}h^2\nabla^2 T(\mathbf{x}),$$

which integrates at once to

$$h(\mathbf{x},t)=\frac{h(\mathbf{x},0)}{1-\left[\frac{t}{2\eta}\right]h(\mathbf{x},0)\nabla^2 T(\mathbf{x})},$$

the desired solution. We note that a trough in T, for $\nabla^2 T>0$, causes steady accretion of cellular matter at the point in question, as anticipated. Of course, if the denominator gets too small, the thin film assumption is not valid, and the approximation fails.

7.5.2. Cell Shape Change and Dissipative Flow. The above was essentially tissue rheology, the study of cellular material. But the same viewpoint may also be appropriate for the contents of a cell, resulting, e.g., in shape change with larger scale effects, including those above. Now we will attend to what might be called cell membrane mechanics and cytoplasmic rheology. First, consider membrane statics in a two-dimensional context. For a nod in the direction of realism, we start with an idealized columnar cell, which maintains its rectangular shape, the side lengths adjusting to minimize the environmental tension energy.

The membrane tension energy is denoted by τ per unit length, and the tension energy due to the substrate on the underside between x_1 and x_2 is denoted by $\sigma(x)$ per unit length. The cell area is fixed at A, and so the total tension energy is

$$T=\tau\left(x_2-x_1+\frac{2A}{x_2-x_1}\right)+\int_{x_1}^{x_2}\sigma(x)dx.$$

We ignore any energy associated with intracellular strain. Then, minimizing with respect to x_1 and x_2, we have

$$0=\tau\left[-1+\frac{2A}{(x_2-x_1)^2}\right]-\sigma(x_1), \quad 0=\tau\left[1-\frac{2A}{(x_2-x_1)^2}\right]+\sigma(x_2).$$

Hence,

$$\sigma(x_1)=\sigma(x_2)\equiv 0, \quad x_2-x_1=\left(\frac{2A\tau}{\tau+\sigma}\right)^{\frac{1}{2}}, \quad h=\left[\frac{(\tau+\sigma)A}{2\tau}\right]^{\frac{1}{2}}.$$

Now we proceed to dynamics. Component biological systems cannot be isolated, so that interaction with the environment, represented in some few-parameter model form, is the norm. In particular, the intracellular "fluid" is highly viscous, so its motion requires "external" energy, later to be dissipated—e.g., as heat—to the outside world. For the enclosing membrane and the intracellular fluid, we have to balance the tension energy change and the motional dissipation in the interior. We know how to model the tension energy T, and so we have to attend to the internal rate of dissipation Φ, entering into $\frac{dT}{dt} + \Phi = 0$. For a fluid of shear viscosity η, a force is produced on a fluid element of size dy:

$$F_x = -(dx)^2\eta\left[\frac{\partial u_x}{\partial y}(y+dy) - \frac{\partial u_x}{\partial y}(y)\right] = -\eta(dx)^2 dy\frac{\partial^2 u_x}{\partial y^2},$$

carried at rate u_x and hence dissipated at rate $-\eta u_x(\partial^2 u_x/\partial y^2)$ per unit volume. For the whole fluid, then,

$$\Phi_{\text{shear}} = -\eta\int\sum_{i\neq j} u_i\frac{\partial^2 u_i}{\partial r_j^2}\,d^3\mathbf{r} = \eta\int\sum_{i\neq j}\left(\frac{\partial u_i}{\partial r_j}\right)^2 d^3\mathbf{r}.$$

But there is also bulk viscosity, acting due to the stretching of fluid elements. If the fluid is isotropic, the total Φ must be rotationally invariant, so the above extends at once to

$$\Phi = \Phi_{\text{shear}} + \Phi_{\text{bulk}} = \eta\int\sum_{i,j}\left(\frac{\partial u_i}{\partial r_j}\right)^2 d^3\mathbf{r}.$$

The condition $\frac{dT}{dt} + \Phi = 0$ does not at all fully determine the local motion. So we assume the principle of maximization of rate of decrease of T at fixed Φ at each moment of time. Inserting this condition by the Lagrange "parameter" $\lambda(t)$, we therefore need

$$\delta\left(\frac{dT}{dt} + \lambda(t)\Phi\right) = 0$$

for any variation of the velocity field, with $\lambda(t)$ then adjusted so that $\frac{dT}{dt} + \Phi = 0$. Now $\dot{r}_i = u_i$, so

$$\frac{dT}{dt} = \sum u_i\frac{\partial T}{\partial r_i},$$

and taking $\delta = \partial/\partial u_i$, it follows that

$$\frac{\partial T}{\partial r_i} + \lambda(t)\frac{\partial\Phi}{\partial u_i} = 0.$$

Typically, and in the case we have examined, Φ is homogeneous of degree two in the velocities. It follows from Euler's equation that

$$\sum u_i\frac{\partial\Phi}{\partial u_i} = 2\Phi,$$

and therefore

$$0 = \sum u_i\frac{\partial T}{\partial r_i} + \lambda(t)\sum u_i\frac{\partial\Phi}{\partial u_i} = \frac{dT}{dt} + 2\lambda(t)\Phi.$$

Comparing with $\Phi = -\frac{dT}{dt}$, we see that $\lambda(t) = \frac{1}{2}$, so that the full dynamics is encompassed by

$$\delta\left(\frac{dT}{dt} + \frac{1}{2}\Phi\right) = 0.$$

We can return to the dynamics of the idealized columnar cell on an adhesive substrate. Assuming that the membranes stretch uniformly, and using that $h = A/(x_2 - x_1)$ so $v = -A(u_2 - u_1)/(x_2 - x_1)^2$, we guess that the full velocity field can be gotten by linear interpolation:

$$u_x(x, y) = u_1\left(\frac{x_2 - x}{x_2 - x_1}\right) + u_2\left(\frac{x - x_1}{x_2 - x_1}\right),$$
$$u_y(x, y) = v\left(\frac{y}{h}\right) = -(u_2 - u_1)\frac{y}{x_2 - x_1}.$$

In addition to satisfying boundary conditions, these also obey incompressible, irrotational flow, which in two dimensions reduces to

$$\frac{\partial u_x}{\partial x} + \frac{\partial u_y}{\partial y} = 0, \qquad \frac{\partial u_x}{\partial y} - \frac{\partial u_y}{\partial x} = 0.$$

Ignoring substrate friction and graded tension, we now have

$$T = (\tau + \sigma)(x_2 - x_1) + \frac{2\tau A}{x_2 - x_1}, \qquad \Phi = 2\eta A\left(\frac{u_2 - u_1}{x_2 - x_1}\right)^2,$$

and from

$$\frac{\partial T}{\partial x_1} + \frac{1}{2}\frac{\partial \Phi}{\partial u_1} = 0 = \frac{\partial T}{\partial x_2} + \frac{1}{2}\frac{\partial \Phi}{\partial u_2},$$

it follows that

$$u_1\left[-(\tau + \sigma) + \frac{2\tau A}{(x_2 - x_1)^2}\right] - \frac{1}{2}\left[\frac{4\eta A(u_2 - u_1)}{(x_2 - x_1)^2}\right] = 0,$$
$$u_2\left[\tau + \sigma - \frac{2\tau A}{(x_2 - x_1)^2}\right] + \frac{1}{2}\left[\frac{4\eta A(u_2 - u_1)}{(x_2 - x_1)^2}\right] = 0.$$

Hence $u_2 = -u_1$, so that if $x_2 = -x_1$ initially, this persists, and

$$u_2 = \frac{dx_2}{dt} = \frac{2\tau A - 4(\tau + \sigma)x_2^2}{8\eta A},$$

with the equilibrating solution

$$x_2 = \left[\frac{\tau A}{2(\tau + \sigma)}\right]^{\frac{1}{2}} \tanh\left(\frac{\tau}{4\eta}t + C\right).$$

A more substantial application is to the bending of the "spherical" sheet of epithelial cells of the amphibian blastula, here in its two-dimensional version. If all of the cell opening angles $\theta_i \ll 1$, the area A_i is fixed, and the intercell tension is e_i per unit length, we have

$$T = \sum_i \left[\left(\xi_i - \frac{1}{2} h_i \theta_i \right) \tau_i + \left(\xi_i + \frac{1}{2} h_i \theta_i \right) \sigma_i + l_i h_i \right],$$

where $\xi_i = \frac{A_i}{h_i}$, and

$$\Phi = \sum A_i \left[\frac{\upsilon (u_i - u_{i-1})^2}{(x_i - x_{i-1})^2} + \frac{4\eta h_i^2 \dot{\theta}_i^2}{(x_i - x_{i-1})^2} \right],$$

where υ denotes substrate friction. The condition that $\{h_i\}$ is periodic around the ring is imposed, while $\sum \theta_i = 2\pi$ must be applied with a Lagrange parameter. This system must be solved numerically. Doing so and applying enhanced outer tension on a group of cells, one does find a typical gastrulation dynamics (which must be manually stopped because cell-cell exclusion has not been included).

7.6. Turing Dynamical Resolution

7.6.1. Dynamics at Low Resolution. From a very general viewpoint, gradient-flow dynamics, i.e.,

$$\dot{x}_i = -\frac{\partial E}{\partial x_i}(\mathbf{x}, \mathbf{a})$$

for parameter set $\mathbf{a}$, can incorporate a great deal of detail via the complexity of E, always "aiming" at minimizing E. But at least at first, phenomenology tends to be qualitative rather than quantitative. Catastrophe (singularity) theory, a once-popular vogue in developmental modeling, is one possible direction. Another is that of concentrating on the basic logical form: $x_i = 0$ is "off" and $x_i = 1$ is "on," and we want to know the transition sequence made by the Boolean vector $\mathbf{x} = \{x_i\}$ in time. Even a system as simple as the Lotka-Volterra predator-prey model, restricted by its constant of motion

$$K = \lambda_1 h + \lambda_2 a - \lambda_1 h_0 \log h - \lambda_2 a_0 \log a,$$

has substantial character in its continuous domain. But its roster of behaviors is really very slim: $(0, 0)$ is an unstable singular point, (a_0, h_0) metastable, and there is a set of closed trajectories, i.e., primitive limit cycles.

Let us instead construct a minimal discrete version: $a \equiv 0$ if small, $a \equiv 1$ if large, and the same for h; time t is an integer; and a (and h) follows $\dot{a}$ if it can:

$$a(t+1) = \begin{cases} 0, & \dot{a} < 0, \\ 1, & \dot{a} > 0. \end{cases}$$

Here, we need a reference (a_0, h_0) as intermediate, say $\left(\frac{1}{2}, \frac{1}{2}\right)$. But also $\dot{a} = 0$ and $\dot{h} = 0$ are marginal, a possibility that is avoided by the replacement

$$\dot{a} = (a + \epsilon)\left(h - \frac{1}{2}\right), \quad \dot{h} = (h + \epsilon)\left(\frac{1}{2} - a\right),$$

giving the table:

$a(t)$	$h(t)$	$a(t+1)$	$h(t+1)$
0	0	0	1
0	1	1	1
1	0	0	0
1	1	1	0

Note that if a prime denotes the next state, $(a, h) \mapsto (a', h')$, the dynamics is encompassed by the simple logical equation

$$a' = h, \quad h' = \bar{a};$$

here the only constant is the trivial $K = a(1-a) + h(1-h)$.

But the ϵ-modification is not unique; we could also write

$$\dot{a} = (a+\epsilon)\left(h - \frac{1}{2}\right), \quad \dot{h} = (h-\epsilon)\left(\frac{1}{2} - a\right),$$

so that:

a	h	a'	h'
0	0	0	0
0	1	1	1
1	0	0	1
1	1	1	0

with $a' = h$ and $h' = \bar{a}h + \bar{h}$, with the nontrivial constant $K = \bar{a}\bar{h}$. The fixed point $(0, 0)$ is recognized as unstable because it disappears under an ϵ-change.

We can also incorporate the information into a single tableau by placing a bar over those entries that change at the next step:

a	h
0	$\bar{0}$
$\bar{0}$	1
$\bar{1}$	0
1	$\bar{1}$

and

a	h
0	0
$\bar{0}$	1
$\bar{1}$	$\bar{0}$
1	$\bar{1}$

Now our main interest, with pattern formation in mind, is in PDEs. Since amplitude and time have already been discretized, we now discretize space as well, say on a unit grid. For a finite system, there are many more concentrations to consider, but that is all. For a sensibly infinite space, conceptual problems arise as well. Consider, for example, basic pure diffusion on a one-dimensional unit grid, thereby taking the form

$$\dot{a}_n \propto a_{n+1} - 2a_n + a_{n-1}.$$

If we append the rule $a'_n = a_n$ when $\dot{a} = 0$, the table (eight entries) of a'_n in terms of a_{n-1}, a_n, and a_{n+1} then looks as follows:

$a_n = 0$:

a_{n-1} \ a_{n+1}	0	1
0	0	1
1	1	1

and $a_n = 1$:

a_{n-1} \ a_{n+1}	0	1
0	0	0
1	0	1

equivalent to the Boolean transformation

$$a'_n = a_{n-1}a_na_{n+1} + \bar{a}_n(1 - \bar{a}_{n-1}\bar{a}_{n+1})$$

with a typical temporal development:

0	0	0	1	1	1	1	0	0	0
0	0	1	0	1	1	0	1	0	0
0	1	0	1	0	0	1	0	1	0
1	0	1	0	1	1	0	1	0	1

Diffusion thus appears as a wave front, but with decreasing coarse-grained density. In this realization, total content is not conserved, and so has to be persistently normalized. A grey level format, say $a = 0, 1$, or 2 would pick up more detail but would contradict the philosophy of the representation.

A more complex dynamics occurs in the Lindenmayer type *cellular automata* (see Wolfram as well), used, e.g., to model one- and higher-dimensional growth patterns. There is a formal model of plant shoot apex in which the apical growth region—say, on the left—maintains its form but leaves a changing pattern on the right. This would result from a table of next states (here, cells really mean biological cells):

	left neighbor	
cell type	0	1
0	0	1
1	1	0

giving rise to

0	1					
0	1	1				
0	1	1	0			
0	1	1	0	1		
0	1	1	0	1	1	0

but even here, the formula for the rightmost cell as a function of time is not known.

7.6.2. Intermediate Resolution. A utility of this construction is that it may also be used in reverse order. Two forms have received the most attention. Suppose we start with a "logical structure" that represents the phenomenology—but then convert this to a different form to permit quantitative analysis. Here is an elementary "biochemical" example in which species a inhibits b, which stimulates c, which inhibits a, with steady states $(0, 1, 1)$ and $(1, 0, 0)$, and two 3-cycles. We first write down the symbolic dynamics

$$a' = \bar{c}, \quad b' = \bar{a}, \quad c' = b,$$

and then replace the $0 \mapsto 1$ and $1 \mapsto 0$ discontinuities by the sigmoidal Hill functions

$$F^+_\lambda(x) = \frac{x^n}{\theta^n_\lambda + x^n}, \quad F^-_\lambda(x) = \frac{\theta^n_\lambda}{\theta^n_\lambda + x^n},$$

for suitable n, together with intrinsic decay rates, so that a species will be driven downward (rate 0) if it is not stimulated (rate 1). The above then becomes

$$\dot{a} = kF_a^-(c) - \alpha a, \quad \dot{b} = k'F_b^-(a) - \beta b, \quad \dot{c} = k''F_c^+(b) - \gamma c.$$

More generally, if one has, e.g., a symbolic dynamics

$$a' = a \cdot \bar{b} + c,$$

where $\cdot$ and $+$ indicate logical product (intersection) and sum (union), this would translate simply to

$$\dot{a} = kF^+(a)F^-(b) + k'F^+(c) - \alpha a.$$

Much more complex immunological situations have been analyzed in this fashion.

Of course, continuous dynamics does not require continuous functions. An interesting format is the piecewise linear kinetics of [**73**], probably the simplest extension of the state transition dynamics that one can imagine. In the somewhat specialized version most heavily studied, the amplitudes x_i (which can be *relative* concentrations) are accompanied by a Boolean characterization

$$\tilde{x}_i = \begin{cases} 0, & x_i < 0, \\ 1, & x_i \geq 0, \end{cases}$$

and a corresponding discrete state vector $\tilde{\mathbf{x}} = \{\tilde{x}_i\}$. The dynamics is then given by

$$\frac{dx_i}{dt} = \Lambda_i(\tilde{\mathbf{x}}) - x_i,$$

where Λ_i is piecewise constant, a function only of the state $\tilde{\mathbf{x}}$, and can of course be taken as multilinear in the $\tilde{x}_i$ (since $\tilde{x}_i^2 = \tilde{x}_i$). Further conditions of convenience are that Λ_i is nowhere zero, and that the sign of $\Lambda_i(\tilde{\mathbf{x}})$ is independent of $\tilde{x}_i$ itself. The dynamics here are such that in each state $\tilde{\mathbf{x}}$, the trajectory aims at the point $\Lambda_i(\tilde{\mathbf{x}})$ belonging to another state, shifting direction when it leaves the current state, signaled by some x_j passing through the value zero. Until this shift, the trajectory is given by solving the above trivial ODE

$$x_i(t) = \lambda_i + (c_i - \lambda_i)e^{-t},$$

where $x_i(0) = c_i$ and $\lambda_i = \Lambda_i(\mathbf{c})$. The process can then be described by the passage from one two-state boundary to another, and, if cyclical, can be characterized by the successive intersections of a given two-state boundary (Poincaré returns).

CHAPTER 8

Somite Formation in Vertebrates

For complex organisms to develop robustly and efficiently, many subdynamic processes must fit together (if not, one has pathologies, such as cleft palate), so timing is of the essence. This means that "clocks" may be expected to play important roles in the process of development, and we start with what has become prototypical: somitogenesis (studied in detail in zebrafish—with transparent eggs—and chicks). This occurs early in embryonic development, where the axial region (featuring the notochord scaffolding) has been laid down between head and tail, and growth is continuing at the tail bud. Somites—compact bundles of cells—eventually develop not only into vertebrae regions but also into many other mesodermal organs. They are laid down with temporal (every 30 minutes in zebrafish) and consequently spatial precision. Their appearance is closely associated with the notch signaling network, in particular correlating with oscillations of two genes, her1 and her7, associated with this network, as well as with an omnipresent messenger, deltaC. We must attend to two basic questions:

(1) How do genes oscillate?
(2) How does this give rise to spatial patterns?

8.1. Elementary Oscillations

"Traditional" models of oscillators stem from physical systems—e.g., pendula—and involve one second-order or two first-order ODEs. A single ODE of the form

$$\dot{x}(t) = f(x(t))$$

will not do, since a given value x_0 will be hit both on the way up and on the way down; thus f could not be single-valued. The first biologically motivated oscillator to be analyzed in detail was the Lotka-Volterra or predator-prey model; see, e.g., [**15**]. In Lotka's case, two populations, $H(t)$ of hares and $L(t)$ of lynx, were involved. The hares had plentiful food and a low natural death rate, so an isolated population would grow exponentially according to $\dot{H}(t) = aH(t)$. But the hares served as (ideally, the only) food for the lynx, who thereby inhibited the hare population: $\dot{H} = aH - bLH$. The lynx produced reproductively via the hare supply and had a substantial natural death rate. Hence, the two populations satisfied the dynamics

$$\dot{H} = aH - bLH, \quad \dot{L} = cHL - dL,$$

and numerical solutions indeed showed the two populations oscillating out of phase.

A biochemical version is easily constructed by including all the players in the dynamics, suitably weighted. To start, a hare is equivalent in bulk to a fraction of a lynx, so let us rescale by setting $H = (b/c)K$, obtaining

$$\dot{K} = aK - bLK, \quad \dot{L} = bKL - dL.$$

Then we include the limitless food supply, A, that the hares live on, and finally the poor lynx, on dying, becomes an endpoint, the corpse P. In chemical kinetics terms, this gives us the reaction set

$$A + K \xrightarrow{a} 2K, \quad L + K \xrightarrow{b} 2L, \quad L \xrightarrow{d} P,$$

with precisely the above dynamics.

The usual analysis of such a nonlinear system of ODEs proceeds by first looking for the stationary points, $\dot{H} = \dot{L} = 0$, or here

$$0 = aH_0 - bL_0H_0, \quad 0 = cH_0L_0 - dL_0,$$

so that

(i) $H_0 = \frac{d}{c}$, $L_0 = \frac{a}{b}$, or
(ii) $H_0 = 0$, $L_0 = 0$,

and finding the nature of the nearby solutions

$$H(t) = H_0 + \delta H(t), \quad L(t) = L_0 + \delta L(t).$$

This results, to first order in δH and δL, in the linear dynamics

$$\delta\dot{H} = a\delta H - bL_0\delta H - bH_0\delta L, \quad \delta\dot{L} = cL_0\delta H + cH_0\delta L - d\delta L.$$

As linear constant-coefficient differential equations, the solution space is spanned by solutions of the form

$$\delta H(t) = \delta He^{\lambda t}, \quad \delta L(t) = \delta Le^{\lambda t}$$

so that

$$\begin{pmatrix} a - bL_0 & -bH_0 \\ cL_0 & cH_0 - d \end{pmatrix} \begin{pmatrix} \delta H \\ \delta L \end{pmatrix} = \lambda \begin{pmatrix} \delta H \\ \delta L \end{pmatrix}.$$

The eigenvalues λ are then given by the consistency conditions

(i) $\begin{vmatrix} -\lambda & -\frac{bd}{c} \\ \frac{ac}{b} & -\lambda \end{vmatrix} = 0$: $\lambda = \pm\imath\sqrt{ad}$ and $\delta L = -\lambda\frac{c}{bd}\delta H$, or

(ii) $\begin{vmatrix} a - \lambda & 0 \\ 0 & -\lambda - d \end{vmatrix} = 0$: $\lambda = a$ and d, with $\delta L = 0$ and $\delta H = 0$.

Case (ii) is a saddle point; the $\lambda = a$ solution for $(\delta H, \delta L)$ grows exponentially in time, while that for $\lambda = d$ decays: there are no nearby solutions. Case (i) says that we have periodic orbits of frequency $\omega = \sqrt{ad}$, but with arbitrary amplitude, surrounding the stationary point, i.e.,

$$\delta H(t) = \alpha \cos\left(t\sqrt{ad}\right) + \beta \sin\left(t\sqrt{ad}\right),$$

and it follows that

$$\delta L(t) = \left[\alpha \sin\left(t\sqrt{ad}\right) - \beta \cos\left(t\sqrt{ad}\right)\right]\frac{c}{b}\sqrt{\frac{a}{d}}.$$

Of course, the nested elliptical orbits distort as they expand, but they remain closed orbits. This would be obvious if we could find an $f(H, L)$ that doesn't depend on time: $(d/dt)f(H(t), L(t)) = 0$; then we would have $f(H(t), L(t)) = m$, a constant, and given $L(t)$ we could solve (it might be multivalued) for $H(t)$. But this requires

$$\dot{H}\frac{\partial f}{\partial H} + \dot{L}\frac{\partial f}{\partial L} = 0$$

or

$$(a - bL)H\frac{\partial f}{\partial H} + L(cH - d)\frac{\partial f}{\partial L} = 0,$$

so that

$$\left(\frac{H}{cH - d}\right)\frac{\partial f}{\partial H} + \left(\frac{L}{a - bL}\right)\frac{\partial f}{\partial L} = 0,$$

clearly satisfied by $f = g(H) + h(L)$ if

$$\frac{\partial g}{\partial H} = \frac{1}{m}\left(c - \frac{d}{H}\right), \quad \frac{\partial h}{\partial L} = -\frac{1}{m}\left(\frac{a}{L} - b\right),$$

for some constant m. So,

$$f(H(t), L(t)) = cH(t) + bL(t) - d\log H(t) - a\log L(t) = m.$$

What is wrong with this as a model for our purposes is that, at fixed parameters a, b, c, d, the oscillating solution is far from unique, and also that, even close to the stationary point where all frequencies are the same, its value depends sensitively on the parameters. Inventing a model without these difficulties is easy enough; see, e.g., [**20**]. We cheat by first going to polar coordinates (r, θ) and require

$$\dot{r} = \frac{1 - r^2}{r}, \quad \dot{\theta} = \omega. \tag{8.1}$$

Then

$$\frac{\dot{\left(r^2\right)}}{1 - r^2} = 2,$$

so that

$$r = (1 - Ke^{-2t})^{\frac{1}{2}}, \quad \theta = \omega t + \theta_0;$$

i.e., (r, θ) spirals in or out to the limit cycle $r = 1$, no matter where it starts, while the frequency is precisely ω. In xy-coordinates ($x = r\cos\theta$, $y = r\sin\theta$), the dynamics read

$$\dot{x} = \left(\frac{1 - x^2 - y^2}{x^2 + y^2}\right)x - \omega y, \quad \dot{y} = \left(\frac{1 - x^2 - y^2}{x^2 + y^2}\right)y + \omega x.$$

So the right type of dynamics is available, but is rather ad hoc.

8.2. Time-Delay Oscillators

The above are artificial, nonbiological, etc. There is of course a fair industry devoted to models of dynamics in the abstract form

$$\dot{x} = f(x, y), \quad \dot{y} = g(x, y),$$

and we will meet examples repeatedly in the future. But the basic temporal locality assumed is often not valid at the level of resolution of biological processes: there are delays due to processes occurring below the level of resolution of the phenomenon to be described. In particular, there are at the gene expression level mechanical motions coupled with the biochemical reactions: a time T_m for the transcription of DNA to mRNA, and a time T_p for the transcription of mRNA to protein. What sort of biochemical kinetics is now appropriate?

Suppose that m is the total mRNA of the assertedly relevant gene (say her1) in a cell and p the total protein stemming from its activity. Since the protein is degraded fairly fast, we would have

$$\dot{p} = am - bp.$$

Now the mRNA, which is produced by the DNA as a source, may be inhibited by its own product, p, which we model by a net source

$$f(p) = \frac{k}{1 + (p/\bar{p})^2}.$$

Since the mRNA is also degraded, we then have

$$\dot{m} = f(p) - cm.$$

At low p, m will settle down to an amount given by $0 = k - cm$, or $m = k/c$; at higher p, there is a p^2 connection to the source strength, consistent with p being a dimer; and at very high p, m will simply decay as $\dot{m} = -cm$, or $m \propto e^{-ct}$.

The general behavior of this small circuit, not yet including delays, is readily seen. First, let us nondimensionalize: set $\tilde{p} = p/\bar{p}$, so that

$$\dot{\tilde{p}} = \left(\frac{a}{\bar{p}}\right) m - b\tilde{p}, \quad \dot{m} = \frac{k}{1 + \tilde{p}^2} - cm;$$

then $t = \frac{T}{c}$ (and $\frac{d}{dT} \equiv {}'$), giving

$$m' = \frac{k/c}{1 + \tilde{p}^2} - m, \quad \tilde{p}' = \left(\frac{a}{c\bar{p}}\right) m - \left(\frac{b}{c}\right)\tilde{p};$$

and finally $m = \frac{k}{c}\tilde{m}$, resulting in

$$\tilde{m}' = \frac{1}{1 + \tilde{p}^2} - \tilde{m}, \quad \tilde{p}' = \alpha\tilde{m} - \beta\tilde{p}.$$

Here, only the parameters $\alpha = ak/(c^2\bar{p})$, $\beta = b/c$ remain as controls. The quickest way to see the behavior is to look at the nullclines

$$\tilde{m}' = 0: \quad \tilde{m} = \frac{1}{1+\tilde{p}^2},$$
$$\tilde{p}' = 0: \quad \tilde{m} = \frac{\alpha}{\beta}\tilde{p}.$$

We see that the component of flow at any point on a nullcline in this system is directed at the stationary point, so that the flow will end up at the stationary point. This can be verified as well by looking at the linearized dynamics near the stationary point.

This is not very interesting. What then is the effect of the delays? Formally, the dynamical equations now become [**109**]

$$\dot{p}(t) = am(t-T_p) - bp(t), \quad \dot{m}(t) = f(p(t-T_m)) - cm(t),$$

which we will analyze in a moment. But even simpler would be a one-step process in which the product directly controls its source strength:

$$\dot{x}(t) = f(x(t-T)) - cx(t), \quad f(x) = \frac{k}{1+x^2}.$$

As a real extreme, suppose that the decay rate is very fast, $c \to \infty$, but that $k \to \infty$ too, with $K = k/c$ fixed. Then we have

$$\frac{1}{c}\dot{x}(t) = \frac{K}{1+[x(t-T)]^2} - x(t), \tag{8.2}$$

and letting $c \to \infty$, see that either $x(t)$ is discontinuous or that it satisfies

$$x(t) = \frac{K}{1+[x(t-T)]^2}.$$

The solution appears to be highly nonunique, since it breaks up into chains $x_0(t)$, $x_0(t+T), x_0(t+2T), x_0(t+3T), \dots$, each with its own x_0.

The simplest solution, corresponding to initial constancy of x, is one in which x stays at x_0 for a period T, switches to x_1 and stays for T, then to x_2 and stays for T, and so forth. When does this become an oscillator? Clearly, if x_2 returns to the value x_0, in which case

$$x_1 = \frac{K}{1+x_0^2}, \quad x_2 = \frac{K}{1+x_1^2} = x_0.$$

It follows on eliminating K that

$$x_1 + \frac{1}{x_1} = x_0 + \frac{1}{x_0},$$

giving just two possibilities: $x_1 = x_0$, in which case there is no oscillation, just steady state, and $x_1 = 1/x_0$ with continued oscillation. Note that then

$$K = \frac{1}{x_0}\left(1+x_0^2\right) \geq 2,$$

which is required for oscillation. If $K < 2$, the oscillation will be damped, eventually arriving at the steady state value given by $\bar{x} = K/(1 + \bar{x}^2)$, or

$$\bar{x}^3 + \bar{x} = K.$$

Suppose that c is very large but not infinity. This is an example of a singular perturbation: raising $1/c$ in (8.2) from 0 to ϵ changes the differential order of the system. What will change of course is what happens at the discontinuity in $x(t)$. To see this, we look closely: set $\tau = ct$, so that with $x(t) = x(\tau/c) = X(\tau)$ we now have

$$X'(\tau) = \frac{K}{1 + [X(\tau - cT)]^2} - X(\tau) \to \frac{K}{1 + x_0^2} - X(\tau),$$

so that

$$X(\tau) = \frac{K}{1 + x_0^2} + Ae^{-\tau};$$

since $X(0) = x_0$ and $\frac{K}{1+x_0^2} = x_1$,

$$X(\tau) = x_0 e^{-\tau} + x_1(1 - e^{-\tau}).$$

How much harder is the two-delay case? To see, we simplify by defining the "advanced" protein population $p_a(t) \equiv p(t+T_p)$, in which case the pair of equations becomes

$$\dot{m}(t) = f(p_a(t - T_m - T_p)) - cm(t), \quad \dot{p}_a(t) = am(t) - bp_a(t).$$

Then introduce $T = T_m + T_p$, and nondimensionalize by setting $\tau = \frac{t}{T}$, $\beta = bT$, $\gamma = cT$, and $\kappa = \frac{ak}{bc\bar{p}}$, as well as

$$\psi(\tau) = \frac{p_a(T\tau)}{\bar{p}}, \quad \mu(\tau) = \frac{c}{k}m(T\tau),$$

so that

$$\mu'(\tau) = \frac{\gamma}{1 + [\psi(\tau - 1)]^2} - \gamma\mu(\tau), \quad \psi'(\tau) = \kappa\beta\mu(\tau) - \beta\psi(\tau).$$

Note that in the absence of inhibition by p, we will have $m_{\max} = k/c$ and $p_{\max} = ak/(bc)$, so that we have actually carried out the rescaling $\mu = m/m_{\max}$, $\kappa = p_{\max}/\bar{p}$.

If we start with large β and γ, corresponding to short lifetimes for m and p, these reduce to

$$\mu(\tau) = \frac{1}{1 + [\psi(\tau - 1)]^2}, \quad \psi(\tau) = \kappa\mu(\tau),$$

so

$$\psi(\tau) = \frac{\kappa}{1 + [\psi(\tau - 1)]^2},$$

which is identical with the single delay system we have already analyzed, and so with the same requirement, $\kappa \geq 2$, for periodicity, with decaying oscillations for

$\kappa < 2$. Without this assumption, we can still eliminate $\mu(\tau)$ to obtain the second-order equation

$$\psi(\tau) + \left(\frac{1}{\beta} + \frac{1}{\gamma}\right)\psi'(\tau) + \frac{1}{\beta\gamma}\psi''(\tau) = \frac{\kappa}{1 + [\psi(\tau - 1)]^2},$$

which will oscillate for β, γ, and κ big enough.

Can we estimate the period? To do so, define

$$\Psi(\tau) = \psi(\tau) + \left(\frac{1}{\beta} + \frac{1}{\gamma}\right)\psi'(\tau) + \frac{1}{\beta\gamma}\psi''(\tau),$$

so $\Psi(\tau)$ is maximized when $\psi(\tau - 1)$ is minimized, and vice versa. We then look at the interval between extrema, which, for large β and γ, would be close to 1. Suppose that $\psi(\tau_0)$ is a maximum, hence $\psi(\tau_0 - 1 - u)$ is a minimum for some small u, implying that $\Psi(\tau_0 - u)$ is a maximum. So

$$\begin{aligned} 0 = \Psi'(\tau_0 - u) &= \psi'(\tau_0 - u) + \left(\frac{1}{\beta} + \frac{1}{\gamma}\right)\psi''(\tau_0 - u) + \frac{1}{\beta\gamma}\psi'''(\tau_0 - u) \\ &= \psi'(\tau_0) + \left(\frac{1}{\beta} + \frac{1}{\gamma} - u\right)\psi''(\tau_0) + \cdots \\ &= \left(\frac{1}{\beta} + \frac{1}{\gamma} - u\right)\psi''(\tau_0) + \cdots, \end{aligned}$$

implying that $u = 1/\beta + 1/\gamma$. We conclude that there is a half period of $1 + 1/\beta + 1/\gamma$ or a full period in real time of $2(T + 1/b + 1/c)$ with a very weak dependence on the system parameters.

8.3. Biochemical Entrainment and Biochemical Waves

8.3.1. Synchronization. So far, we have dealt with individual cells, but the phenomenology of interest is that of cell populations. In somite formation, there are three essential steps:

(1) establishing the presomitic mesoderm (PSM) that will be converted to the somite structure,
(2) synchronizing the behavior of the cell population, and
(3) carrying out the condensation process.

In this section, we look first at (2), a matter of entrainment, and then at (1), the delineation of the PSM.

In an extended system paced by a clock, it is crucial that different parts of the system synchronize their clocks. The PSM appears to be running on a clock characterized by variation of the proteins her1 and her7, but it was observed some time ago that deltaC, an important messenger in the notch signaling system, oscillates in unison. Experiments—relying mainly on inputs to the system—made it clear that deltaC is not necessary for the oscillation of the protein expression and suggested that its role is that of synchronizing the oscillations of nearby cells by its transmission from cell to cell. How would this work? We need an effective general way of analyzing nonlinear oscillations, and a standard one is the "equivalent linearization" of Kyrlov and Bogoliubov.

We start with a very basic example, the Van der Pol oscillator, which on scaling takes the form

$$\ddot{x}(t) - \epsilon[1 - (x(t))^2]\dot{x}(t) + x(t) = 0.$$

If ϵ were to vanish, the solution, to within phase, would be simply

$$x(t) = A\cos t, \quad \dot{x}(t) = -A\sin t,$$

which is neutrally stable, as in the predator-prey case, or for that matter, as in any linear system. If $\epsilon \neq 0$, we could still assume the form

$$x(t) = A(t)\cos\theta(t), \quad \dot{x}(t) = -A(t)\sin\theta(t),$$

but would then have the consistency condition (this is also the familiar variation-of-constants method)

$$\dot{A}(t)\cos\theta(t) - A(t)\dot{\theta}(t)\sin\theta(t) = -A(t)\sin\theta(t).$$

The Van der Pol equation itself would become

$$\begin{aligned}-\dot{A}(t)\sin\theta(t) - A(t)\dot{\theta}(t)\cos\theta(t) = -\epsilon A(t)\sin\theta(t)[1 - (A(t))^2\cos^2\theta(t)]\\ - A(t)\cos\theta(t),\end{aligned}$$

and we could solve at once for $\dot{A}$ and $\dot{\theta}$:

$$\begin{aligned}\dot{A}(t) &= \epsilon A(t)\sin^2\theta(t)[1 - (A(t))^2\cos^2\theta(t)],\\ \dot{\theta}(t) &= \epsilon\sin\theta(t)\cos\theta(t)[1 - (A(t))^2\cos^2\theta(t)] + 1.\end{aligned}$$

If ϵ is small, $A(t)$ and $\theta(t)$ would vary slowly with time, and so we could average the above over a cycle of $\theta(t)$, changing nothing. Doing so, and using $\langle\sin^2\theta\rangle = \frac{1}{2}$, $\langle\sin^2\theta\cos^2\theta\rangle = \frac{1}{8}$, and $\langle\sin\theta\cos\theta\rangle = \langle\sin\theta\cos^3\theta\rangle = 0$, we then obtain

$$\dot{A} = \frac{\epsilon}{8}A(2 - A)(2 + A), \quad \dot{\theta} = 1,$$

almost identical with the model of (8.1) and with the solution (limit cycle)

$$A(t) = \frac{2}{(1 + Ce^{-\epsilon t})^{1/2}}, \quad \theta(t) = t + C'.$$

Of course, one can do better; e.g., systematically, take $A(t)$ as a Fourier series in $\theta(t)$ and take the Fourier coefficients of the result to rewrite the series coefficients. But for a quick idea of oscillating behavior, even for complex systems, this technique is a great first step. For example, let's look at the reduced form of the her1 oscillator (period 2π):

$$x(t) = \frac{K}{1 + [x(t - \pi)]^2}.$$

For steady state, we would choose the form

$$x(t) = a + b\cos t,$$

yielding at once

$$a + b\cos t = \frac{K}{1 + (a - b\cos t)^2},$$

whose first two Fourier coefficients would give (essentially the Galerkin approach to an overdetermined system)

$$a = \frac{1}{2\pi}\int_0^{2\pi} \frac{K}{1 + (a - b\cos t)^2}\,dt, \quad b = \frac{1}{\pi}\int_0^{2\pi} \frac{K(a + b\cos t)}{1 + (a - b\cos t)^2}\,dt,$$

a pair of algebraic equations for a and b, solvable only within a range of K.

At this simple level, then, the time-delay oscillator and the usual, i.e., Van der Pol, oscillator show many characteristics, and we will generally use oscillator (and other) models in a local time format for a semiquantitative overview (with suitable effective parameters, to be sure). We can now proceed to the synchronization of cellular oscillators, e.g., via deltaC signaling.

In fact, we will examine a prior phenomenon, that of entrainment of frequency. Suppose that a Van der Pol oscillator is driven by an external source, for example, a cell that has previously been set into oscillation at not quite the same frequency,

$$\ddot{x} - \epsilon(1 - x^2)\dot{x} + x = K\sin\omega t.$$

If ϵ were zero, the periodic solution would simply be

$$x_0(t) = \frac{K}{1 - \omega^2}\sin\omega t,$$

so let us set

$$x(t) = y(t) + \frac{K}{1 - \omega^2}\sin\omega t,$$

giving us instead

$$\ddot{y} - \epsilon[1 - (y + \gamma\sin\omega t)^2](\dot{y} + \omega\gamma\cos\omega t) + y = 0,$$

where $\gamma = K/(1 - \omega^2)$. Assuming noncommensurability of y and ωt, we average over a cycle of ωt, obtaining

$$\ddot{y} - \epsilon\left(1 - y^2 - \frac{\gamma^2}{2}\right)\dot{y} + y = 0,$$

and tackle this by standard equivalent linearization: set

$$y = a\cos\theta, \quad \dot{y} = -a\sin\theta,$$

requiring

$$\dot{a}\cos\theta - a\dot{\theta}\sin\theta + a\sin\theta = 0,$$

while the equation becomes

$$-\dot{a}\sin\theta - a\dot{\theta}\cos\theta + \epsilon\left(1 - a^2\cos^2\theta - \frac{\gamma^2}{2}\right)a\sin\theta + a\cos\theta = 0,$$

so that

$$\dot{a} - \epsilon a \sin^2\theta\left(1 - a^2\cos^2\theta - \frac{\gamma^2}{2}\right) = 0,$$

$$a\dot{\theta} - \epsilon a \sin\theta\cos\theta\left(1 - a^2\cos^2\theta - \frac{\gamma^2}{2}\right) - a = 0.$$

Taking the zeroth Fourier coefficient, as before, we obtain

$$\dot{a} = \frac{\epsilon a}{2}\left(1 - \frac{\gamma^2}{2} - \frac{a^2}{4}\right), \quad \dot{\theta} = 1.$$

We therefore have a solution

$$x = \gamma \sin \omega t + a \sin(t + \theta_0),$$

where, if $\gamma^2/2 < 1$, a converges to $2(1 - \gamma^2/2)^{1/2}$ and a mixed oscillating state results. But if $\gamma^2 > 2$, then $a \to 0$, and so the self-oscillation is entrained by the external force: oscillations are synchronized if the coupling is large enough, i.e., from nearby cells.

So we can imagine a locally synchronized oscillating population whenever the cells have been activated to an her1/7 oscillating mode. The activity appears to go as a wave. What is the nature of this biochemical wave?

8.3.2. Biochemical Waves. There seems to be no difficulty in biologically reasonable "differential delay" models of oscillating genes and their products, and we know how to synchronize a population. Consequently, the protein production serves at the same time to identify the phase of the gene cycle, a biological clock. For the corresponding spatial pattern, one needs a biological ruler or spatial marker, most directly the concentration of a molecular species. As we have indicated before, the tail bud growth appears to be accompanied by an adjoining region of PSM moving at the same rate. The region is that bounded by high FGF8 activity on the tail side and high RA (FGF8 is a growth factor, RA is retinoic acid) on the head side. The presomite population induced by the FGF8 halts at the position, time, and phase at which the FGF8 is turned off, and the RA turned on. As we will show, this determines the biochemical repertoire of each cell and ultimately the interfaces identifying the somites.

One can imagine two processes contributing to the FGF8 PSM zone next to the tail [**37, 132**]. The first is obvious: FGF8 production at the tail that decays as one moves away. This is presumably sufficient to create the spatial context we need but is not very sharp and is subject to large fluctuations. The second is the FGF8-RA interaction serving to stabilize a sharp boundary (this dichotomy, sufficiency/stabilization, is a theme running through biological development, invagination accompanied by bottle cell production in gastrulation being a dramatic example), and this boundary must be moving, a biochemical wave front.

Suppose that a cell is characterized by the absence or presence of a molecular species A (standing as well for its concentration). If we start at $A = \epsilon$ but have a nondepleting source of A, say of concentration 1, that is processed autocatalytically,

$$S + A \underset{k'}{\overset{k}{\rightleftharpoons}} 2A$$

or

$$\dot{A} = kA - k'A^2$$

(the "logistic" equation), then the A concentration will saturate at $A_0 = \frac{k}{k'}$, in fact with the time dependence

$$A(t) = \frac{A_0}{1 + (A_0/\epsilon - 1)e^{-kt}}.$$

This system has two stationary states; the state $A = 0$ is unstable, while $A = A_0$ is stable.

Now consider a closely packed cell population, so that A can be transferred between cells—in one dimension to start. The concentration will depend on which cell and hence which spatial location we are referring to: $A(x, t)$. The transfer can be regarded as a monomolecular reaction

$$A(x - \Delta x) \underset{\gamma}{\overset{\gamma}{\rightleftharpoons}} A(x) \underset{\gamma}{\overset{\gamma}{\rightleftharpoons}} A(x + \Delta x),$$

so that now

$$\dot{A}(x,t) = kA(x,t) - k'[A(x,t)]^2 + \gamma[A(x + \Delta x) + A(x - \Delta x) - 2A(x,t)].$$

If A changes slowly from cell to cell, then

$$A(x + \Delta x) + A(x - \Delta x) - 2A(x) = (\Delta x)^2 A''(x) + \cdots,$$

so in terms of the diffusion constant $D = \gamma(\Delta x)^2$, we have

$$\frac{\partial}{\partial t}A(x,t) = kA(x,t) - k'[A(x,t)]^2 + D\frac{\partial^2}{\partial x^2}A(x,t).$$

This is an example of a reaction-diffusion equation, which, like most, is not exactly solvable due to its nonlinearity, but a semiquantitative analysis is usually not hard.

Is this relevant in the case of somitogenesis? A little bit. The tail end is the source of growth and the differentiation that can go with it. An A-producing cell, state $A = A_0$, can then transfer A to the next cell that has been resting at $A = 0$, get it up to A_0, and so on down the line. In other words, we may expect to see a wave front of "A-activation." So we should look for a solution of the form (reading away from the tail)

$$A(x,t) = A(x - ct).$$

To make this clearer, we start by nondimensionalizing

$$\dot{A} = kA - k'A^2 + DA''.$$

The variables t, A, and x are all available for scaling: $t = \alpha\tau$, $A = \beta a$, and $x = \gamma X$, so

$$\frac{\beta}{\alpha}\frac{\partial a}{\partial \tau} = \beta k a - \beta^2 k' a^2 + \frac{D}{\gamma^2}\frac{\partial^2 a}{\partial X^2},$$

and we can set

$$\frac{\beta}{\alpha} = \beta k = \beta^2 k' = \frac{D}{\gamma^2},$$

or

$$\beta = \frac{k}{k'} = A_0, \quad \gamma^2 = \frac{D}{\beta^2 k'}, \quad \alpha = \frac{1}{k},$$

yielding

$$\frac{\partial a}{\partial \tau} = a - a^2 + \frac{\partial^2 a}{\partial X^2}$$

(Fisher's equation; see, e.g., [**22**]).

Now if $a(X, \tau) = a(X - c\tau) \equiv a(\xi)$, we have

$$a''(\xi) + ca'(\xi) + a(\xi)[1 - a(\xi)] = 0.$$

Since the two uniform steady states are $a = 0, 1$, we can presumably make a transition between them. Look at the $a \sim 0$ part:

$$a'' + ca' + a = 0,$$

so

$$a(\xi) = k_1 \exp\left(-\frac{c}{2} + \sqrt{\frac{c^2}{4} - 1}\right)\xi + k_2 \exp\left(-\frac{c}{2} - \sqrt{\frac{c^2}{4} - 1}\right)\xi.$$

Since $a \geq 0$, it can't oscillate below zero, giving us the condition $c \geq 2$.

To actually find c, let us start at

$$a(X, 0) = \begin{cases} 1, & X < 0, \\ 0, & X > 0, \end{cases}$$

and again look at $a \sim 0$: $\dot{a} = a + a''$. We then solve by Laplace transform

$$\hat{a}(X, p) = \int_0^\infty e^{-p\tau} a(X, \tau) d\tau, \quad p > 1,$$

so

$$\int_0^\infty e^{-p\tau} \dot{a}(X, \tau) d\tau = -a(X, 0) + p\hat{a}(X, p),$$

giving

$$\hat{a}'' + (1 - p)\hat{a} = \begin{cases} -1, & X < 0, \\ 0, & X > 0. \end{cases}$$

For a finite solution at $X = \pm\infty$, the solution must take the form

$$\hat{a}(X, p) = \begin{cases} \frac{1}{p-1} + k_3 e^{X\sqrt{p-1}}, & X < 0, \\ k_4 e^{-X\sqrt{p-1}}, & X > 0, \end{cases}$$

or by continuity of a and a' at $X = 0$,

$$\hat{a}(X, p) = \begin{cases} \frac{1}{p-1} - \frac{1}{2(p-1)} e^{X\sqrt{p-1}}, & X < 0, \\ \frac{1}{2(p-1)} e^{-X\sqrt{p-1}}, & X > 0. \end{cases}$$

Since a is small at large X, this tells us that then

$$\hat{a}'(X, p) = -\frac{1}{2\sqrt{p-1}} e^{-X\sqrt{p-1}},$$

or by inverse Laplace transforming,

$$a'(X, \tau) = -\frac{1}{\sqrt{4\pi\tau}} e^{\tau - \frac{x^2}{4\tau}}.$$

At fixed amplitude $a' = -u$, then

$$e^{\tau - x^2/(4\tau)} = u\sqrt{4\pi\tau}$$

or

$$x = 2\tau\left(1 - \frac{1}{\tau}\log u\sqrt{4\pi\tau}\right)^{\frac{1}{2}}$$

with a wave velocity

$$\dot{x} = \left(2 + \frac{1}{\tau}\log u\sqrt{4\pi\tau} - \frac{1}{2\tau}\right)\left(1 - \frac{1}{\tau}\log u\sqrt{4\pi\tau}\right)^{\frac{1}{2}},$$

allowing us to conclude that $c = 2$ for large τ [**22**].

8.4. Clock, Wavefront, and Somite Condensation

8.4.1. Wave Front. Our discussion of biochemical waves has been rather abstract. And, in the case of somite production, there certainly appear to be two interacting molecular species (morphogens), determining the "wave front," F for FGF8 and A for retinoic acid RA. It is easy to derive a one-dimensional model to represent this that does not violate obvious biological information. Let us adopt a frame in which the tail end is fixed and is a concentrated source of F production. Denoting the F decay rate by γ, with α the coefficient of production inhibition by A, and including diffusion, we have—outside the source region $x = 0$ at which F is held to the value F_0,

$$\dot{F} = -\gamma F - \alpha AF + DF''.$$

As for A, its production is imagined to follow the (suitably scaled) $A(1 - A)$ model, likewise inhibited by F—but we assume a very low diffusion constant, so

$$\dot{A} = A(1 - A) - \beta AF.$$

We have a pair of coupled reaction-diffusion equations, a format we will meet on many occasions. In the absence of A, F would settle down ($\dot{F} = 0$) to the profile

$$0 = -\gamma F + DF'',$$

so that

$$F(x) = F_0 e^{-(\frac{\gamma}{D})^{\frac{1}{2}} x},$$

not at all a sharp boundary. How does the RA modify this?

In the absence of source F_0 and diffusion, the dynamics has two nonnegative stationary states, $F = 0$, $A = 0$ and $F = 0$, $A = 1$. The first, for which

$$\delta \dot{F} = -\gamma \delta F, \quad \delta \dot{A} = \delta A,$$

is an unstable saddle point, leading to the second,

$$\delta \dot{F} = -(\alpha + \gamma)\delta F, \quad \delta \dot{A} = -\delta A - \beta \delta F,$$

which indeed is stable but uninteresting: A rises to its maximum throughout. Diffusion, however, lets F spread from its source, and we want to see the nature of the spread. In steady state,

$$A(1 - A) = \beta A F.$$

If $\beta F > 1$, then only $A = 0$ is possible, with the above exponential falloff of F. This ceases when βF falls to 1, and then

$$A = 1 - \beta F,$$

so we have

$$DF'' - (\alpha + \gamma)F + \alpha\beta F^2 = 0.$$

To solve, we set up an "energy integral." Multiply by F' and integrate from x to infinity, where $F = F' = 0$:

$$\frac{D}{2}F'^2 + V(F) = 0, \quad V(F) = -\left(\frac{\alpha + \gamma}{2}\right)F^2 + \frac{\alpha\beta}{3}F^3.$$

As initial condition: if $\beta F_0 > 1$, then

$$\beta F(x_0) = 1 = \beta F_0 e^{-(2/D)^{\frac{1}{2}} x_0},$$

so that

$$\beta F(x_0) = 1 \quad \text{at } x_0 = \left(\frac{D}{2}\right)^{\frac{1}{2}} \log \beta F_0.$$

We then have the deep well in $V(F)$, indicating that F falls rapidly to its "potential minimum" value of $(\alpha + \gamma)/(\alpha\beta)$, staying there for a while, independently of F_0, and then falling rapidly towards zero. In other words, there is a sizable, roughly constant region of F attached to the tail, its head end comprising the wave front in question—here only because the origin is moving.

Cooke and Zeeman emphasized in their original work that the segmentation process has as its basic elements smooth oscillation, a slow wave front, and rapid cellular change. The details have only gradually emerged and are still a source of controversy, in part because the biological messengers involved never seem to come singly but rather in clusters of similar species. The most effective models are therefore those in which species with similar properties are amalgamated, providing what in evolutionary terms might be called quasi-species. Even then, it is not trivial to distinguish between a causative species and one that follows in unison. And finally, the genesis of the omnipresent diffusion constant D is not that sharply defined. It arose in our treatment from transfer between cells, with the cell as unit, implying a mesoscopic viewpoint. But any process, such as thermal or Brownian motion, that moves a molecule randomly from a spatial point to one nearby will give rise to the same qualitative phenomenon in the same format. The distinction is not obvious without detailed examination, although quantitative values of D may often give a hint.

The most detailed analytical models of somatogenesis are those of the Oxford group (Collier, Maini, Baker, etc.). We will not study these, but what pretty much all models have in common is a tail-associated region of high concentration of FGF8, which is required to cause oscillations in the cells of the underlying presomitic mesoderm. As the tail end grows, cells exit this region at a gene expression time in their cycle that determines their further function, subsequent to which the determined cells aggregate to form functional somites. We now turn our attention to this aggregation process.

8.4.2. Somite Condensation. The claim is that cells leave their FGF8 environment at various protein expression stages in their her cycle, and it is imagined that the next stage of their fate is then determined, such as whether they will have increased adhesivity and therefore be associated with a tighter cluster of cells. But there will be large quantitative fluctuations, and so a boundary sharpening process is called for. Perhaps the simplest model—with details very much not known—would posit a rapidly diffusing activator species $a(x, t)$, externally secreted, which biases the adhesive expression level of a cell but is balanced by a locally produced inhibitor $h(x, t)$ serving to sharpen the pattern of a. The obvious question, of course, is whether this will be compatible with a periodic somite arrangement. One can carry relevant general remarks quite far in this respect.

Suppose one has a coupled two-morphogen reaction-diffusion system, still effectively in one dimension,

$$\dot{a} = f(a, h) + Da'', \quad \dot{h} = g(a, h) + D'h'',$$

which, in the absence of diffusion, would settle down to an uninteresting spatially uniform steady state

$$f(a_0, h_0) = 0, \quad g(a_0, h_0) = 0.$$

Settling down means that a virtual perturbation, satisfying

$$\delta\dot{a} = f'_a(a_0, h_0)\delta a + f'_h(a_0, h_0)\delta h, \quad \delta\dot{h} = g'_a(a_0, h_0)\delta a + g'_h(a_0, h_0)\delta h,$$

when tested by

$$\delta a(t) = e^{\lambda t}\delta a, \quad \delta h(t) = e^{\lambda t}\delta h,$$

would have only λ with negative real part:

$$\mathrm{Re}\left(\lambda\left\{\begin{pmatrix} f'_a & f'_h \\ g'_a & g'_h \end{pmatrix}\right\}\right) < 0.$$

Now introduce the diffusion mechanism and choose

$$\delta a(x,t) = e^{\imath kx}e^{\lambda t}\delta a, \quad \delta h(x,t) = e^{\imath kx}e^{\lambda t}\delta h$$

as the deviations from space-time uniformity. The controlling matrix now becomes

$$M_k = \begin{pmatrix} f'_a - Dk^2 & f'_h \\ g'_a & g'_h - D'k^2 \end{pmatrix},$$

and we ask whether decreasing the diagonal elements of M_0 could provoke an instability (the Turing effect) by raising the real part of at least one eigenvalue of M_k to positivity. This certainly could not happen if M_0 is symmetric, but that will frequently not be the case.

An example from [**156**] is as follows:

$$a + h \longrightarrow 2a, \quad \varnothing \longrightarrow h, \quad a \longrightarrow \varnothing,$$

which scaled becomes

$$\dot{a} = \frac{1}{16}a(h-1) - \frac{3}{4} + \epsilon a'', \quad \dot{h} = -\frac{1}{16}ah + 1 + \nu h''.$$

The diffusionless uniform steady state is at $a_0 = h_0 = 4$, and then

$$\delta\dot{a} = \frac{3}{16}\delta a + \frac{1}{4}\delta h, \quad \delta\dot{h} = -\frac{1}{4}\delta a - \frac{1}{4}\delta h.$$

The matrix

$$M_0 = \begin{pmatrix} \frac{3}{16} & \frac{1}{4} \\ -\frac{1}{4} & -\frac{1}{4} \end{pmatrix}$$

has eigenvalues $\lambda = \frac{1}{32}(-1 \pm \imath\sqrt{15})$ and so this state (a_0, h_0) is stable. But now introduce diffusion

$$M_k = \begin{pmatrix} \frac{3}{16} - \epsilon k^2 & \frac{1}{4} \\ -\frac{1}{4} & -\frac{1}{4} - \nu k^2 \end{pmatrix},$$

with eigenvalues given by

$$0 = \lambda^2 + \left[\frac{1}{16} + (\epsilon + \nu)k^2\right]\lambda + \frac{1}{16} - \left(\frac{3}{16} - \epsilon k^2\right)\left(\frac{1}{4} + \nu k^2\right).$$

One direction from the uniform steady state will be unstable if

$$\left(\frac{3}{16} - \epsilon k^2\right)\left(\frac{1}{4} + \nu k^2\right) > \frac{1}{16}$$

or

$$-\frac{1}{64} + \left(\frac{3}{16}\nu - \frac{1}{4}\epsilon\right)k^2 - \epsilon\nu k^4 > 0,$$

readily seen to be solvable if $\frac{\epsilon}{\nu} < \frac{1}{4}$.

While the stability of the uniform steady state is mathematically convenient, it has no particular biological justification—there is no reason to imagine that diffusion is just "turned on." A model that corresponds to the activator-inhibitor setup that is appropriate to somite condensation is due to [**68**] and (with altered notation) goes as

$$\dot{a} = \mu - h^2 a + Da'', \quad \dot{h} = ch^2 a - \lambda h,$$

corresponding to

$$a + 2h \longrightarrow 3h, \quad \varnothing \longrightarrow a, \quad h \longrightarrow \varnothing$$

(a is produced, h inhibits a, h is degraded). Here, the $D = 0$ uniform steady state $h = 0$ has neutral stability, but we can solve directly for the $D \neq 0$ nonuniform steady state. Eliminating $h = \lambda/(ca)$ in steady state, we have

$$Da'' - \frac{\lambda^2}{ca} + \mu = 0,$$

again solved by the "energy integral"

$$\frac{D}{2}(a')^2 + V(a) = K, \quad V(a) = \mu a - \left(\frac{\lambda^2}{c}\right) \log a.$$

The solution (by mechanical analogy $a \mapsto x$, $x \mapsto t$) is clearly periodic in x. Since the minimum

$$V_m = \frac{\lambda^2}{c}\left(1 - \log \frac{\lambda^2}{c\mu}\right)$$

at

$$a_m = \frac{\lambda^2}{c\mu}, \quad V''(a_m) = \frac{\lambda^2}{a_m^2 c} = \frac{\mu^2 c}{\lambda^2},$$

the profile in the vicinity of a_m is determined by

$$Da'' + \frac{\mu^2 c}{2\lambda^2}(a - a_m) = 0$$

or

$$a = a_m + A \cos\left(\frac{\mu}{\lambda}\sqrt{\frac{c}{2}}\right)(x + \alpha),$$

a spatial oscillator independent of K, but with amplitude determined by a more complete analysis.

It seems clear that the "rate-limiting step" is not the mathematical analysis—the concepts and details are comfortably universal—but the data that calls for mathematical organization, and that is accumulating at high speed!

CHAPTER 9

Compartments

In the last chapter we saw how cell proliferation, under suitable control, provides a local mechanism for morphogenesis. Our next task will be to investigate to what extent growth might influence the *global* organization and determination of cell type, by causing the tissue to evolve into a state where pattern-forming instabilities are possible, for example; this is a second means by which cell proliferation can affect the fate of developing tissue. At the present time the role of size-dependent diffusive instabilities in early development is largely conjectural, but recent experiments with insects have revealed some remarkable properties of cell proliferation that strongly suggest some kind of global organization. In the present section we shall examine this evidence and formulate the Kauffman-Shymko-Trabert (KST) model as a dynamic reaction-diffusion model of pattern formation [**102**].

9.1. The Evidence and Its Implications

We first summarize the discoveries of Garcia-Bellido and his coworkers [**66**] (see also Crick and Lawrence [**53**]) concerning the development of clones in imaginal discs of *Drosophila*. In this insect the adult is formed by a process involving a number of stages. In the first stage the nucleus of the fertilized egg divides a number of times, followed by migration of the nuclei to the inner surface of the egg where cell membranes are formed; further cell proliferation produces a monolayer of cells within the egg, comprising the blastoderm. In the second stage the blastoderm becomes folded up, segmentation occurs, and eventually the egg hatches the first of three larval stages; the final larva then pupates, and, following metamorphosis, the adult fly emerges.

During the larval stages the cells which are eventually to form the adult are carried as passive groups of cells, some of which are the *imaginal discs*. During the larval stages these discs grow considerably, from on the order of 100 cells to about 10^5 cells in the case of the wing disc. The latter produces, in the adult, one wing and a piece of the thorax. It is a flattened structure and for our purposes may be taken as a monolayer of cells.

What Garcia-Bellido has done is to examine, by a process of marking individual cells (through irradiation with X-rays), the progeny of single cells in a developing disc. Any such group of cells makes up a marked *clone*, associated with a particular cell marked at a specific time. That is, a clone is the group of descendants of a given cell. A typical observation of marked clones is then the following: suppose that a certain cell is marked at time zero, divides into two daughter cells at time t_1, and each of these divides at time t_{21} and t_{22}, etc., eventually forming the clone C

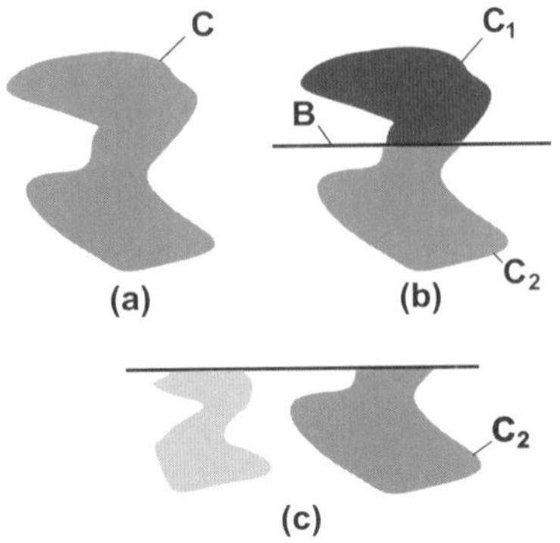

FIGURE 9.1. Formation of clones and compartments from cell progeny.

(see Figure 9.1(a)). Now suppose that we are able to separately mark the two cells formed at t_1 and thereby follow the two associated clones separately. In the situation we want to examine, it is found that the two clones respect a common boundary B, almost as if a fence had been created beyond which cells of a given clone could not pass. The union of the two clones C_1 and C_2 is of course the original clone C, as indicated in Figure 9.1(b). Finally, suppose that after $t = t_1$ one of the cells is marked, and simultaneously any other cell on the same side of the presumptive boundary is marked. The two clones again respect the border B as in Figure 9.1(c).

The results of many such experiments lead to the conclusion that such borders are irreversibly formed and once present limit the territory accessible to neighboring clones. In the example it could be concluded that the boundary was established between times 0 and t_1. Once established, the boundary is said to divide the disc into two *compartments*. At any subsequent time the compartment is a union of clones, the founder cells being those cells present on one side of the boundary at the time of its formation. By the same token a compartment is in general not in itself a clone, since cells on both sides of the boundary may share a common ancestor.

Crick and Lawrence [**53**] introduce the term *polyclone* for the cells making up a given compartment. A polyclone can then be characterized as consisting of *all descendants* of a certain (generally small) group of founder cells (but not generally a single founder cell). If we stipulate that the founder cells were present simultaneously, then all we can say is that founder cells were, as above, those cells present when the boundary was formed. If founder cells are not so restricted then cell lineages may be traced backward with the result that for each founder cell we have the following:

(i) The progeny of the cell form a clone lying entirely within the compartment.
(ii) Property (i) does not apply to the mother cell; i.e., there is a sister cell having progeny not in the compartment.

In general such cells form a much smaller group than the cells present when the boundary is formed.

We may now summarize the main thrust of the argument put forward by Kauffman, Shymko, and Trabert [**102**]. As a disc evolves, its properties will change as a function of time, with the result that a diffusive instability might be realized at some critical time; there may even be a succession of critical times and instabilities. For a planar geometry the spatial gradients associated with an instability are decided by eigenfunctions of a linearized reaction-diffusion system, which can be characterized by nodal lines. The latter form a possible explanation for the appearance of compartmental boundaries, provided that the integrity of a compartment can be explained by cell properties established by chemical gradients. Such a theory might be termed a "tympanic" theory because of the analogy with the eigenfunctions of a vibrating membrane. It is an interesting realization of the irreversible "decisions" that are, in a general way, the essence of development. We shall presently look more closely at this model but first examine a one-dimensional problem that introduces the dynamic process underlying successive instabilities in a reaction-diffusion system.

9.1.1. Time-Dependent Diffusivities in a Linear System. Let us consider one-dimensional tissue characterized by two reaction-diffusion equations with time-dependent diffusion coefficients. We assume first that cell proliferation is by *cleavage*, the length of the domain remaining constant. A diffusion coefficient μ may be thought of as measuring $(\Delta x)^2/\Delta t$ for typical point-to-point motions. If the path is broken up by N cell membranes per unit length, then approximately $\Delta x \to \Delta x/\sqrt{N}$, $\Delta t \to \Delta t/\sqrt{N}$, and $\mu \to \mu/\sqrt{N}$. An exponential decrease as e^{-at} is probably not unreasonable over a prolonged period of cleavage.

Linearized about a constant equilibrium, the system we are considering has the form

$$\frac{\partial u}{\partial t} - \mu \frac{\partial^2 u}{\partial x^2} = a_{11}u + a_{12}v,$$
$$\frac{\partial v}{\partial t} - \nu \frac{\partial^2 v}{\partial x^2} = a_{21}u + a_{22}v,$$

where the a_{ij} are constants. Assuming for the moment that μ and ν are constant, conditions for diffusive instability for a mode varying as $\cos(kx)$ are then (see Chapter 3)

$$a_{11} + a_{22} < 0, \quad D = a_{11}a_{22} - a_{21}a_{12} > 0,$$
$$\nu a_{11} + \mu a_{22} > 0, \qquad \frac{(\nu a_{11} + \mu a_{22})^2}{4\mu\nu} > D.$$

For slow variation of μ and ν it is useful to assume that the ratio $r = \nu/\mu$ remains constant, in which case the last conditions remain true if satisfied initially. The growth rate σ associated with an instability is given by

$$2\sigma = a_{11} + a_{22} - (\mu + \nu)k^2 + \sqrt{(a_{11} + a_{22} - (\mu + \nu)k^2)^2 + 4Q(k^2)},$$
$$Q(k^2) = -D + (\nu a_{11} + \mu a_{22})k^2 - \mu\nu k^4,$$

TABLE 9.1

m	0	1	2	3	4	5	6	7	8	9
$A\theta^{-m}$	0.80	1.13	1.60	2.26	3.20	4.52	6.40	9.04	12.80	18.80
$B\theta^{-m}$	1.20	1.70	2.40	3.39	4.80	6.80	9.60	13.60	19.20	27.20
n^*	1	-	-	-	2	-	3	-	4	5

and is indeed positive provided $Q > 0$. For an interval of unit length, the eigenvalues are $k_n^2 = n^2\pi^2$. For slowly varying diffusivities the corresponding eigenfunctions are excited sequentially as the positive part of the Q function passes over the k_n^2. Here the sequence of maxima of Q are identical, and as diffusivities decrease the maximum growth rate approaches a fixed positive value, so the amplitude obtained during excitation depends primarily on the duration of the "window" of positive σ. The latter steadily lengthens, since the two zeros of Q are inversely proportional to diffusivity, as is the position of the maximum. For an exponential decrease in μ and ν, eventually several eigenvalues will lie in the same window, since its duration increases faster than the time between capture of eigenvalues.

The last points may be demonstrated most succinctly by considering synchronized cleavage and diffusivities obeying

$$(\mu, \nu) = (\mu_0, \nu_0)\theta^m, \quad 0 < \theta < 1,$$

where m is a generation number. Then there will be two constants A and $B > A$ such that the nth node is unstable provided that

$$A < n^2\theta^m < B,$$

and this will be the case for many m if n is sufficiently large. Nevertheless, it is surprisingly easy to filter out all but one growing mode through a large number of generations, at least in one dimension. Table 9.1 is constructed by taking $A = 0.8$, $B = 1.2$, and $\theta = 2^{-1/2}$ (the star indicates the number of unstable modes).

Here it is only at the 13th generation that more than one mode ($n = 9, 10$) is excited during one generation. Note that the eigenvalues are not always centered in the interval of positive Q, and so growth rates vary considerably. Also higher-dimensional examples will have considerably less separation of excited states with respect to generation.

Now let us suppose that cell size stays the same but that the size of the tissue increases. Since chemicals then have further to diffuse, the effective diffusion coefficients are again reduced. Indeed, if the length of the interval under consideration is now $L(t)$, and the variation with t is suitably slow, we may again treat the parameters as constants in the stability problem. Eigenfunctions proportional to $\cos(n\pi x/L)$ are introduced, and we see that p and v, and μ, ν, and L, always occur in the combinations μ/L^2 and ν/L^2. The problem thus reverts to the one already studied with the time dependence of the effective diffusion coefficients given by the factor L^{-2}. (This dependence is of course an immediate consequence of dimensional arguments.) For synchronized division L^{-2} decreases by a factor $\frac{1}{2}$ with

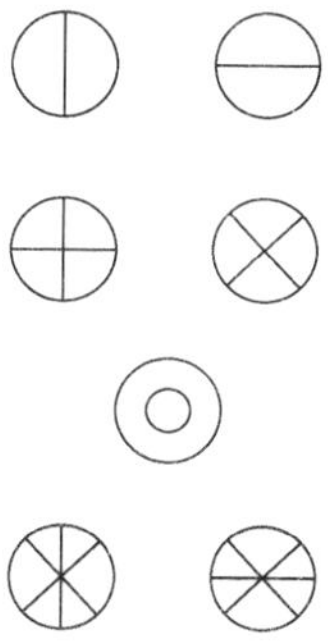

FIGURE 9.2. Modes corresponding to the eigenfunctions of Table 9.2.

each generation in a plane monolayer, and by a factor $2^{-2/3} \approx 0.63$ for a three-dimensional region.

It is a simple matter to extend these arguments to higher dimension. We consider first a planar monolayer filling a domain D, which, by the above normalization in terms of time-dependent diffusivities, may be taken to be independent of t. We consider the solutions of the eigenvalue problem

$$\nabla^2\phi + \lambda\phi = 0 \quad \text{in } D, \qquad \frac{\partial\phi}{\partial n} = 0 \quad \text{on boundary of } D.$$

In the case of a disc of radius unity, we have the following sequence of eigenvalues and eigenfunctions, with nodal lines indicated in Figure 9.2:

TABLE 9.2

$\lambda = k^2$	Eigenfunctions
3.4	$J_1(kr)(\cos\phi, \sin\phi)$
9.6	$J_2(kr)(\cos 2\phi, \sin 2\phi)$
14.4	$J_0(kr)$
17.6	$J_3(kr)(\cos 3\phi, \sin 3\phi)$

The multiple eigenvalues tend to split under deformation of the boundary. For the case of an ellipse with major axis $2a$ and minor axis $2b$, the mode with the minor axis as its nodal set is excited first, and there are changes in the nodal structure of the various other modes for the disc. For eccentricity $e = (1 - b^2/a^2)^{1/2} = 0.8$, for example, one obtains the sequence of Figure 9.3. Here mode 3 is a perturbation of $J_2(kr)\sin 2\phi$, which can be excited more easily than mode 4 as a higher "one-dimensional" harmonic similar to mode 1. The same applies to mode 5, which comes from $J_3(kr)\sin 3\phi$; for sufficiently large eccentricity it is excited more easily than the continuation of the J_0 mode represented by the final entry. In the KST theory the elliptical domain is taken as typical of domains where separation of the first few eigenfunctions, with respect to the magnitude of the eigenvalue, is complete. Mode 4, by the way, could be discarded as a recapitulation of modes 1 and 2, and part of 5 recapitulates mode 1.

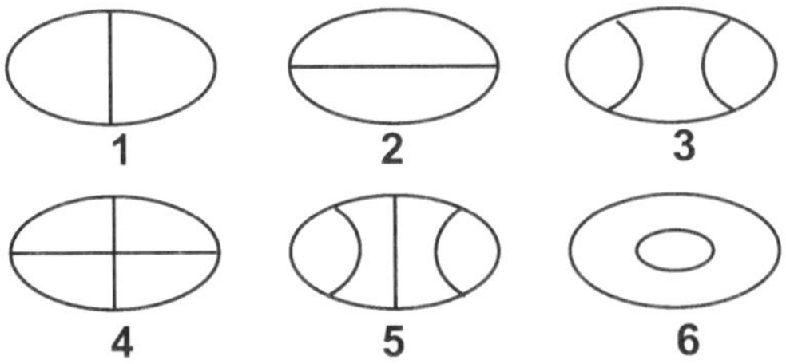

FIGURE 9.3. Planar modes in an ellipse of eccentricity $e = 0.8$.

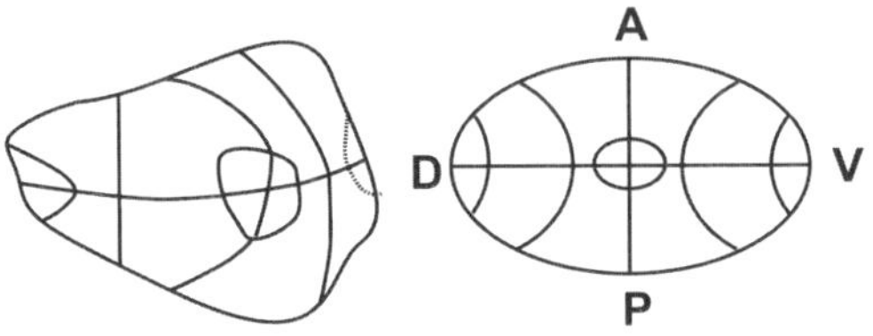

FIGURE 9.4. The fate map of the wing disc, left, and the elliptical representation, right.

We note in passing the corresponding results for a spherical shell blastula. In this case ∇^2 is replaced by part of the Laplacian in spherical coordinates involving differentiation with respect to angles, so the eigenvalues are multiple. For a suitable ellipsoidal shell the nodal surfaces of the first two eigenvalues can be regarded as establishing the anterior-posterior and dorsal-ventral axes of the blastula.

9.1.2. Evidence of Compartmentalization of the Wing Disc. The larva of *Drosophila* carries 19 imaginal discs, 9 pairs plus a single genital disc. The pair of dorsal mesothoracic or wing discs form, in the adult, the wings and a portion of the thorax to which they are attached. In the first larval stage the wing disc consists of only a few dozen cells. Growth is exponential during the larval stages, involving 10 or 11 divisions, with a mean division of about 8 hours.

The techniques used by Garcia-Bellido et al. to mark clones rest upon mitotic recombination produced by X-rays. Recombinants that are phenotypically different in the adult mark the clone, an example being cells that produce multiple wing hairs in the final epithelium. For compartments formed late in development prior to pupation it is also necessary to utilize mutant cells that have a smaller division time, so that clones may grow rapidly enough to delineate a compartmental boundary. These faster-growing clones in a slower-growing background do not, by the way, cause any change of size or shape of the final structure, meaning that cell sizes are regulated to accommodate the different rates of division.

The compartmental boundaries observed in the adult wing can be traced back to the disc just prior to pupation to obtain a fate map for clonal restriction. The remarkable fact noted by Kauffman et al. [**102**] is that the superimposed boundaries compare well with the sequence of nodal lines for an ellipse; see Figure 9.4. The sequence is formed as follows: the anterior-posterior (AP) line is drawn first.

This line is always drawn second for the growing elliptical domain, but the latter is believed to be only a recapitulation of a basic anterior-posterior polarity already induced in the disc by the first segmentation of the ellipsoidal blastoderm (see below). The next boundary to form is the dorsal-ventral (DV). It should be noted that the disc eventually is folded over to form a sort of bag, which then collapses to form both sides of the adult wing. The third and fourth boundaries (dtw and vtw) separate the dorsal and ventral thoracic compartments from the wing compartment. The next two divide the dorsal and ventral thorax into two parts (dsp, vsp), the ventral member being an as yet unobserved prediction of KST. The final restriction (dpw) is to the distal and proximal parts of the wing. The sequence thus repeats that obtained for an ellipse with $e = 0.8$, as well as for a path of growth (in the ξq-plane), which roughly approximates that of the disc (see [**102**]).

9.1.3. Coding of Compartmentalization in the Blastoderm. The KST model can also be applied to segmentation of the blastoderm prior to formation of the discs. The blastoderm is roughly ellipsoidal with axis ratio about 3.7 to 1.0. In the context of a Turing model, the easiest modes to excite will be similar to one-dimensional waves down the long axis. After a few such restrictions the eigenvalue associated with the dorsal-ventral compartment will intervene. The observed fate map of the blastula accords with the model if the dorsal-ventral restriction follows three distinct longitudinal restrictions. The sequence must here be regarded as distinguished by the reduction of diffusivities associated with membrane formation.

These four developmental events can be postulated to act upon an ordered sequence of four binary switches, so that at the end cell types are specified by binary numbers $a_1a_2a_3a_4$, $a_i = 0, 1$. For one choice of code the sequence of restrictions can be diagrammed as shown in Figure 9.5. In the KST model this compartmentalization and the above code agree quite well with observation.

The method of testing the code involves the phenomenon of *transdetermination*, as well as the ability of certain mutant cells to transform one type of tissue to another. A transdetermination of a disc can be induced by culturing it in the abdomen of an adult, then implanting it in a larva. Following metamorphosis one observes, with a certain frequency, adult tissue normally contributed by another disc. The set of transdetermination pathways and frequencies might now be regarded as indicative of the number of binary switches that must be reset to pass between tissue types. The predictions that can be made on this basis agree remarkably well with observed frequencies, thus providing evidence of a quantitative measure of cell differentiation along a developmental pathway.

These discoveries raise a number of questions. An interesting mathematical problem is the extraction of the probability distribution over the set of possible codes from data such as transdetermination frequencies. All codes have certain common features; e.g., adjacent compartments differ by only one letter, but they will differ in the predicted "distance" between highly dissimilar cell types. On the other hand, transdetermination between two such types would presumably be a rare event. Thus the most revealing data may be the least accurate. It is conceivable that

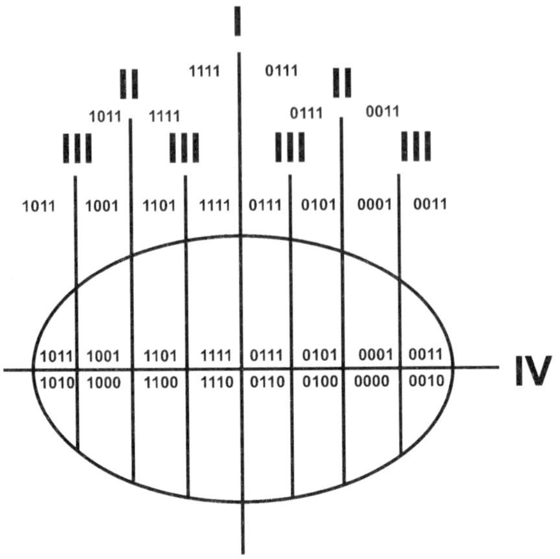

FIGURE 9.5. Coding of compartments of the blastoderm.

chemical models of the switches postulated here (cf. Chapter 3) will prove helpful in assigning a code purely on the basis of the compartmental sequence.

What process creates and maintains the demarcation between compartments? One possibility is that mitotic division along the boundary is oriented. Another interesting suggestion is that differential adhesion maintains a sharp transition between compartments. Both of these mechanisms lose some of their force when compartmental boundaries are identified with nodal lines, since the chemical gradients themselves are not necessarily sharp; indeed the linear theory deals with smooth eigenfunctions. This suggests that nonlinear processes may be quite significant in models of compartmentalization based on diffusion-reaction equations.

9.2. Nonlinear Theory of Compartmentalization

We take up, then, the manner by which sharp compartmental boundaries might be established by reactive-diffusive chemicals. Our approach will be to determine the consequences of nonlinear behavior on the progress of the diffusive instability and to study this as a mathematical problem of *dynamic bifurcation*.

We begin by considering how eigenfunctions are excited under nonlinear control. For simplicity we assume synchronized mitosis and a system of two equations, although the results are easily extended to more complicated systems. We further suppose that, immediately following one division, the domain D of the tissue (coordinate x) admits a sequence of simple eigensolutions $\{\lambda_i, \phi_i(x) : i = 0, 1, 2, \dots\}$ where

$$\lambda_0 = 0, \quad \phi_0 = 1, \quad \frac{1}{L}\int \phi_i \phi_j \, dx = \delta_{ij}, \quad L = \text{length of } D.$$

We write the nonlinear Turing system in the form

$$\frac{\partial u}{\partial t} - \mu\frac{\partial^2 u}{\partial x^2} - (a_{11}u + a_{12}v) = f(u,v) = P(u,v) + M(u,v),$$
$$\frac{\partial v}{\partial t} - \nu\frac{\partial^2 v}{\partial x^2} - (a_{21}u + a_{22}v) = g(u,v) = Q(u,v) + N(u,v),$$

where P and Q are homogeneous of degree 2 in u and v, M and N then containing terms of degree 3 or higher.

A single eigenfunction is to be selected by a diffusive instability, which we take to be ϕ_1. To create an asymptotic theory, we take the positive growth rate to be $\epsilon \ll 1$, this being a result of linearized theory. All other linear growth rates (for 1 as well as the other eigenfunctions) will have negative real parts of order unity.

Under this last assumption, consider the projection of the solution of the linearized initial value problem onto the subspace of the eigenfunction ϕ_1:

$$\Pi_1(u,v) = \left(\alpha^{(1)}(t;\epsilon), \beta^{(1)}(t;\epsilon)\right), \quad \Pi_i(\cdot) = \frac{1}{L}\int (\cdot)\phi_i\, dx.$$

We have, introducing the matrix

$$B = \{b_{ij}\} = A - \lambda_1\begin{pmatrix}\mu & 0\\ 0 & \nu\end{pmatrix},$$

the system

$$\frac{d\alpha^{(1)}}{dt} - \left(b_{11}\alpha^{(1)} + b_{12}\beta^{(1)}\right) = 0,$$
$$\frac{d\beta^{(1)}}{dt} - \left(b_{21}\alpha^{(1)} + b_{22}\beta^{(1)}\right) = 0.$$

The general solution has the form

$$(\alpha^{(1)}, \beta^{(1)}) = C_1(p_{11}(\epsilon), p_{12}(\epsilon))e^{\epsilon t} + C_2(p_{21}(\epsilon), p_{22}(\epsilon))e^{\sigma t},$$

where $\mathbf{p}_1$ and $\mathbf{p}_2$ are the eigenvectors of B associated with ϵ and σ, with σ approaching a negative number as $\epsilon \to 0$. For small ϵ it follows that the second term decays to 0 on the time scale $\frac{1}{\epsilon}$, for which the first term grows. Thus after an initial transient, $(\alpha^{(1)}, \beta^{(1)})$ will be orthogonal to $\mathbf{q}_2$, where $\mathbf{q}_1$ and $\mathbf{q}_2$ are the corresponding eigenvectors of the transpose of B. The component proportional to $\mathbf{p}_2$ is therefore ultimately determined by nonlinear effects. At the same time we see that the natural time scale for the dominant linear terms suggests the variable $\tau = \epsilon t$. Finally, given that nonlinear quadratic terms are present and that these will determine the equilibrium amplitude, the latter should be proportional to ϵ in general.

For the other eigenspaces the results are similar, but then the nonlinear terms will in every case determine equilibrium amplitudes. Therefore, with the exception of the unstable mode, all contributions to (u, v) are of order ϵ^2.

Therefore, now adding nonlinear terms, following the initial transients the behavior of (u, v) on the time scale $\tau \sim O(1)$ is given by the following expansion:

$$(u,v) = \epsilon\mathbf{p}_1(0)m(\tau)\phi_1(x) + O(\epsilon^2),$$

where

$$\frac{dm}{dt} - m = m^2[\mathbf{q}_1(0)\cdot\mathbf{p}_1(0)]^{-1}\big[q_{11}(0)P(p_{11}(0), p_{12}(0)) \\ + q_{12}(0)Q(p_{11}(0), p_{12}(0))\big]\Pi_1(\phi_1^2) \equiv -\frac{m^2}{m_0}.$$

In addition, it can be deduced that all leading (i.e., $O(\epsilon^2)$) contributions to modes orthogonal to the growing one come from the quadratic terms of f and g evaluated at the $O(\epsilon)$ term. Thus m grows to an equilibrium amplitude m_0 of the same (undetermined) sign, while all orthogonal components have a passive dependence on $m^2(\tau)$ and equilibrate at values $O(\epsilon^2)$.

The quantity m plays the role of a Turing morphogen, and our assumptions have reduced the problem to a single such quantity. In a sense the case considered is the most general, since only in special circumstances will eigenvalues not be simple or will the quadratic terms not give the dominant nonlinear effect. (In addition, it might be desirable to consider at most quadratic nonlinearity in the chemical kinetics.)

The preceding calculation will ultimately be useful to us in a very simple way, applicable to systems of any size. The suggestion is that when there is preferential excitation of a chemical mode λ, ϕ, the quadratic nonlinear terms can be limiting on amplitude, and the nature of the instability, which may involve only a few chemicals in a large system, is modulated in space and time by the predominant eigenfunction $\phi(x,t)$ (allowing now also a slow dependence upon time, as might occur as a result of change of shape of the domain). If the morphogen concentration is "read" by cells we expect that there will be induced a gradient in the underlying chemical kinetics of cells. The important assumption is that gradients will be developed in the linearization of the kinetics about equilibrium. In the nonlinear terms the principal modulation in space and time comes from $\phi(x,t)$, however, and this makes nodal lines exceptional in nonlinear theory.

We examine next what happens following a second division, which we may assume to select a second growing mode (with again a small growth rate). Because of the nonlinearity there will be a residual component of the growing mode left over from the previous step, of order ϵ^2. This mode will then grow to saturation (amplitude $\sim \epsilon$) in a time t of order $-\ln\epsilon/\epsilon$, which gives an estimate of the division time needed in our theory. Generally, with continuous growth involving a change of shape as well as size, during a period of excitation the relevant mode will be a function of both x and t, so that in our simplification of the Turing system the modulation ϕ of the dynamic has temporal as well as spatial parametric dependence.

9.2.1. Multiple Eigenvalues. This can be an important case since we do not use this term in the strict sense; all that is required is that there be more than one growing mode with nearby growth rates. We now allow the problem to be two- or three-dimensional as well, over a domain D. To take a definite case, however, we assume a multidimensional eigenspace S for some eigenvalue λ. From the above calculation it follows directly that equilibrium spatial organization is obtained by solving a nonlinear eigenvalue problem of the form

$$\eta\phi = \mathcal{P}(\phi^2),$$

where $\mathcal{P}$ is the projection operator onto the eigenspace S, subject to

$$\frac{1}{|D|}\int \phi^2\, dD = 1, \quad |D| = \text{content of } D.$$

There will generally be more than one solution, hence more than one possible nodal set.[1] The evolved nodal set will, however, generally be unique as it minimizes a *Liapunov function*; the exceptional nodal set corresponds to the minimum modulus $|\eta|$ of the eigenvalue.

To see this, suppose S has dimension s and orthonormal basis $\phi_1, \dots, \phi_s$. Writing $\phi = \alpha_1(\tau)\phi_1 + \cdots + \alpha_s(\tau)\phi_s$, the evolution equation establishing the gradient has the form

$$\frac{d\alpha_i}{d\tau} = \alpha_i - c\Pi_i(\phi^2) = \alpha_i - c\sum_{j,k=1}^{s} c_{ijk}\alpha_j\alpha_k \equiv -\frac{\partial V}{\partial \alpha_i},$$

where c is a constant and

$$c_{ijk} = \frac{1}{|D|}\int \phi_i\phi_j\phi_k\, dD, \quad V = \frac{c}{3}\sum_{i,j,k=1}^{s} c_{ijk}\alpha_i\alpha_j\alpha_k - \frac{1}{2}\sum_{i=1}^{s}\alpha_i^2.$$

The existence of the potential V is the key point. Since

$$\frac{dV}{d\tau} = \sum_{i=1}^{s}\frac{d\alpha_i}{d\tau}\frac{\partial V}{\partial \alpha_i} = -\sum_{i=1}^{s}\left(\frac{d\alpha_i}{d\tau}\right)^2 \le 0,$$

it follows that V is a Liapunov function for the system and that locally stable equilibria are local minima of V. Global stability is reached when, at equilibrium, V is smallest. Since $-\sum_{i=1}^{2}\alpha_i^2 = 6V$ at equilibrium, V is smallest when $-\sum_{i=1}^{2}\alpha_i^2$ is smallest and hence when $|\eta| = [c^2\sum_{i=1}^{s}\alpha_i^2]^{-1/2}$ is smallest. But note that sufficiently small $|\eta|$ will generally bring cubic and higher-order nonlinearity into the problem of equilibrium.

9.2.2. Compartmental Boundaries as Shocks. We turn now to the question of nonlinear control of the formation of compartmental boundaries, given a parametrization of the chemical dynamic of the kind just obtained. The time variation of the dynamic will be assumed to be slow relative to the typical equilibration time for reactions; nevertheless, the nonlinear phenomena we will study will occur on an intermediate time scale, fast relative to the variation of the dynamic. We shall therefore not explicitly consider diffusive processes, although these are partially contained in our approximate equation (for **U** below). Our investigation will reveal conditions on the chemical kinetics determining the compartmental boundaries which can be sharpened by nonlinear effects.

[1]This eigenvalue problem was introduced by Turing in unpublished work, and applied to the symmetry of Radiolaria. Reference to this theory may be found in the biography by Sara Turing (W. Heffer, 1959, chap. XIV).

We restrict attention again to a one-dimensional spatial parametrization. Let $\mathbf{u}(x,t)$ be an N-vector of chemical concentrations, satisfying

$$\epsilon\frac{d\mathbf{u}}{dt} = \mathbf{f}(\mathbf{u}, x, t)$$

for some N-vector f, ϵ as always a small parameter. We are interested in the bifurcation away from an equilibrium $\mathbf{u}_0$ in the neighborhood of a nodal set, the latter being here a point x_0. Creation of the boundary will occur near $t = t_0$.

The $N + 2$ numbers determining $\mathbf{u}_0$, x_0, and t_0 are obtained in principle as follows: (i) since $\mathbf{u}_0$ is an equilibrium we have

$$\mathbf{f}^0 = \mathbf{f}(\mathbf{u}_0, x_0, t_0) = 0 \quad (N \text{ equations});$$

also, (ii) we set

$$\det \mathbf{A}^0 = 0, \quad \mathbf{A}^0 = \nabla_{\mathbf{u}}\mathbf{f}(\mathbf{u}_0, x_0, t_0) \quad (1 \text{ equation}).$$

A final determing equation will be obtained below by requiring that quadratic contributions to the dominant nonlinear effect vanish. This condition will serve to select those cells on the nodal set from the large group of cells participating in the diffusive instability. We will, by the way, now allow $\mathbf{f}$ to have a general dependence upon $\mathbf{u}$, not necessarily stopping at quadratic terms, although the theory does not depend on these higher-order terms (cf. the parameter λ below).

To study locally the evolution of u, we expand $\mathbf{f}$ in a Taylor's series in $N + 2$ variables, about the (as yet not fully specified) point $(\mathbf{u}_0, x_0, t_0)$, where we use dyadic notation for the double and triple index sums:

$$\begin{aligned}\epsilon\frac{d\mathbf{u}}{dt} &= \mathbf{f}^0 + \mathbf{f}^0_x \cdot (x - x_0) + f^0_t \cdot (t - t_0) + \mathbf{A}_0 \cdot (\mathbf{u} - \mathbf{u}_0)\\ &+ \left[\mathbf{A}^0_a \cdot (x - x_0) + \mathbf{A}^0_t \cdot (t - t_0)\right] \cdot (\mathbf{u} - \mathbf{u}_0)\\ &+ \mathbf{B}^0 : (\mathbf{u} - \mathbf{u}_0)^2 + \mathbf{C}^0 \vdots (\mathbf{u} - \mathbf{u}_0)^3 + \mathbf{E}.\end{aligned}$$

Here $\mathbf{B}^0 = \frac{1}{2}(\nabla_{\mathbf{u}}\nabla_{\mathbf{u}}\mathbf{f})^0$, $\mathbf{C}^0 = \frac{1}{6}(\nabla_{\mathbf{u}}\nabla_{\mathbf{u}}\nabla_{\mathbf{u}}\mathbf{f})^0$, e.g.,

$$\mathbf{C} \vdots \mathbf{u}^3 = \sum_{j,k,l=1}^{N} C_{jkl}u_ju_ku_l.$$

With $\mathbf{u} - \mathbf{u}_0 = \Delta\mathbf{u}$, etc., the error is

$$\mathbf{E} = O(\Delta x^2) + O(\Delta t)^2 + O(\Delta x\Delta\mathbf{u}^2) + O(\Delta t\Delta\mathbf{u}^2) + O(\Delta\mathbf{u}^4).$$

We now add two postulates: (iii) the null space of $\mathbf{A}^0$ is one-dimensional, 0 being an eigenvalue of multiplicity 1; and (iv) if $\mathbf{A}_0 \cdot \mathbf{p} = 0$ and $(\mathbf{A}_0)^t \cdot \mathbf{q} = 0$, then $(\mathbf{q}, \mathbf{B}^0 : \mathbf{p}^2) = 0$, where $(\cdot\,,\cdot)$ is the inner product on $\mathbb{R}^N$. Postulate (iv) thus provides the final equation for the bifurcation (nodal) point. This postulate anticipates a compatibility condition on the local system that will have the effect of eliminating a fold catastrophe in favor of a cusp (Chapter 2), since the fold is degenerate at a node. This will lead to steepening of a morphogen gradient near the compartmental boundary; the boundary is in fact completed as a shock.

9.2.3. Expansion for Small ϵ. We now attempt to order the terms of the expanded equation by appropriate stretchings. Let $\Delta\mathbf{u} = \epsilon^{\alpha}\mathbf{U}$, $\Delta t = \epsilon^{\beta}\tau$, and $\Delta x = \epsilon^{\gamma}\xi$. We then have

$$\begin{aligned}\epsilon^{1+\alpha-\beta}\mathbf{U}_{\tau} &= \epsilon^{\gamma}\mathbf{f}_x^0\xi + \epsilon^{\beta}\mathbf{f}_t^0\tau + \epsilon^{\gamma}\mathbf{A}^0\cdot\mathbf{U}\\ &+ \epsilon^{\alpha+\gamma}\mathbf{A}_x^0\cdot\mathbf{U}\xi + \epsilon^{\alpha+\beta}\mathbf{A}_t^0\cdot\mathbf{U}\tau + \epsilon^{2\alpha}\mathbf{B}^0 : \mathbf{U}^2\\ &+ \epsilon^{3\alpha}\mathbf{C}_0 \,\vdots\, \mathbf{U}^3 + O(\epsilon^{2\beta} + \epsilon^{2\gamma} + \epsilon^{\beta+2\alpha} + \epsilon^{\gamma+2\alpha} + \epsilon^{4\alpha}),\end{aligned}$$

At this point we must distinguish between two cases. For Case (1), suppose first $(\mathrm{v})_a$: $(\mathbf{q},\mathbf{f}_t^0) = 0$. Then, taking the left product with the adjoint eigenvector $\mathbf{q}$, we attempt to make all terms other than those involving $\mathbf{f}_x^0$ and $\mathbf{f}_t^0$ of the same order. Thus $1+\alpha-\beta = \alpha+\gamma = 3\alpha = \alpha+\beta$ or $\beta = \gamma = 1-2\alpha$, $\alpha = \frac{1}{4}$. Using $(\mathrm{v})_a$ we have left only $\mathbf{f}_x^0$, a term of order ϵ^{γ}, so we must take $\gamma = 3\alpha$, which contradicts $\alpha+\gamma = 3\alpha$. Thus we cannot retain the $\mathbf{A}_x^0$ term, and we impose instead $1+\alpha-\beta = \alpha+\beta = 3\alpha = \gamma$, so that $\beta = 2\alpha$, $\gamma = 3\alpha$, $\alpha = \frac{1}{4}$. The equation for $\mathbf{U}$ then has the form

$$\begin{aligned}\epsilon^{3/4}\mathbf{u}_{\tau} &= \epsilon^{3/4}\mathbf{f}_x^0\xi + \epsilon^{1/2}\mathbf{f}_t^0\tau + \epsilon^{1/4}\mathbf{A}^0\cdot\mathbf{U}\\ &+ \epsilon\mathbf{A}_x^0\cdot\mathbf{U}\xi + \epsilon^{3/4}\mathbf{A}_t^0\cdot\mathbf{U}\tau + \epsilon^{1/2}\mathbf{B}^0 : \mathbf{U}^2 + \epsilon^{3/4}\mathbf{C}^0 \,\vdots\, \mathbf{U}^3 + O(\epsilon).\end{aligned}$$

Letting

$$\mathbf{U}(\xi,\tau;\epsilon) = \mathbf{U}_0(\xi,\tau) + \epsilon^{1/4}\mathbf{U}_1(\xi,\tau) + \epsilon^{1/2}\mathbf{U}_2(\xi,\tau) + \cdots,$$

we obtain from terms of like order a sequence of equations, the first three being

$$\begin{aligned}&\mathbf{A}^0\cdot\mathbf{U}_0 = 0,\\ &\mathbf{A}^0\cdot\mathbf{U}_1 = -\mathbf{f}_t^0\tau - \mathbf{B}^0 : \mathbf{U}_0^2,\\ &\mathbf{A}^0\mathbf{U}_2 = -\mathbf{f}_x^0\xi - \mathbf{A}_t^0\cdot\mathbf{U}_0\tau - \mathbf{B}^0 : (\mathbf{U}_0\mathbf{U}_1 + \mathbf{U}_1\mathbf{U}_0) - \mathbf{C}^0 \,\vdots\, \mathbf{U}_0^3 + \mathbf{U}_{0\tau}.\end{aligned}$$

In view of (iv) and $(\mathrm{v})_a$ the equation for $\mathbf{U}_1$ can be solved and from the form of $\mathbf{U}_0$ we may write

$$\mathbf{U}_1 = \mathbf{h}_1 + \mathbf{g}_1\phi^2$$

where $\mathbf{g}_1$ and $\mathbf{h}_1$ are N-vectors. The compatibility equation for the equation for $\mathbf{U}_2$ is then

$$\begin{aligned}(\mathbf{q},\mathbf{p})\phi_{\tau} &= (\mathbf{q},\mathbf{f}_x^0)\xi + \big[(\mathbf{q},\mathbf{A}_t^0\cdot\mathbf{p})\tau + (\mathbf{q},\mathbf{B}^0 : (\mathbf{p}\mathbf{h}_1 + \mathbf{h}_1\mathbf{p}))\big]\phi\\ &+ \big[(\mathbf{q},\mathbf{B}^0 : (\mathbf{p}\mathbf{g}_1\mathbf{g}_1\mathbf{p})) + (\mathbf{q},\mathbf{C}^0 \,\vdots\, \mathbf{p}^3)\big]\phi^3.\end{aligned}$$

Suppose finally that $(\mathrm{vi})_a$: $(\mathbf{q},\mathbf{A}_t^0,\mathbf{p}) \neq 0$. With τ replaced by

$$\theta = \tau + \big(\mathbf{q},\mathbf{A}_t^0,\mathbf{p}\big)^{-1}\big(\mathbf{q},\mathbf{B}^0 : (\mathbf{p}\mathbf{h}_1 + \mathbf{h}_1\mathbf{p})\big),$$

we then have

$$(\mathbf{q},\mathbf{p})\phi_{\theta} = \sigma + \big(\mathbf{q},\mathbf{A}_t^0,\mathbf{p}\big)\theta\phi + \lambda\phi^3,$$

where $\sigma = (\mathbf{q},\mathbf{f}_x^0)\xi$, $\lambda = (\mathbf{q},\mathbf{B}^0 : (\mathbf{p}\mathbf{g}_1\mathbf{g}_1\mathbf{p})) + (\mathbf{q},\mathbf{C}^0 \,\vdots\, \mathbf{p}^3)$. This ordinary differential equation for ϕ is the main result of this first case. From (iii) we may take $(\mathbf{q},\mathbf{p}) = 1$ and it is seen that if $(\mathrm{vii})_a$ $(\mathbf{q},\mathbf{A}_t^0\cdot\mathbf{p}) > 0$, and $\lambda < 0$, solution curves for

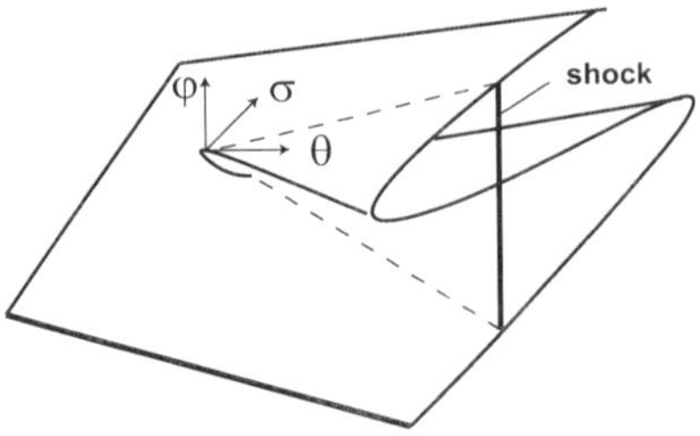

FIGURE 9.6. Formation of a stationary shock.

fixed σ converge to the cusp surface obtained by setting the right-hand side equal to 0. Developmental paths on either side of $\sigma = 0$ diverge, and a stationary shock is formed at $\sigma = 0$; see Figure 9.6.

We shall not discuss this equation further and proceed to case (2), our second, probably more representative case: $(\mathrm{v})_b$: $(\mathbf{q}, \mathbf{f}_t^0) \neq 0$. For this case a modification of the ordering of variables is needed. To see this, consider once more the equation for $\mathbf{U}$ with arbitrary orders, but because of $(\mathrm{v})_b$ we impose the additional condition $1 + \alpha - \beta = \beta$. With $\alpha = \frac{1}{4}$ we obtain $\beta = \frac{5}{8}$, contradicting the value of $\frac{1}{2}$ obtained from the remaining conditions. To resolve this difficulty we focus attention on the term

$$\sigma^* = (\mathbf{q}, \mathbf{f}_x^0)(x - x_0) + (\mathbf{q}, \mathbf{f}_t^0)(t - t_0),$$

and note that σ^* and x can be used in place of t and x as coordinates in space-time. We also set $\sigma^* = \epsilon^\delta \sigma$, $\delta > 0$. Attempting to retain all remaining terms in the equation for $\mathbf{U}$, we set $\alpha + \gamma = \alpha + \beta = 3\alpha$, yielding $\beta = \gamma = 2\alpha$. Tentatively take $\delta > 2\alpha$ and utilize a "two-time" formalism with

$$\left(\frac{\partial}{\partial t}\right)_x = \left(\frac{\partial}{\partial t}\right)_{\sigma^*} + (\mathbf{q}, \mathbf{f}_t^0)\frac{\partial}{\partial \sigma^*} \sim O(\epsilon^{-\delta}).$$

We should then impose $1 + \alpha - \delta = \delta = 3\alpha$, yielding $\delta = \frac{1+\alpha}{2} = 3\alpha$ or $\alpha = \frac{1}{5}$. Thus

$$\alpha = \frac{1}{5}, \quad \beta = \gamma = \frac{2}{5}, \quad \delta = \frac{3}{5} > 2\alpha,$$

as required. With this ordering we have

$$\begin{aligned}\epsilon^{3/5}\frac{\partial \mathbf{U}}{\partial \sigma} &= \epsilon^{2/5}(\mathbf{f}_t^0 \tau + \mathbf{f}_x^0 \xi) + \epsilon^{1/5}\mathbf{A}^0 \cdot \mathbf{U} + \epsilon^{3/5}(\mathbf{A}_x^0 \xi + \mathbf{A}_t^0 \tau) \cdot \mathbf{U} \\ &\quad + \epsilon^{2/5}\mathbf{B}^0 : \mathbf{U}^2 + \epsilon^{3/5}\mathbf{C}^0 \vdots \mathbf{U}^3 + O(\epsilon^{4/5}).\end{aligned}$$

Now for this ordering to be possible we must have $(\mathrm{vi})_b$: $(\mathbf{q}, \mathbf{F}_x^0) \neq 0$. This allows elimination of ξ in favor of τ by

$$\xi = \frac{\epsilon^{1/5}\sigma - (\mathbf{q}, \mathbf{f}_t^0)\tau}{(\mathbf{q}, \mathbf{f}_x^0)}.$$

Using this and expanding,

$$\mathbf{U}(\sigma, \tau; \epsilon) = \mathbf{U}_0(\sigma, \tau) + \epsilon^{1/5}\mathbf{U}_1(\sigma, \tau) + \epsilon^{2/5}\mathbf{U}_2(\sigma, \tau) + \cdots$$

we obtain as before the sequence

$$\begin{aligned}
\mathbf{A}^0\mathbf{U}_0 &= 0, \\
\mathbf{A}^0\mathbf{U}_1 &= \left[-\mathbf{f}_t^0 + \frac{(\mathbf{q}, \mathbf{f}_t^0)\mathbf{f}_x^0}{(\mathbf{q}, \mathbf{f}_x^0)}\right]\tau - \mathbf{B}^0 : \mathbf{U}_0^2, \\
\mathbf{A}^0\mathbf{U}_2 &= -(\mathbf{q}, \mathbf{f}_t^0)\frac{\partial \mathbf{U}_0}{\partial \sigma} + \frac{\mathbf{f}_x^0 \sigma}{(\mathbf{q}, \mathbf{f}_x^0)} + \mathbf{B}^0 : (\mathbf{U}_1\mathbf{U}_0 + \mathbf{U}_0\mathbf{U}_1) \\
&\quad + \left[\mathbf{A}_t^0 - \frac{(\mathbf{q}, \mathbf{f}_t^0)}{(\mathbf{q}, \mathbf{f}_x^0)}\mathbf{A}_x^0\right] \cdot \mathbf{U}_0 \tau + \mathbf{C}^0 : \mathbf{U}_0^3,
\end{aligned}$$

etc. Then and $\mathbf{U}_0$ and $\mathbf{U}_1$ have the previous forms, and compatibility for $\mathbf{U}_2$ gives

$$(\mathbf{q}, \mathbf{f}_t^0)\phi_\sigma = \sigma + \gamma(\tau - \tau_0)\phi + \lambda\phi^3,$$

which should be compared with the corresponding result in the first case. Here

$$\begin{aligned}
(\mathbf{q}, \mathbf{p}) = 1, \quad \tau_0 = -\gamma^{-1}(\mathbf{q}, \mathbf{B}^0 : (\mathbf{p}\mathbf{h}_1 + \mathbf{h}_1\mathbf{p})), \\
\gamma = (\mathbf{q}, \mathbf{f}_x^0)^{-1}\left[(\mathbf{q}, \mathbf{A}_t \cdot \mathbf{p})(\mathbf{q}, \mathbf{f}_x^0) - (\mathbf{q}, \mathbf{A}_x^0 \cdot \mathbf{p})(\mathbf{q}, \mathbf{f}_t^0)\right],
\end{aligned}$$

and λ is as defined previously.

In what follows we must have $(\text{vii})_b$: $\gamma \neq 0$, $\lambda \neq 0$, $\operatorname{sgn}(\gamma/\lambda) = -1$.

9.2.4. Analysis of the Equation for ϕ in Case 2. We consider the various subcases:

		$\operatorname{sgn}((\mathbf{q}, \mathbf{f}_t^0))$	$\operatorname{sgn}(\gamma)$	$\operatorname{sgn}(\lambda)$
Case (2a)	a	+1	+1	+1
	b	+1	+1	−1
	c	+1	−1	+1
	d	+1	−1	−1
Case (2b)	a	−1	+1	+1
	b	−1	+1	−1
	c	−1	−1	+1
	d	−1	−1	−1

Case (2a). If $(\mathbf{q}, \mathbf{f}_x^0) > 0$ (< 0), then $\sigma^* = \text{const}$ is a line of positive (negative) slope in the $\xi\tau$-plane, and so the catastrophe set of the cusp on the right (the projection onto the $\xi\tau$-plane of the folds of the cusp) is as shown in Figures 9.7 and 9.8(a). The dotted line in Figure 9.8(a) indicates an integral curve asymptotic for large $|\sigma|$.

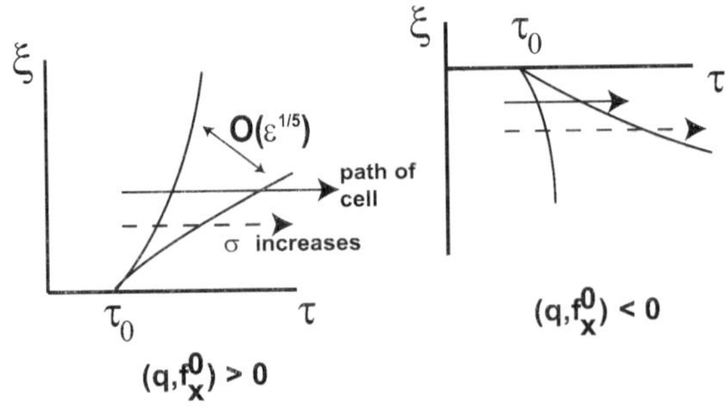

FIGURE 9.7. Cusp formation in Case (2a).

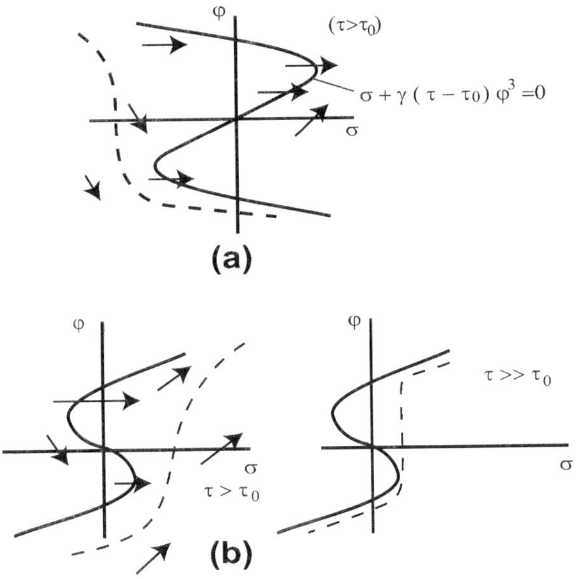

FIGURE 9.8. Phase portraits in Cases (2a) and (2b).

This solution does not, however, represent stable development with respect to the fast time variable σ since if $\phi = \phi_0 + \delta\phi$ for $\tau - \tau_0 \gg 1$, we have approximately

$$\frac{d\delta\phi}{d\sigma} \approx -2\lambda\phi_0^2\delta\phi \approx 2\gamma(\tau - \tau_0)\delta\phi.$$

There is thus instability for increasing σ, i.e., along the path of development. This excludes case (2a) and implies that that $\operatorname{sgn}(\gamma) = +1$ is necessary for case (2a).

The phase portrait for the remaining case, (2b), has the form shown in Figure 9.8(b). The transition is thus from lower to higher values of the morphogen ϕ.

Case (2b). This case may be reduced to the previous one by replacing σ by $-\sigma$. Then ϕ will *decrease* as a compartmental boundary is crossed. Again, only subcase (2b) is allowed.

Physically, cases (1) and (2) differ essentially only in the orientation of the axis of the cusp with respect to the time axis. In the bifurcation represented by case (2), each cell "differentiates" along a transition profile of ϕ as shown above, but the process steepens to a shock as $\tau - \tau_0$ increases. Different cells encounter the transition as different times τ. In the catastrophe literature the wave of transitions in case (2) is known as a *primary wave*. In effect the use of σ and τ has allowed us to follow the path of the wave on one time scale, but to study the transition on another, smaller one by a factor $\epsilon^{1/5}$. In the original variables the path of the wave may be shown to be given by

$$\left(\frac{dx}{dt}\right)_{\text{wave}} = -\frac{(\mathbf{q}, \mathbf{f}_t^0)}{(\mathbf{q}, \mathbf{f}_x^0)} + \frac{\gamma \operatorname{sgn}(\mathbf{q}, \mathbf{f}_t^0)}{(\mathbf{q}, \mathbf{f}_x^0)} \left|\frac{\gamma}{3\lambda}\right|^{\frac{1}{2}} (t - t_0)^{\frac{1}{2}}$$

to first order.

Case (2) thus represents incomplete compartmentalization in that the compartment boundary is in motion relative to the tissue. One can only consider a compartment to have been established when all waves of differentiation have stabilized. However, the theory would seem to imply that, under the restrictions on $\mathbf{f}$ detailed above, once compartments are established the boundaries are abrupt transitions.

To sum up, the question we have addressed is, given that a Turing instability produces moderate gradients in morphogens, how can sharp compartmental boundaries be formed? The present proposal is that nonlinear chemical kinetics provides a natural steepening process. We have effectively assumed a weak space and/or time parametrization of the kinetics, since this is consistent with the ability of cells to "read" morphogen levels. The results show that steepening to shocks occurs at nodal lines, where quadratic nonlinearities are negligible relative to linear ones (when projected onto the subspace of the morphogen). It is noteworthy that the analytic description does not require nonlinearity higher than quadratic.

CHAPTER 10

Segmentation of Insect Embryos

One of the developmental systems most heavily studied in a quantitatively respectable fashion is that of the early insect embryo. The story begins with the maternal oocyte to which nurse cells have transported a variety of mRNAs. Principal actors at this stage are, with varying concentrations along the anterior-posterior axis: bicoid and nanos, which are transcription factors, eventually organizing other genes in this part of the embryo; and hunchback (hb) and caudal, which are regulatory proteins. As the embryo cleaves without cell membranes being formed, termed a syncytium, allowing for easy molecular diffusion, the mRNAs produce proteins that diffuse to create a pattern quite different from their sources, with reactions as well to help shape their profiles—compare bcd mRNA and protein at the beginning of the process. Actually, the total axial concentration (integrated over the cross section) does not tell the whole story, as a two-dimensional projection of bicoid indicates.

As cleavage proceeds and corresponding cell nucleus number increases, a number of eventual structural genes are activated by the bicoid gradient and the already expressed genes: in all cases, the proteins control the production reactions while diffusing, and it is the collection of these reaction-diffusion processes that is responsible for expression patterns that determine the qualitative fate of the associated cells.

10.1. Prototypical Reaction-Diffusion: One Component

We have previously met with diffusion in the form of random transfer of molecules from one of a linear array of cells to its neighbors. The molecules need not be attached to cells, but may move by thermal or Brownian motion, in which case we have transport from one infinitesimal volume of space to its neighbors. It will be useful to repeat the "derivation" on another version, emphasizing the role of current density. If $A(\mathbf{r})$ is the concentration of A at $\mathbf{r}$ in molecules per unit volume, and $J_A(\mathbf{r})$ the (vectorial) rate of flow of A across a unit surface, then for a source $f(A)$ of production of A per unit volume, we have the usual conservation equation for A:

$$\dot{A}(\mathbf{r})(dx)^3 = f(A(\mathbf{r}))(dx)^3 + \left[J_{A,x}\left(\mathbf{r} - \frac{1}{2}\mathbf{e}_x dx\right) - J_{A,x}\left(\mathbf{r} + \frac{1}{2}\mathbf{e}_x dx\right)\right](dx)^2$$
$$+ \left[J_{A,y}\left(\mathbf{r} - \frac{1}{2}\mathbf{e}_y dx\right) - J_{A,y}\left(\mathbf{r} + \frac{1}{2}\mathbf{e}_y dx\right)\right](dx)^2 +$$

$$+\left[J_{A,z}\left(\mathbf{r}-\frac{1}{2}\mathbf{e}_z dx\right)-J_{A,z}\left(\mathbf{r}+\frac{1}{2}\mathbf{e}_z dx\right)\right](dx)^2,$$

where $\mathbf{e}_x$ is the unit vector in direction x, or, to $O(dx)$,

$$\dot{A}=f(A)-\nabla\cdot J.$$

But if the transport is represented by the "reactions"

$$A(\mathbf{r})\underset{\gamma}{\overset{\gamma}{\rightleftharpoons}}A(\mathbf{r}+\mathbf{dr}),$$

then clearly

$$J_{A,x}(\mathbf{r})(dx)^2=\gamma\left[A(\mathbf{r})(dx)^3-A\left(\mathbf{r}+\mathbf{e}_x dx\right)(dx)^3\right],$$

and setting $D=\gamma(dx)^2$, we can conclude that, including y- and z-directions,

$$J_A(\mathbf{r})=-D\nabla A(\mathbf{r}),$$

known as Fick's law. And of course, coupled with conservation, we have

$$\dot{A}=f(A)+D\nabla^2 A.$$

The above applies to an open space. If the space is closed, we need boundary conditions. A standard representation imagines a species B on the other side of the boundary to which A is converted,

$$A\overset{\gamma'}{\longrightarrow}B,$$

so that

$$J_{A,x}(\mathbf{r})(dx)^2=\gamma' A(\mathbf{r})(dx)^3.$$

Defining $\alpha=(\gamma'/D)dx$, Fick's law tells us that $\partial A/\partial x=-\alpha A$ at the boundary, or, generalizing to a boundary with unit normal $\hat{n}$,

$$\hat{n}\cdot\nabla A=-\alpha A.$$

In particular, if $\alpha=0$, no absorption or "no flux," we have the Neumann boundary condition $\hat{n}\cdot\nabla A=0$.

Our basic closed bounded reaction-diffusion system in the volume Ω is then (scaled so that $D=1$)

$$\dot{A}=f(A)+\nabla^2 A,$$

with

$$\hat{n}\cdot\nabla A=-\alpha A$$

on the boundary $\partial\Omega$. The general properties of the solution are best assessed by setting up the Lyapunov function (time dependence is implicit)

$$\mathcal{L}(A)=\frac{1}{2}\int_{\Omega}|\nabla A(\mathbf{r})|^2\,d\mathbf{r}+\frac{\alpha}{2}\int_{\partial\Omega}[A(\mathbf{r})]^2\,dS-\int_{\Omega}\varphi(A(r))d\mathbf{r},$$

where $\varphi'(A) = f(A)$. Let us assume that $\varphi(A)$ is bounded from above, so that for finite Ω, $\mathcal{L}(A)$ will be bounded from below. We now have, on time-differentiating,

$$\begin{aligned}\dot{\mathcal{L}}(A) &= \int_\Omega \nabla \dot{A}(\mathbf{r}) \cdot \nabla A(\mathbf{r}) d\mathbf{r} + \alpha \int_{\partial\Omega} \dot{A}(\mathbf{r}) A(\mathbf{r}) dS - \int_\Omega \dot{A}(\mathbf{r}) \varphi'(A(\mathbf{r})) d\mathbf{r} \\ &= \int_{\partial\Omega} \dot{A}(\mathbf{r}) \nabla A(\mathbf{r}) dS - \int_\Omega \dot{A}(\mathbf{r}) \nabla^2 A(\mathbf{r}) d\mathbf{r} \\ &\quad + \alpha \int_{\partial\Omega} \dot{A}(\mathbf{r}) A(\mathbf{r}) dS - \int_\Omega \dot{A}(\mathbf{r}) f(A(\mathbf{r})) d\mathbf{r},\end{aligned}$$

or with a little bit of algebra,

$$\dot{\mathcal{L}}(A) = -\int_\Omega \dot{A}(\mathbf{r})[f(A(\mathbf{r})) + \nabla^2 A(\mathbf{r})] d\mathbf{r} = -\int_\Omega [f(A(\mathbf{r})) + \nabla^2 A(\mathbf{r})]^2 \, d\mathbf{r}.$$

In other words, $\mathcal{L}$ is a decreasing function of time, becoming stationary (generally as $t \to \infty$) when

$$f(A(\mathbf{r})) + \nabla^2 A(\mathbf{r}) = 0$$

and boundary conditions are satisfied. Thus, if the solution is unique, then a closed, bounded system will necessarily reach this steady state. In contrast, if the domain is not bounded, e.g., an open system, then $\mathcal{L}(A)$ need not be bounded from below, and the conclusion need not hold. Indeed, we have already found traveling wave solutions under these circumstances.

But when the domain is bounded, the nature of the solution depends very much on the boundary conditions. In particular, for the simplest conceptual model, an isolated system with no-flux Neumann boundary conditions, there are typically at least two steady state solutions: one in which $A_0(\mathbf{r}) = A_0$ is constant, where $f(A_0) = 0$, i.e., no pattern at all, and one or more with $A_1(\mathbf{r})$ not constant, satisfying

$$f(A_1(\mathbf{r})) + \nabla^2 A_1(\mathbf{r}) = 0$$

in Ω and

$$\hat{n} \cdot \nabla A_1(\mathbf{r}) = 0$$

on $\partial\Omega$. Before focusing on the latter pattern-forming case, we must check that it is stable, and this is typically not true. This is easy to see in one-dimensional space, where

$$A_1(x,t) = f(A_1(x,t)) + A_1''(x,t), \quad A_1'(0) = A_1'(L) = 0.$$

Here, for a perturbation from steady state,

$$\dot{A}_1(x,t) \mapsto A_1(x) + \epsilon g(x) e^{\lambda t};$$

we have

$$\lambda g(x) = f'(A_1(x)) g(x) + g''(x), \quad g'(0) = g'(L) = 0,$$

and we require that all possible λ be negative for stability. Now it is readily shown that

$$\lambda_{\max} = \max_g \left\{ \frac{1}{2} \int_0^L \left[f'(A_1(x))(g(x))^2 - (g'(x))^2 \right] dx \right\},$$

where we maximize over all g normalized by

$$\int_0^L [g(x)]^2 \, dx = 1,$$

and that the maximizing g_0 automatically satisfies $g_0'(0) = g_0'(L) = 0$.

Let us choose a special $g_1(x)$ that does *not* satisfy $g_1'(0) = g_1'(L) = 0$, so that we will have

$$\lambda_{\max} > \frac{1}{2} \int_0^L \left[f'(A_1(x))(g_1(x))^2 - \left(g_1'(x)\right)^2 \right] dx.$$

To do so, observe that from

$$f(A_1(x)) + A_1''(x) = 0,$$

we have

$$f'(A_1(x))A_1'(x) + A_1'''(x) = 0, \quad A_1'(0) = A_1'(L) = 0.$$

Hence, if

$$g_1(x) = \frac{A_1'(x)}{\left[\int_0^L \left(A_1'(x)\right)^2 dx\right]^{1/2}},$$

then

$$f'(A_1(x))g_1(x) + g_1''(x) = 0, \quad g_1(0) = g_1(L) = 0.$$

It is now trivial to show that

$$\int_0^L \left[f'\left(A_1(x)\right)\left(g_1(x)\right)^2 - \left(g_1'(x)\right)^2 \right] dx = 0,$$

and we conclude that $\lambda_{\max} > 0$, so that $A_1(x)$ is unstable.

On the other hand, for the solution $A_0(x) = A_0$, we need only check

$$\lambda g(x) = f'(A_0)g(x) + g''(x), \quad g'(0) = g'(L) = 0,$$

which is solvable with

$$g(x) \propto \cos\left[\frac{\pi}{L} \left(f'(A_0) - \lambda\right)^{\frac{1}{2}} x \right]$$

if $\lambda < f'(A_0)$, and so the constant solution is stable if $f'(A_0) < 0$. The conclusion that only the uniform solution is stable under Neumann boundary conditions holds as well in three dimensions if the boundary is convex, but bona fide patterns can occur for nonconvex boundaries.

In general, then, one must allow for "leakage" ($\alpha > 0$) at the boundary to establish a nontrivial concentration profile. And of course, $\hat{n} \cdot \nabla A = -\alpha A$ will do for any $\alpha > 0$ since this precludes a constant solution. But then, we may take α arbitrarily close to 0—it only results in steady state taking a long time to establish. Another option is to break the homogeneity of space in that $f(A(\mathbf{r}), \mathbf{r})$ depends explicitly on $\mathbf{r}$ as well. We will take advantage of both paths to pattern formation.

Suppose now that we have a one-dimensional caricature of our one-species steady state reaction-diffusion with leaky boundary, in the form

$$A''(x) + \varphi'(A(x)) = 0, \quad A'(0) = \alpha A(0), \quad A'(L) = -\alpha A(L).$$

The solution is simple but informative. We form the energy integral

$$\frac{1}{2}[A'(x)]^2 + \varphi(A(x)) = K \tag{10.1}$$

and integrate from $x = 0$ with $dx = |dA(x)|/A'(x)$, where the positive value of $A'(x)$ is chosen, to obtain

$$x = \int_{A(0)}^{A} \frac{|dA|}{[2(K - \varphi(A))]^{1/2}}, \tag{10.2}$$

determining K in principle from

$$L = \int_{A(0)}^{A(L)} \frac{|dA|}{[2(K - \varphi(A))]^{1/2}} \tag{10.3}$$

together with the boundary conditions, which now read

$$\frac{\alpha^2}{2}[A(0)]^2 + \varphi(A(0)) = K = \frac{\alpha^2}{2}[A(L)]^2 + \varphi(A(L)).$$

If $\alpha \to 0$, so that $A'(0) = A'(L) = 0$, there is a smooth dependence of $A(x)$ on α, and the only difference is that the boundary conditions simplify to

$$K - \varphi(A(0)) = K - \varphi(A(L)) = 0. \tag{10.4}$$

Expressions (10.1), (10.3), and (10.4) give rise to a standard pictorial representation: the allowed region for A is given by $\varphi(A) \geq K$ and is terminated by $\varphi^{-1}(K)$. The "path" $A(x)$ follows is given by (10.2), and the value of K is determined by (10.3); any number of transits of the "half-cycle" between the boundaries $\varphi_1^{-1}(K)$ and $\varphi_2^{-1}(K)$ of possible A can occur in the resulting concentration pattern.

The prototypical case is that in which $\varphi(A)$ has a local minimum,

$$\varphi(A) = \varphi_0 + \frac{\omega^2}{2}(A - A_0)^2.$$

In this vicinity, the profile equation

$$A'' + \omega^2(A - A_0) = 0, \quad A'(0) = A'(L) = 0,$$

has the immediate solution

$$A(x) = A_0 + C\cos\omega x$$

if $L = n\pi/\omega$. The required quantization of L is relaxed when deviations from constant curvature of $\varphi(A)$ are taken into account, but no matter what the value of L, the basic $\cos \omega x$ dependence of the pattern does not change. This is absolute regulation so that organisms of the same species with a distribution of sizes would nonetheless have the same absolute pattern of concentration. In real life, this is rarely the case—biochemical patterns universally scale with the size of the organism. How would this relative size regulation come about?

There are a number of possible mechanisms that we will attend to, but even with this class of models, there is a very simple one that was introduced back in Chapter 3, which we will later analyze in further detail. Suppose that $\varphi'(A)$ is discontinuous at A_0, say,

$$\varphi(A) = \begin{cases} \varphi_0 + \omega_1(A - A_0), & x \geq 0, \\ \varphi_0 - \omega_2(A - A_0), & x \leq 0, \end{cases}$$

with $A(0) = A_0$. For convenience, take $\varphi_0 = 0$. Then, on solving

$$\varphi(A) + \frac{1}{2}(A')^2 = K^2,$$

we find

$$A = \begin{cases} A_0 + K^2\left[\frac{\sqrt{2}}{K}x - \frac{\omega_1}{2}\left(\frac{x}{K}\right)^2\right], & x \geq 0, \\ A_0 + K^2\left[\frac{\sqrt{2}}{K}x + \frac{\omega_2}{2}\left(\frac{x}{K}\right)^2\right], & x \leq 0. \end{cases}$$

The x-excursion from $A'(x_m) = 0$ to $A'(x_n) = 0$ therefore becomes

$$L = x_m - x_n = \left(\frac{1}{\omega_1} + \frac{1}{\omega_2}\right)K\sqrt{2},$$

and we see that the pattern regulates perfectly under change of L and hence of K.

Let us return briefly to the one-dimensional caricature. When is it valid? We confine our attention to Neumann boundary conditions. Suppose that the diffusion constant is independent of location (it does not have to be if the randomizing species is not uniform). Then the surfaces of constant ρ ($A \mapsto \rho$) being normal to the boundary means that for a slowly changing cross section, they are nearly parallel and hence longitudinally (here, x) directed. Define the aggregate one-dimensional concentration

$$P(x) = \int_{a(x)} \rho(\mathbf{r})da(x)$$

over a cross section $a(x)$ (also denoting the area of the cross section) and likewise an aggregate current density

$$J(x) = \int_{a(x)} j_x(\mathbf{r})da(x)$$

for three-dimensional current density $j(\mathbf{r})$. Then the dynamics

$$\dot{\rho} + \nabla \cdot j = S(\rho),$$

with $j = -D\nabla P$ and $\hat{n}\cdot j = 0$ on the boundary, implies at once (using the divergence theorem)

$$\dot{P} + \frac{\partial J}{\partial x} = \int_{a(x)} S(\rho(\mathbf{r}))da(x).$$

But since ρ is constant on $a(x)$, we have

$$P(x) = a(x)\rho(x, y, z), \quad J(x) = a(x)j_x(x, y, z) = -Da(x)\frac{\partial \rho}{\partial x},$$

$$\int_{a(x)} S(\rho(\mathbf{r}))da(x) = a(x)S\left(\frac{P(x)}{a(x)}\right),$$

giving us the slightly modified (Fick-Jacobs) reaction-diffusion equation

$$\dot{P}(x) = \frac{\partial}{\partial x}\left[Da(x)\frac{\partial}{\partial x}\frac{P(x)}{a(x)}\right]$$

for $S \equiv 0$.

10.2. Prepattern Activation

In the insect embryo, a short-time anterior source of bicoid rapidly forms a pattern by diffusion, getting "used up" by binding to nuclei and then getting distributed as the nuclei divide. There results a bcd gradient that organizes the future structure of the embryo. What is its form? A boundary source is the opposite of an absorbing sink, and in this kind of model, we would have $J_{A,x}(dx)^2 = kS(dx)^3$ for S the source. Scaling suitably, this means that at the boundary,

$$\hat{n} \cdot \nabla A = \beta,$$

a constant. In other words, the basic one-dimensional reaction-diffusion would now be

$$A''(x) + \varphi'(A(x)) = 0, \quad A'(0) = -\beta, \quad A'(L) = -\alpha A(L).$$

The energy integral would still read

$$\frac{1}{2}[A'(x)]^2 + \varphi(A(x)) = K,$$

and the boundary conditions (now take $\alpha = 0$) become

$$\varphi(A(0)) = K - \frac{1}{2}\beta^2, \quad \varphi(A(L)) = K.$$

The resulting profile is again given by

$$x = \int_{A(0)}^{A} \frac{|dA|}{[2(K - \varphi(A))]^{1/2}},$$

but we should separate out the binding contribution $A + B \xrightarrow{\gamma} C$. This is given by $\dot{A}_b = f_b = -\gamma BA$, or $\varphi_b(A) = -(\gamma B/2)A^2$, so

$$\varphi(A) = \Phi(A) - \frac{\gamma B}{2} A^2,$$

and $L = x_m - x_n$ determines K, as usual.

The dominant effect is of course the binding, and if $\Phi(A) = 0$, we would simply have (with D unscaled)

$$\dot{A} = DA'' - \gamma BA, \quad A'(0) = -\beta, \quad A'(L) \sim 0,$$

with the steady state exponential profile

$$A(x) = \beta \sqrt{\frac{D}{\gamma B}} \frac{\cosh\left[\sqrt{\gamma B/D}(x - L)\right]}{\sinh\left(\sqrt{\gamma B/D L}\right)}.$$

Suppose then that the bicoid gradient has been established. How would that be converted into location-dependent ("positional information" by Wolpert) activation of several genes? Such a concentration—sensitive activation—is not hard to imagine, in principle. We want a multilevel switch—but first a two-level switch in which an activator a turns on a target gene when a exceeds some threshold, and the gene stays on when a is removed. This is based on a derepressor g that is produced autocatalytically but is also degraded:

$$\dot{g} = \frac{g^2}{1 + g^2} - \frac{g}{2K}, \qquad K \geq 0.$$

Hence, $\dot{g} = 0$ at $\bar{g} = 0, K \pm \sqrt{K^2 - 1}$, of which only 0 and $K + \sqrt{K^2 - 1}$ are stable, since

$$\delta\dot{g} = \left[\frac{2\bar{g}}{(1 + \bar{g}^2)^2} - \frac{1}{2K}\right]\delta g$$

with $\lambda < 0$, $\lambda > 0$, and $\lambda < 0$ for $\bar{g} = 0$, $K - \sqrt{K^2 - 1}$, and $K + \sqrt{K^2 - 1}$, respectively. But now let a be an activator concentration acting as a source:

$$\dot{g} = \gamma a + \frac{g^2}{1 + g^2} - \frac{g}{2K}.$$

Then if

$$a > \tilde{a} = \frac{|\dot{g}_{\min}|}{\gamma},$$

there is only one stationary state of g, and it is stable. It is close to the $a = 0$ stable state, and so if a is switched off, the system remains activated.

Now for a related multilevel switch, in which the states correspond to activation of one gene with repression of the others, as determined by a common repressor, we start with

$$\dot{g}_i = \frac{c_i g_i^2}{r} - \gamma g_i, \quad \dot{r} = \sum_i c_i g_i^2 - \beta r,$$

with $\beta > \gamma$, so that the g_i are autocatalytically produced, with a common repressor r. Equilibrium requires either $g_i = 0$ or $g_i = \gamma r / c_i$ for each i. By examining the matrix

$$\mathbf{J} = \begin{pmatrix} \frac{\partial \dot{g}_i}{\partial g_j} & \frac{\partial \dot{r}}{\partial g_j} \\ \frac{\partial \dot{g}_i}{\partial r} & \frac{\partial \dot{r}}{\partial r} \end{pmatrix} = \begin{pmatrix} \left(\frac{2c_i g_i}{r} - \gamma\right)\delta_{ij} & 2c_j g_j \\ -\frac{c_i g_i^2}{r^2} & -\beta \end{pmatrix},$$

we see that the ith stable state satisfies $g_j = 0$ for $j \neq i$, $g_i = \frac{\gamma r}{c_i} = \frac{\beta}{\gamma}$, and $r = \frac{\beta c_i}{\gamma^2}$.

Next, introduce the activator a and the corresponding replacement

$$c_i \mapsto \bar{c}_i = c_i\left(1 + \frac{a}{r}\right).$$

It can be shown that as a rises to its final value, successive g_i reach their maxima and then shut off, leaving just one activated g_i at each value of a. This is perhaps the simplest mechanism for a genetic interpretation of spatially structured activator.

Once the gap genes are producing their respective proteins, how do they set up a concentration pattern to lead to the next stage? This must be a matter of at least two-species reaction-diffusion, and so we will continue along the lines of the pair

$$\dot{a} = f(a,h) + D\nabla^2 a, \quad \dot{h} = g(a,h) + D'\nabla^2 h,$$

and when necessary we will explicitly assume Neumann boundary conditions

$$\hat{n} \cdot \nabla a = \hat{n} \cdot \nabla h = 0.$$

The basic activator concentration is produced by having the activator reproduce by autocatalysis, controlled by an inhibitor (or the equivalent) produced by cross catalysis. It is localized by giving the activator a small diffusion constant, while the inhibitor diffuses rapidly, causing lateral inhibition.

Stability analysis then proceeds, as it has on several previous occasions, by introducing the static uniform solution $\bar{a}$, $\bar{h}$ satisfying

$$f(\bar{a}, \bar{h}) = g(\bar{a}, \bar{h}) = 0$$

and expanding a and h around $\bar{a}$ and $\bar{h}$: $a = \bar{a} + \delta a$ and $h = \bar{h} + \delta h$, so that

$$\delta\dot{a} = f_a \delta a + f_h \delta h + D\nabla^2 \delta a, \quad \delta\dot{h} = g_a \delta a + g_h \delta h + D'\nabla^2 \delta h,$$

where the coefficients are evaluated at $\bar{a}$, $\bar{h}$. The solution space is spanned by

$$\delta a = A_k \psi_k(\mathbf{r}) e^{\lambda t}, \quad \delta h = H_k \psi_k(\mathbf{r}) e^{\lambda t},$$

where $\nabla^2 \psi_k(\mathbf{r}) = -k^2 \psi_k(\mathbf{r})$ with boundary conditions enforced yields the set of possible k's. (For a large region, k will be continuous, and we can ignore boundary conditions.)

In any event, we now have, suppressing index k,

$$\left(f_a - Dk^2\right)A + f_h H = \lambda A, \quad g_a A + \left(g_h - D'k^2\right)H = \lambda H,$$

and solvability requires the eigenvalues λ to satisfy

$$\lambda^2 - \left[f_a + g_h - (D + D')k^2\right]\lambda + \left[f_a g_h - f_h g_a - \left(D g_h + D' f_a\right)k^2 + DD'k^4\right] = 0.$$

In the Turing scenario, the nondiffusing state $k = 0$ is stable, so that

$$\lambda^2 - (f_a + g_h)\lambda + (f_a g_h - f_h g_a)$$

has only roots of negative real part:

$$f_a + g_h < 0, \quad f_a g_h - f_h g_a > 0.$$

On the other hand, for large k, with $\lambda = \Lambda k^2$,

$$\Lambda^2 + (D + D')\Lambda + DD' = 0,$$

so $\Lambda = -D$ or $-D'$ is also stable. Since the linear terms in λ, i.e., $-(f_a + g_h) + (D + D')k^2$, remain positive for all k, instability occurs if and only if the constant term

$$DD'k^4 - \left(D g_h + D' f_a\right)k^2 + f_a g_h - f_h g_a$$

passes through zero for some k^2. But $f_a g_h - f_h g_a > 0$, so this quadratic in $k^2 = x$ having a positive root is equivalent ($Ax^2 - Bx + C = 0$ gives $x > 0$ if $B > \sqrt{4AC}$) to

$$\left(\frac{D'}{D}\right)^{\frac{1}{2}} f_a + \left(\frac{D}{D'}\right)^{\frac{1}{2}} g_h > 2(f_a g_h - f_h g_a)^{\frac{1}{2}}.$$

In one-dimensional space, it is, in a way, too easy to create models that yield stably a given monotonic activity profile. Suppose that $\tilde{a}(x)$ is the desired pattern, and we are willing to produce $\tilde{h}(x) \equiv \tilde{a}(x)$ ($\tilde{h}(x) = \alpha\tilde{a}(x) + \beta$ is no harder). Define the single-valued function $F(\tilde{a})$ to satisfy

$$F(\tilde{a}(x)) = \tilde{a}''(x)$$

and construct the model

$$\dot{a} = -F(a) + \sigma(a - h) + a'', \quad \dot{h} = -\nu F(h) + \nu\sigma(a - h) + \nu h'',$$

for $\nu > 1$ and large enough σ. Checking stability requires analysis of the matrix

$$\mathbf{J} = \begin{pmatrix} -F'(\tilde{a}) + \frac{d^2}{dx^2} + \sigma & -\sigma \\ \nu\sigma & -\nu F'(\tilde{a}) + \nu\frac{d^2}{dx^2} - \nu\sigma \end{pmatrix}.$$

Since the eigenvalues of

$$\left[-F'(\tilde{a}(x)) + \frac{d^2}{dx^2}\right]\psi_l(x) = l\psi_l(x)$$

with no-flux boundary conditions are bounded from above (under mild conditions on F), and the $\psi_l(x)$ span function space, we only need to look at the eigenvalues of

$$\begin{pmatrix} l+\sigma & -\sigma \\ \nu\sigma & \nu l - \nu\sigma \end{pmatrix}, \quad \lambda^2 + [(\nu-1)\sigma - (\nu+1)l]\lambda + \nu l^2 = 0.$$

For $l \neq 0$ ($l = 0$ is a special case, disappearing under a shift of container size), λ necessarily has a negative real part if $(\nu-1)\sigma - (\nu+1)l > 0$, i.e., if $\nu > 1$ and $\sigma > [(\nu+1)/(\nu-1)]l_{\max}$, as desired. One consequence is that something like a bicoid pattern is trivially arranged without any external source—but it doesn't regulate properly (nor do real patterns have to, as in limb bud expansion).

The Turing mechanism, in which a system that would be stable in a uniform state is destabilized by diffusion, is behind most of the current reaction-diffusion models. It restricts the possibilities for models, which is good, and it allows for a cleaner description of the dynamics, which is better. The reason for the ubiquity of this mechanism is that it does *not* arise from the transition $0 \to D$. At $D = 0$, reactions would proceed independently at all points of space. Rather, the idealized initial state has complete spatial correlation, effectively $D = \infty$, and then as more random interactions appear in the environment, D becomes finite. The uniform system instability requirement can, however, allow for more than one mode when diffusivity becomes finite, which has an unknown amplitude, and is of course invariant under the transformation $(\delta a, \delta h) \mapsto (-\delta a, -\delta h)$. It is the source prepattern, together with initial fluctuations, that play supporting roles in deciding which mode is chosen to propagate and at what strength.

An orthogonal viewpoint is that, while a graded source prepattern can indeed switch on or off a localized biochemical reaction, the reactants will diffuse, so that the pattern results from internal transport as well. We can carry out a relatively model-free analysis of this situation. Let us look at a three-dimensional system with a weak graded activation source $\mu(\mathbf{r})$, but which is isolated in the sense of having Neumann reflective boundary conditions:

$$\dot{a} = f(a,h) + \mu(\mathbf{r}) + D\nabla^2 a, \quad \dot{h} = g(a,h) + D'\nabla^2 h,$$

in Ω with

$$\hat{n} \cdot \nabla a = \hat{n} \cdot \nabla h = 0$$

on $\partial\Omega$. We imagine a reference system $(\bar{a}, \bar{h})$, uniform and sourcefree, deviations from which will then satisfy

$$\begin{pmatrix} \delta\dot{a} \\ \delta\dot{h} \end{pmatrix} = \begin{pmatrix} f_a + D\nabla^2 & f_h \\ g_a & g_h + D'\nabla^2 \end{pmatrix} \begin{pmatrix} \delta a \\ \delta h \end{pmatrix} + \mu(\mathbf{r}) \begin{pmatrix} 1 \\ 0 \end{pmatrix}$$

with

$$\hat{n} \cdot \begin{pmatrix} \delta a \\ \delta h \end{pmatrix} = \mathbf{0}$$

on $\partial\Omega$ and $(\delta a(\mathbf{r},0), \delta h(\mathbf{r},0)) = \mathbf{0}$. Now introduce the orthonormal "free space" modes

$$\nabla^2 \psi_k(\mathbf{r}) + k^2 \psi_k(\mathbf{r}) = 0$$

in Ω with

$$\hat{n} \cdot \nabla \psi_k(\mathbf{r}) = 0$$

on $\partial\Omega$, and expand $\delta a(\mathbf{r},t) = \sum \delta a_k(t)\psi_k(\mathbf{r})$, and similarly for δh and μ. Taking the coefficient of $\psi_k(\mathbf{r})$, it follows that

$$\frac{\partial}{\partial t}\begin{pmatrix} \delta a_k \\ \delta h_k \end{pmatrix} = \begin{pmatrix} f_a - Dk^2 & f_h \\ g_a & g_h - D'k^2 \end{pmatrix}\begin{pmatrix} \delta a_k \\ \delta h_k \end{pmatrix} + \mu_k \begin{pmatrix} 1 \\ 0 \end{pmatrix},$$

which we write as

$$\frac{\partial}{\partial t}\delta c_k = M_k \delta c_k + \mu_k \begin{pmatrix} 1 \\ 0 \end{pmatrix}$$

and solve formally as

$$\delta c_k(t) = e^{tM_k}\delta c_k(0) + (e^{tM_k} - 1)M_k^{-1}\mu_k \begin{pmatrix} 1 \\ 0 \end{pmatrix}.$$

But if M_k, a 2×2 matrix, has the eigenvalues λ_k^+ and λ_k^-, where $\mathrm{Re}_e(\lambda_k^+) \geq \mathrm{Re}_e(\lambda_k^-)$, then for any function $p(M)$, with index k understood,

$$p(M) = \frac{M - \lambda^+}{\lambda^- - \lambda^+}p(\lambda^-) + \frac{M - \lambda^-}{\lambda^+ - \lambda^-}p(\lambda^+)$$

and so

$$\begin{aligned} \delta c_k(t) = &\left[\left(\frac{M_k - \lambda_k^+}{\lambda_k^- - \lambda_k^+}\right)e^{t\lambda_k^-} + \left(\frac{M_k - \lambda_k^-}{\lambda_k^+ - \lambda_k^-}\right)e^{t\lambda_k^+}\right]\delta c_k(0) \\ &+ \left[\left(\frac{M_k - \lambda_k^+}{\lambda_k^- - \lambda_k^+}\right)\frac{e^{t\lambda_k^-} - 1}{\lambda_k^-} + \left(\frac{M_k - \lambda_k^-}{\lambda_k^+ - \lambda_k^-}\right)\frac{e^{t\lambda_k^+} - 1}{\lambda_k^+}\right]\begin{pmatrix} \mu_k \\ 0 \end{pmatrix}. \end{aligned}$$

We have previously analyzed the eigenvalues of M_k and found the instability region for k. It appears that the λ of largest real part will grow fastest, so that, approximately for this k,

$$\delta c_k(t) \sim \left(\frac{M_k - \lambda_k^-}{\lambda_k^+ - \lambda_k^-}\right)\left[\delta c_k(0) + \frac{\mu_k}{\lambda_k^+}\begin{pmatrix} 1 \\ 0 \end{pmatrix}\right]\psi_k(\mathbf{r})e^{t\lambda_k^+}.$$

However, it is the combination $\delta a_k(0) + \mu_k/\lambda_k^+$ that serves as an effective amplitude, so that if $\mathrm{Re}(\lambda^+)$ is nearly degenerate, it is the k of largest occupancy in $\mu(\mathbf{r})$ that will at first be preferentially chosen to grow, even if $\mathrm{Re}_e(\lambda_k^+) > \mathrm{Re}_e(\lambda_k^-)$.

10.3. Further Aspects of Reaction-Diffusion

10.3.1. Intermediate Dynamics. We have seen how a pattern can emerge, and how it appears in steady state. But, as we have previously noted, biologically developing systems usually do not have the luxury of reaching even temporary steady state. What then is their dynamical path past very short time? Let us look again at the two-species situation in three dimensions,

$$\dot{a} = f(a,h) + D\nabla^2 a, \quad \dot{h} = g(a,h) + D'\nabla^2 h,$$

in Ω with $\hat{n}\cdot\nabla a = \hat{n}\cdot\nabla h = 0$ on $\partial\Omega$, and see how the dynamics proceeds from an unstable uniform initial state that satisfies

$$f(\bar{a},\bar{h}) = g(\bar{a},\bar{h}) = 0.$$

For the initial dynamics, we proceed as usual: set $a = \bar{a} + \delta a$ and $h = \bar{h} + \delta h$, so that, to leading order,

$$\frac{\partial}{\partial t}\begin{pmatrix}\delta a\\ \delta h\end{pmatrix} = \begin{pmatrix} f_a + D\nabla^2 & f_h \\ g_a & g_h + D'\nabla^2\end{pmatrix}\begin{pmatrix}\delta a\\ \delta h\end{pmatrix}$$

in Ω with $\hat{n}\cdot\nabla\delta a = \hat{n}\cdot\nabla\delta h = 0$ on $\partial\Omega$, and expand the space of suitably bounded functions in terms of the set

$$\begin{pmatrix}\delta a\\ \delta h\end{pmatrix} = \begin{pmatrix}\tilde{a}\\ \tilde{h}\end{pmatrix} e^{\lambda t}\psi_k(\mathbf{r}),$$

where

$$\nabla^2\psi_k(\mathbf{r}) + k^2\psi_k(\mathbf{r}) = 0$$

in Ω with $\hat{n}\cdot\nabla\psi_k(\mathbf{r}) = 0$ on $\partial\Omega$, therefore requiring that

$$\begin{pmatrix} f_a - Dk^2 & f_h \\ g_a & g_h - D'k^2\end{pmatrix}\begin{pmatrix}\tilde{a}\\ \tilde{h}\end{pmatrix} = \lambda\begin{pmatrix}\tilde{a}\\ \tilde{h}\end{pmatrix}.$$

For finite (closed, bounded) Ω, the set of k's is discrete, and the dynamics is dominated by the mode of maximum $\mathrm{Re}_e(\lambda)$ or by that biased by a prepattern. The corresponding mode—call it $\psi_0(\mathbf{r})$—may continue to dominate the form of the pattern, but we want to know its amplitude in time. Let us then define

$$q(\mathbf{r},t) = \begin{pmatrix} a(\mathbf{r},t) - \bar{a}\\ h(\mathbf{r},t) - \bar{h}\end{pmatrix}, \qquad M = \begin{pmatrix} f_{\bar{a}} & f_{\bar{h}}\\ g_{\bar{a}} & g_{\bar{h}}\end{pmatrix},$$

$$\mathbf{D} = \begin{pmatrix} D & 0\\ 0 & D'\end{pmatrix}, \qquad \Delta(q) = \begin{pmatrix} f(a,h) - (a-\bar{a})f_{\bar{a}} - \left(h-\bar{h}\right)f_{\bar{h}}\\ g(a,h) - (a-\bar{a})g_{\bar{a}} - \left(h-\bar{h}\right)g_{\bar{h}}\end{pmatrix},$$

so that

$$\dot{q} = Mq + \Delta(q) + \mathbf{D}\nabla^2 q$$

in Ω with $\hat{n}\cdot\nabla q = 0$ on $\partial\Omega$, where $\Delta(q)$ starts at second order in q. For the initial unstable mode, we of course have

$$(M + \mathbf{D}\nabla^2)\psi_0 = \lambda_0\psi_0.$$

Next, write

$$q(\mathbf{r},t) = q_0(t)\psi_0(\mathbf{r}) + Q(\mathbf{r},t),$$

where Q is orthogonal to ψ_0. What this means is that, on introducing the dual eigenfunctions of $M + \mathbf{D}\nabla^2$,

$$(M^+ + \mathbf{D}\nabla^2)\varphi_k = \lambda_k^* \varphi_k,$$

where M^+ is the adjoint, the conjugate transpose of M, we have

$$\int \varphi_k^*(\mathbf{r})\psi_{k'}(\mathbf{r})d\mathbf{r} = \delta_{kk'},$$

allowing us to define the projection

$$P_k f(\mathbf{r}) = \psi_k(\mathbf{r}) \int \varphi_k^*(\mathbf{r}) f(\mathbf{r}) d\mathbf{r}.$$

We then set

$$q_0(t) = \int \varphi_0^*(\mathbf{r}) q(\mathbf{r},t) d\mathbf{r}, \quad Q(\mathbf{r},t) = (I - P_0) q(\mathbf{r},t),$$

and, applying P_0 to the dynamics and removing the common $\psi_0(\mathbf{r})$, there results

$$\dot{q}_0(t) = \lambda_0 q_0(t) + \int \varphi_0^*(\mathbf{r}) \Delta(q_0(t)\psi_0(\mathbf{r}) + Q(\mathbf{r},t)) d\mathbf{r}.$$

But $Q(\mathbf{r},t)$ only contains $\psi_k(\mathbf{r})$ for $k \neq 0$, and so to lowest order,

$$\dot{q}_0(t) = \lambda_0 q_0(t) + \int \varphi_0^*(\mathbf{r}) \Delta(q_0 \psi_0(\mathbf{r})) d\mathbf{r}$$

(recognized as "equivalent linearization" applied to spatial variation).

An example is the basic Gierer-Meinhardt depletion model:

$$C \longrightarrow A, \quad D \longrightarrow H, \quad 2A + H \longrightarrow 3A, \quad A \longrightarrow E,$$

so that in one dimension, with suitable scaling on the interval $(0, L)$,

$$\dot{a} = c + a^2 h - \gamma a + D a'', \quad \dot{h} = c' - a^2 h + D' h'',$$

plus Neumann boundary conditions. Here the production of A is catalyzed by H, which is depleted in the process, and if $D' \gg D$, A production will be confined to a small region. Here,

$$\bar{a} = \frac{c + c'}{\gamma}, \quad \bar{h} = \frac{c'\gamma^2}{(c' + c)^2},$$

from which

$$M = \begin{pmatrix} 2\bar{a}\bar{h} - \gamma & \bar{a}^2 \\ -2\bar{a}\bar{h} & -\bar{a}^2 \end{pmatrix} = \begin{pmatrix} \gamma \frac{c'-c}{c'+c} & \frac{(c'+c)^2}{\gamma^2} \\ -2\gamma \frac{c'}{c'+c} & -\frac{(c'+c)^2}{\gamma^2} \end{pmatrix},$$

$$\Delta = \left(\bar{h}(\delta a)^2 + 2a(\delta a)(\delta h) + (\delta a)^2(\delta h)\right)\begin{pmatrix} 1 \\ -1 \end{pmatrix}.$$

The space dependence of the dominant eigenfunction is $\cos kx$, $k = \frac{\pi K}{L}$ for integral K, and so with $\cos kx$ understood, we have

$$\begin{pmatrix} \gamma\frac{c'-c}{c'+c} - Dk^2 & \frac{(c'+c)^2}{\gamma^2} \\ -2\gamma\frac{c'}{c'+c} & -\frac{(c'+c)^2}{\gamma^2} - D'k^2 \end{pmatrix} \psi_0 = \lambda_0 \psi_0.$$

The eigenvalue λ_0 is a bit messy, but having found it, we can clearly choose

$$\psi_0 = \begin{pmatrix} \frac{(c'+c)^2}{\gamma^2} \\ \lambda_0 + Dk^2 - \gamma\frac{c'-c}{c'+c} \end{pmatrix} \cos kx,$$

as well as

$$\varphi_0 = \begin{pmatrix} -2\gamma\frac{c'}{c'+c} \\ \lambda_0 + Dk^2 - \gamma\frac{c'-c}{c'+c} \end{pmatrix} \frac{2}{NL} \cos kx,$$

where

$$N = \left[\left(\lambda_0 + Dk^2 - \gamma\frac{c'-c}{c'+c}\right)^2 + \left(2\gamma\frac{c'}{c'+c}\right)\right]^{\frac{1}{2}}$$

for normalization. Now $\Delta(q_0\psi_0)$ has only $\cos^2 kx$ and $\cos^3 kx$, with no $\cos kx$ component, and

$$\cos^3 kx = \frac{1}{4}\cos 3kx + \frac{3}{4}\cos kx.$$

Hence, after some algebra,

$$\int \varphi_0^*(\mathbf{r})\Delta(q_0\psi_0)d\mathbf{r} = -\frac{3}{4}\frac{(c'+c)^4}{\gamma^4}\left(\lambda_0 + Dk^2 - \gamma\frac{c'-c}{c'+c}\right)(\lambda_0 + Dk^2 + \gamma)\frac{q_0^3}{N},$$

and the dominant mode dynamics takes the simple form

$$\dot{q}_0 = \lambda_0 q_0 - Bq_0^3.$$

10.3.2. Static Size Regulation. We have seen that there are two generic "potentials" in stationary, one-component, boundary-reflecting reaction-diffusion: (a) smooth, showing absolute size regulation in the vicinity of the minimum, and (b) a slope discontinuity, showing relative size regulation as long as the curvature remains small. Biochemical examples of (a) are typical, but what about (b)? Here, in fact, various limiting two-component systems will suffice.

In almost all models, the activator diffuses slowly. If we take the limit in which $D = 0$, the pair of stationary state equations becomes

$$0 = f(a,h) = g(a,h) + D'h''.$$

In solving $f(a, h) = 0$ for a as a function of h, there can be more than one solution: $a = A_i(h)$. Diffusionless stability would require $f_a < 0$ and $f_h < 0$, so

$$\left.\frac{da}{dh}\right|_{f=0} = \frac{f_h}{f_a} > 0,$$

following an intermediate branch A_1 in ah-space. Whatever branch is followed, the problem reduces to

$$g(A(h), h) + D'h'' = 0$$

on that branch, and the boundary conditions $\hat{n} \cdot \nabla h = 0$ imply $\hat{n} \cdot \nabla a = 0$ as well. Then if the corresponding g_1 (with diffusion) has become unstable, h must jump from g_0 to g_2 at some switch point h_s. Since $A(h)$ jumps from A_0 to A_2, the energy integral [with $\varphi'(h) = g(A(h), h)$] has exactly the same form,

$$\frac{D'}{2}(h')^2 + \varphi(h) = K,$$

needed for relative size regulation.

But what is h_s? The reason that the switch can occur is that D becomes important in the vicinity of h_s, where a is changing very fast. We recall that to focus on this region, we stretch x-space:

$$x = x_s + D^{\frac{1}{2}} X,$$

so that

$$f(a, h_s) + \frac{d^2 a}{dX^2} = 0,$$

whose solution is supposed to yield A_0 and A_2 at large $\pm x$: $a(-\infty) = A_0(h_s)$ and $a(\infty) = A_2(h_s)$. We can solve as usual, setting

$$U(a) = \int_{A_1(h_s)}^{a} f(a, h_s) da$$

so that

$$U(a) + \frac{1}{2}\left(\frac{da}{dX}\right)^2 = U_0.$$

The endpoints $x = \pm\infty$ are at A_0 and A_2, and so we require

$$U(A_0(h_s)) = U(A_2(h_s)),$$

which determines h_s. Geometrically, h_s is to be adjusted so that the area under $f(a, h_s)$ between A_0 and A_2 vanishes.

An example is the basic Gierer-Meinhardt activator-inhibitor model. First, a preliminary: biochemical systems typically have intermediate steps. The classical Michaelis-Menten (Henri) enzymatic step is

$$S + E \underset{k'}{\overset{k}{\rightleftharpoons}} C \xrightarrow{l} E + P,$$

converting substrate S to product P via enzyme E and its complex C. The time scale is imagined to result from $k, k' \to \infty$ at fixed $K_M = k'/k$. Consequently, $S + E \rightleftharpoons C$ is in equilibrium at each time, resulting in $K_M C = ES$. But if $C(0) = 0$ initially and E_0 is the initial enzyme concentration, we have $C + E = E_0$. Thus, $C = E_0 S/(K_M + S)$, from which

$$\dot{P} = lC = lE_0 \frac{S}{K_M + S}.$$

Based on this picture, the GM model takes the form

$$\dot{a} = p_a \frac{a^2}{\left(1 + k_a a^2\right)h} - \gamma_a a + \mu_a + D_a \nabla^2 a, \quad \dot{h} = p_h a^2 - \gamma_h h + \mu_h + D_h \nabla^2 h,$$

which is certainly autocatalytic in a and inhibitory in h. Suitably scaled, and in the absence of sources μ_a, μ_h, we then have

$$f(a,h) = \frac{a^2}{h(1 + a^2)} - \frac{1}{2}a, \quad g(a,h) = a^2 - \lambda h,$$

and in the $D_a \to 0$ limit, $A^2 = hA(1 + A^2)/2$ has three solutions:

$$A_0 = 0, \quad A_1(h) = \frac{1}{h} - \sqrt{\frac{1}{h^2} - 1}, \quad A_2(h) = \frac{1}{h} + \sqrt{\frac{1}{h^2} - 1},$$

precisely of the form previously considered. If $a_s = A_2(h_s)$, then since $A_0(h_s) = 0$, the switch point is determined by

$$0 = \int_0^{a_s} \left[\frac{a^2}{h_s\left(1 + a^2\right)} - \frac{1}{2}a \right] da = \frac{a_s}{h_s} - \frac{a_s^2}{4} - \frac{1}{h_s} \tan^{-1} a_s,$$

but $h_s = 2\frac{a_s}{1+a_s^2}$, so

$$\tan^{-1} a_s = a_s \left(\frac{1 + a_s^2/2}{1 + a_s^2} \right),$$

yielding $a_s \approx 1.525$ and $h_s \approx 0.9171$.

10.4. Pattern Under Growth: Chick Limb Bud

When size change is not a matter of comparison of organisms, but rather of the development process in a single organism, it becomes instead a crucial pattern-changing mechanism. To illustrate this, we turn to a later stage of embryonic development.

Let us start with a situation in which the "size" is effectively reduced in time. This is the avian limb bud, a paddle-shaped and so effectively two-dimensional growing population that we model as a growing rectangle. In the simplest semi-realistic model, only one morphogen A is involved, fibronectin, which (as a prepattern) produces adhesion and hence clumping of cells. It appears to not cross the lateral boundary, but to be homeostatically (fixed current) absorbed at the body-bud interface and produced at the AER (apical ectodermal ridge); thus the boundary conditions are

$$\left.\frac{\partial A}{\partial y}\right|_{y=0,b} = 0, \quad \left.\frac{\partial A}{\partial x}\right|_{x=0} = M_0, \quad \left.\frac{\partial A}{\partial x}\right|_{x=a} = M_1.$$

If the relevant "potential" $\varphi(A)$ is expanded about its minimum, the resulting dynamics of course becomes

$$\dot{A} = r(A - A_0) + D\nabla^2 A,$$

with the separated solution at $\dot{A} = 0$:

$$A(x, y) = A_0 + \frac{aM_0}{\lambda}\cos\left(\frac{n\pi y}{b}\right)e^{\lambda x/a}, \quad \lambda = \log\frac{M_1}{M_0},$$

where we need

$$D\left[\frac{\lambda^2}{a^2} - \left(\frac{n\pi}{b}\right)^2\right] + r = 0,$$

so

$$\frac{n\pi}{b} = \left(\frac{r}{D} + \frac{\lambda^2}{a^2}\right)^{\frac{1}{2}},$$

where n transverse half-cycles fit (or nearly fit when realistic reactions are used). Now the a that is relevant is that from the full body, which has been increasing in size longitudinally as cells are activated, and one sees experimentally a *decreasing* chamber size a, and consequently an increase in the number of condensations, leading, e.g., to digits.

Such a model depends crucially on the identification of reactants and reactions. Another possibility for which there is no evidence would be a growing but isolated chamber—Neumann throughout—so that $\partial A/\partial x = 0$ at $x = 0$ and $x = a$, yielding

$$A(x, y) = A_0 + C\cos\left(\frac{n\pi y}{b}\right)\cos\left(\frac{\pi x}{a}\right),$$

where

$$D\left[-\frac{\pi^2}{a^2} - \left(\frac{n\pi}{b}\right)^2\right] + r = 0,$$

or

$$\frac{n\pi}{b} = \left(\frac{r}{D} - \frac{\pi^2}{a^2}\right)^{\frac{1}{2}},$$

which would even produce a maximum number of digits as a grows, if precondensation is "sealed in."

There is also evidence for an inhibitor (hyaluronate) as a comorphogen. At its most primitive level, this could change the effective diffusion constant, e.g., if diffusion involves passing through channels that can be partially blocked. Suppose that "next grid cell" transmission goes via

$$A^- + C \underset{k'}{\overset{k}{\rightleftharpoons}} B \overset{l}{\rightarrow} C + A^+$$

but is accompanied by inhibitor interference

$$I + C \underset{k'_I}{\overset{k_I}{\rightleftharpoons}} B_I,$$

so that

$$\dot{A}^+ = lB, \quad \dot{B} = kA^-C - (k' + l)B, \quad \dot{B}_I = k_I IC - k'_I B_I,$$

where inhibitor concentration I is fixed. For fast k's, $\dot{B} = \dot{B}_I = 0$, so

$$k'_I B_I = k_I IC, \quad kA^-C = (k' + l)B,$$

and of course by conservation of C,

$$C_0 = B_I + B + C,$$

so on solving,

$$\dot{A}^+ = C_0 l \frac{A^-}{K(1 + K_I I) + A^-},$$

where $K = (k' + l)/k$, $K_I = k_I/k'_I$. In other words,

$$\dot{A}^+ = \gamma A^-, \quad \gamma = \frac{C_0 l}{K(1 + K_I I) + A^-}.$$

Hence $D = \gamma(\Delta x)^2$ decreases as I increases, and so, no matter what the model, $n = (\alpha/D + \beta)^{1/2}$ must increase.

10.5. Insect Imaginal Disc

At this stage, we can also make contact with the material of Chapter 9. The point is that the early but not very early insect embryo, after initial segmentation, blastulation, etc., has created a number of subdomains that will produce via evagination the various major organs. Each such *imaginal disc* expands with the embryo, and as it does so, compartmentalizes into increasingly specific suborgan regions or clones. The nodal lines $a = 0$ for morphogen a denote the boundaries between clones, depending upon the sign of excess a from uniformity, and are binary signals, such as anterior/posterior, dorsal/ventral, etc. If nothing changes but the boundary, we have only to look at the successive patterns, each nodal line remaining after being switched on. The sequence must go according to increased global ramification of what will fit.

For a really toy model, imagine then a single morphogen, and denote its departure from uniform equilibrium by a. Then if not far from a potential minimum, we will have

$$\nabla^2 a + k^2 a = 0$$

with no-flux boundary conditions. Suppose that the imaginal disc is just a circular disc of radius R, so that in polar coordinates,

$$\left(\frac{\partial^2}{\partial r^2}+\frac{1}{r}\frac{\partial}{\partial r}+\frac{1}{r^2}\frac{\partial^2}{\partial \phi^2}\right)a+k^2a=0$$

with $\partial a/\partial r = 0$ at $r = R$. We want to let the radius R expand gradually. For the solution, we proceed as usual with the ansatz $a(r,\phi) = a(r)b(\phi)$, so that

$$\frac{r^2}{a(r)}\left[a''(r)+\frac{1}{r}a'(r)\right]+\frac{b''(\phi)}{b(\phi)}+k^2r^2=0, \quad a'(R)=0.$$

The term $b''(\phi)/b(\phi)$ depends upon neither r nor ϕ, and so must be a constant. Since ϕ is an angle, we must have

$$b(\phi)=\begin{cases}\cos m\phi,\\ \sin m\phi,\end{cases}$$

and then

$$a''(r)+\frac{1}{r}a'(r)-\frac{m^2}{r^2}a(r)+k^2a(r)=0, \quad a'(R)=0,$$

with the solution (regular at the origin)

$$a(r)=J_m(kr), \quad kR=\alpha_m^{1/2},$$

where $\alpha_0 \approx 3.4$, $\alpha_1 \approx 9.6$, $\alpha_2 \approx 14.4$, $\alpha_3 \approx 17.6, \ldots$. The sequence is scaled to a unit circle. As R expands, the compartmental boundaries, the nodal lines that have emerged, become fixed, and the new ones just add by superposition.

To be a little more realistic, let us at least consider an elliptical boundary—eccentricity $e = (1-b^2/a^2)^{1/2} = 0.8$. This in fact is analytically solvable. But more to the point, the observed boundary of the imaginal disc as it grows can be fed in and the Helmholtz equation solved numerically to find the predicted compartmental sequence. The results are remarkably accurate.

Note that the boundary—with Neumann or Dirichlet conditions—can always be converted to a circle or an annulus with the modified Helmholtz equation, making approximate solutions relatively easy to obtain. For example, if one wants

$$\nabla_Z^2 a(X,Y)+K^2a(X,Y)=0$$

on the ellipse

$$\frac{X^2}{1+t^2}+\frac{Y^2}{(1-t)^2}=1,$$

then the conformal map

$$Z=X+\imath Y=\left(z+\frac{1}{z}\right)t^{\frac{1}{2}}$$

converts the boundary to a circular annulus $1 \geq |z| \geq t$. Since

$$\nabla_z^2=\frac{\partial^2}{\partial Z\,\partial Z^*}=\frac{1}{\partial Z/\partial z}\frac{\partial}{\partial z}\frac{1}{\partial Z^*/\partial z^*}\frac{\partial}{\partial z^*}=\frac{1}{|\partial Z/\partial z|^2}\nabla_{z}^2,$$

we then have

$$\nabla_z^2 a(x, y) + K^2 t(1 - z^{-2})^2 a(x, y) = 0,$$

with the stationary principle

$$K^2 = \frac{\iint |\nabla_z a|^2 \, dx \, dy}{t \iint \dfrac{a^2}{|1 - z^{-2}|^2} \, dx \, dy},$$

convenient for approximation.

Supplementary Notes

Chapter 2

After the publication of Thom's influential book [**31**], there was considerable interest in developing his and others' ideas (see, e.g., [**25, 78**]), and there followed a period of intense interest in applications of the theory to a diverse set of problems in the social and biological sciences. In some of this work the models were poorly conceived, the formulations highly speculative, and the conclusions unwarranted; see, e.g., the survey [**57**]. As a result, what is fundamentally a basic mathematical technique for classifying robust geometrical structure came to be viewed in some circles as irrelevant and overblown. This occurred despite the numerous applications, for example to the classification of the asymptotic structure of oscillatory integrals [**60**], where well established formulations exist. Although the applications to development discussed in this chapter are minimal, the possibility that the control parameters of these catastrophes might one day be given biological significance led us to view this material as essential to a course in mathematical models in developmental biology.

Chapter 3

Lecture notes in mathematics can be expected to have a long, if not indefinite, shelf life. The same cannot be said for accounts of mathematical models in biology. The pace of biological science has in the past decades brought forth data at an astonishing rate. And if, as has been said, an experiment is where theories go to die, one must question the relevance of theoretical speculations in developmental biology formulated over three decades ago. This is particularly true of ideas about pattern formation, which encompass many phenomena of early development where there is room for inventive modeling.

The 1970s was a decade of considerable activity in the modeling of pattern-forming processes at the macroscopic, multicellular level. This research owed much to the pioneering study of Alan Turing of the chemical basis of morphogenesis [**156**], although the idea of chemical gradients determining cell properties dates from the early twentieth century [**111**]. Turing's postulated *morphogens* intrigued applied mathematicians, as they provided a link between cell differentiation and morphogenesis generally, and the *processes* by which the morphogens could be created, spread, and evolve. These processes involve the stuff of physics, chemistry, and continuum mechanics—diffusion, kinetic equations, elastic and deformable continua—and the ordinary and partial differential equations that then arise. Much of this work did involve comparison between model and experiment, often with very good agreement; see, e.g., [**67**]. But as a result of this activity, theoretical modeling in biology began to move much closer to the realities of the laboratory. (Any mathematician wishing today to make a contribution would do well to be firmly embedded in the biology and to work closely with a biologist studying the system of interest.)

Nevertheless, it is fair to say that the material of this chapter reflects many of the ideas that have proved useful in subsequent developments. A survey of the mathematical modeling in the field in 2000 may be found in [**17**]. Among biologists, the role of morphogen gradients in cell differentiation is now accepted in many systems, where the relevant morphogen has been isolated. We quote from an excellent survey in 2011 by Rogers and Schier [**135**][1]:

> The study of morphogens has seen remarkable progress in the past 25 years. Morphogens are no longer a theoretical concept but have been identified in many developmental contexts.
>
> A plethora of regulatory interactions have been demonstrated to regulate morphogen gradient formation and interpretation. Although there are many variations on the theme, the general picture that emerges for morphogen gradient formation and interpretation is as follows. Morphogen molecules are released from a local but dynamic source, assemble with themselves and/or other molecules, and move via restricted diffusion through the extracellular milieu. Gradient shape is determined by the flux from the source, the diffusivity of the morphogen, and its clearance kinetics in target tissues.

Another excellent source of papers summarizing the field in 2010 may be found at: `http://cshperspectives.cshlp.org/`.

It would, however, be wrong to suggest that the most naive picture of cell differentiation activated by morphogen gradients has been fully substantiated by observations. For example, in the experiments of Ochoa-Espinosa et al. [**127**], the effect of the Bicoid (bcd) gradient along the anterior-posterior of the Drosophila embryo did not appear to be consistent with the Turing picture. The latter would say that a given cell type would correspond to a certain range of levels of the morphogen within the existing wild-type gradient. However, in these experiments the gradient was flattened and the bcd level adjusted to a certain target range. It was found that the target cells developed at a much smaller level of bcd that existed in the wild-type gradient. This would indicate that cell interpretation of morphogen gradients could be far more complicated than a simple "reading" of the local concentration.

Likewise, there is little biological evidence for the occurrence of really elegant economical solutions to problems of self-organization, such as that of Meinhardt

[1]We thank our colleague Stephen Small for this reference.

[**119**] for insect embryo compartmentalization; rather, a sequence of inhibitors appears to be involved, consistent with the messy trial-and-error process of evolution.

Chapters 4 and 5

In these lectures the modeling of morphogenetic movements was derived from modulation of the adhesive properties of cells. This reflected our interest at the time in the cell-sorting experiments of Holtfreter [**91, 92**], Steinberg [**142**], and others. However, cells do more than simply adhere. They may deform, break contacts, locomote, move relative to other cells, etc. The cell architecture of microtubules provides a source of internal stress, and the excitation and modulation of these stresses must somehow result in the observed properties of cell sorting.

Subsequent to the 1977–78 course some of the ideas discussed here were further explored in [**47**]. Further progress was made in the Ph.D. dissertation of Deborah Sulsky, who studied sorting of two-dimensional cells utilizing a Voronoi construction; see [**145**]. In this construction "cells" are defined by "nuclei." The cell boundaries are determined by the orthogonal bisectors of lines connecting pairs of nuclei. Thus the cell determined by a given nucleus consists of points closer to that nucleus than any other. As nuclei move, boundaries shift, contacts are broken and formed, and a semblance of cell migration results, allowing the energy landscape of the cell-sorting algorithms to be explored.

While such a scheme produces reasonable cell shapes from a configuration space of nuclei, and can provide a discrete geometry suitable for cell-sorting and engulfment, it is very far from providing a realistic representation of cell boundaries. To get an idea of the vast range of cell behaviors, the structure of the cell membrane and its role in forming and breaking attachments with substrate and cells, and other contributors to the morphogenetic movements of cells and cell aggregates, see the provocative monograph of Trinkaus [**32**].

Clearly more is needed to properly treat the cell membrane. One approach, now accessible for computational modeling, is to simply increase the dimension of the configuration space. For example, let a "cell" now be composed of M "subcells." Within the confines of a differential adhesion model, the subcells would have adhesion properties reflecting their internal architecture. Boundary subcells then represent the cell membrane. Models of this type have received considerable attention in resent years; see, e.g., [**81, 162**]. For other more recent assessments of differential-adhesion models see [**65, 151**]. It is fair to say that we remain far from having an adequate model of cell deformation and motility, both in monolayers and within cell aggregates.

Chapter 6

Following on the material presented in Section 6.2.3, the authors considered other aspects of the nonlinear behavior of solutions of the simplest of the Keller-Segel models. The results were presented in [**48, 46**]. The starting point was the conjecture of Nanjundiah concerning the possible collapse of cell aggregates to essentially delta functions [**124**]. The arguments presented by Nanjundiah were independent of the dimension of the space of aggregation, which led to an investigation of the role of dimension on the fate of aggregating cells in the Keller-Segel model with constant coefficients. It was found that the singularity formation conjectured by Nanjundiah could not happen in one dimension within this model. In two dimensions, the result of [**48**] suggested that a certain critical number of cells (or critical mass) had to be isolated in an aggregate before a delta function would form in finite time. In dimensions ≥ 3 self-similar collapsing aggregates of arbitrary size could be found. In subsequent years there has been considerable work elaborating the critical mass and collapse to a singularity in various models of chemotaxis; see, e.g., [**94, 95**].

Chapters 7 and 8

(Note: These chapters have considerable overlap, at the technical level, with previous chapters, which, while not intentional, is not a bad thing pedagogically.)

The clock and wave front mechanism [**52**] is ingenious, and its basic structural components biologically identifiable. This, however, does not mean that, consistent with frequent biological strategy, it is the only process that has been evolutionarily activated. The point is that it is not unusual to have one process that initially produces the desired phenomena, and another that then takes over and supplies detailed reproducibility. For example, as we have discussed in Section 5.3.4, while anisotropic differential adhesion of deformable cells may be sufficient to drive the invagination of gastrulation, its reliability is dwarfed by the consequences of bottle cell production.

A recent attempt to trace an underlying mechanism for somatogenesis [**58**] posits a basic differential adhesion hypothesis and makes a strong case for it being sufficient to reproduce observed data. The extent to which it is actually used by current organisms has not yet been established.

Chapters 9 and 10

The KST model represents an extremely effective extension of the Turing mechanism, at a semi-empirical level, to an intrinsically complex phenomenon. Elegant alternatives (see, e.g., [**19**]) have been proposed as well, but as the tools of molecular biology have increased dramatically in power, attention has shifted to the identification of molecular pathways that are activated in omnipresent processes such as compartmentalization (see, e.g., [**10**]). With this identification increasingly under control, statistical physicists have joined in the activity as well, attending more to the emergent properties of groups of cells in the frequent massive motions of morphogenesis [**42**].

In Chapter 10, we return to an earlier embryonic stage, when the gross spatial characteristics of the developing embryo are being established, and elaborate in greater detail on the responsibility of reaction-diffusion in the process. Particular attention is paid to the role of prepattern in biasing the developing pattern (for a more general context, see [**96**]) and to the dynamics associated with, but not literally caused by, the instability (see also [**20**]).

Bibliography

Selected Reference Texts

[1] Balinsky, B. I. *An introduction to embryology*. W. B. Saunders, Philadelphia-London, 1960.

[2] Batchelor, G. K. *Introduction to fluid dynamics*. Cambridge University Press, Cambridge, 1967.

[3] Berrill, N. J., and Karp, G. *Development*. McGraw-Hill, New York, 1976.

[4] Bonner, J. T. *Morphogenesis*. Princeton University Press, Princeton, N.J., 1952.

[5] Curtis, A. S. G. *The cell surface: its molecular role in morphogenesis*. Academic Press, New York, 1967.

[6] Ebert, J. D., and Sussex, I. M. *Interacting systems in development*. Second edition. Holt, Rinehart and Winston, New York, 1970.

[7] Edelstein-Keshet, L. *Mathematical models in biology*. SIAM, Philadelphia, 2004.

[8] Feller, W. *An introduction to probability theory and its applications: Vol. I*. Wiley, New York, 1950.

[9] ———. *An introduction to probability theory and its applications: Vol. II*. Wiley, New York, 1966.

[10] Forgacs, G., and Newman, S. A. *Biological physics of the developing embryo*. Cambridge University Press, Cambridge, 2005.

[11] Gantmacher, F. R. *Applications of the theory of matrices*. Interscience, New York, 1959.

[12] Gilbert, S. F. *Developmental biology*. Sinauer Associates, Sunderland, Mass., 2006.

[13] Grenander, U., and Szegö, G. *Teoplitz forms and applications*. University of California Press, Berkeley-Los Angeles, 1958.

[14] Keener, J. P., and Sneyd, J. *Mathematical physiology*. Springer, New York, 1998.

[15] Krylov, N. M., and Bogoliubov, N. N. *Introduction to nonlinear mechanics*. Princeton University Press, Princeton, N.J., 1943.

[16] Landau, L. D., and Lifshitz, E. M. *Fluid mechanics*. Second edition. Butterworth-Heinemann, Oxford, 1987.

[17] Maini, P. K., and Othmer, H. G. *Mathematical models for biological pattern formation*. IMA Volumes in Mathematics and Its Applications, 121. Springer, New York, 2001.

[18] McLachlan, N. W. *Theory and application of Mathieu functions*. Clarendon Press, Oxford, 1951.

[19] Meinhardt, H. *Models of biological pattern formation*. Academic Press, London–New York, 1982.

[20] Minorsky, N. *Introduction to non-linear mechanics: topological methods, analytical methods, non-linear resonance, relaxation oscillations*. J. W. Edwards, Ann Arbor, Mich., 1947.

[21] Mostow, G. D. *Mathematical models of cell rearrangement*. Yale University Press, New Haven, Conn., 1975.

[22] Murray, J. D. *Mathematical biology*. Second edition. Springer, Berlin, 1993.

[23] Nelsen, O. E. *Comparative embryology of the vertebrates*. McGraw-Hill, New York, 1953.

[24] Nüsslein-Volhard, C. *Coming to life: how genes drive development*. Kales Press, San Diego, 2006.

[25] Poston, T., and Stewart, I. *Catastrophe theory and its applications*. Surveys and Reference Works in Mathematics. Pitman, London–San Francisco, 1978.

[26] Prandtl, L. *Essentials of fluid dynamics*. Blackie, London, 1952.

[27] Reiner, M. *Deformation, strain, and flow*. Interscience, New York, 1960.

[28] Saunders, J. W. *Patterns and principles of animal development*. Macmillan, New York, 1970.

[29] Slack, J. M. W. *From egg to embryo*. Second edition. Cambridge University Press, Cambridge, 1991.

[30] Sussman, M. *Growth and development*. Second edition. Prentice-Hall, Englewood Cliffs, N.J., 1964.

[31] Thom, R. *Structural stability and morphogenesis: an outline of a general theory of models*. Benjamin-Cummings, Reading, Mass., 1980.

[32] Trinkhaus, J. P. *Cells into organs: the forces that shape the embryo*. Second edition. Prentice-Hall, Englewood Cliffs, N.J., 1984.

[33] Wessels, N. K. *Tissue interactions and development*. W. A. Benjamin, Menlo Park, Calif., 1977.

[34] Wolpert, L., Beddington, R., Jessell, T., Lawrence, P., Meyerowitz, E., and Smith, J. *Principles of development*. Third edition. Oxford University Press, London, 2007.

[35] Woodcock, A. E. R., and Poston, T. *A geometrical study of the elementary catastrophes*. Springer, Berlin, 1974.

Other References

[36] Allen, S. M., and Cahn, J. W. A microscopic theory for antiphase boundary motion and its application to antiphase domain coarsening. *Acta Metall.* 27(6):1085–1095, 1979.

[37] Baker, R. E. Periodic pattern formation in developmental biology: a study of the mechanisms underlying somitogenesis. Thesis, University of Oxford, 2005.

[38] Bard, J., and Lauder, I. How well does Turing's theory of morphogenesis work? *Jour. Theor. Biol.* 45(2): 501–531, 1974.

[39] Bell, G. I. Models of specific adhesion of cells. *Science* 200(4342), 618–627, 1978.

[40] Bellevaux, C., and Maille, M. Applications of finite element methods in fluid dynamics. *AGARD Computational Methods for Inviscid and Viscous Two-and-Three-Dimensional Flow Fields*, 7-1–7-28. AGARD Lecture Series, 73. Techical Editing and Reproduction, London, 1975.

[41] Berrill, N. J. Development and medusa-bud formation in the hydromedusae. *Quart. Rev. Biol.* 25(3): 292–316, 1950.

[42] Bi, D., Lopez, J. H., Schwarz, J. M., and Manning, L. M. Energy barriers and cell migration in densely packed tissues. *Sort Matter* 10(12): 1885–1890, 2014.

[43] Burz, D. S., Rivera-Pomar, R., Jäckle, H., and Hanes, S. D. Cooperative DNA-binding by bicoid provides a mechanism for threshold-dependent gene activation in the Drosophila embryo. *EMBO J.* 17(20): 5998–6009, 1998.

[44] Cahn, J. W., and Hilliard, J. E. Free energy of nonuniform system. I. Interfacial free energy. *J. Chem. Phys.* 28(2): 258–267, 1958.

[45] Chafee, N. Asymptotic behavior for solutions of a one-dimensional parabolic equation with homogeneous Neumann boundary conditions. *J. Diff. Eq.* 18(1): 118–134, 1975.

[46] Childress, S. Chemotactic collapse in two dimensions. *Modelling of patterns in space and time*, 61–66. Lecture Notes in Biomathematics, 55. Springer, Berlin-Heidelberg, 1984.

[47] Childress, S., and Percus, J. K. *Modeling of cell and tissue movements in the developing embryo.* Lectures on Mathematics in the Life Sciences, 14. American Mathematical Society, Providence, R.I., 1981.

[48] ______. Nonlinear aspects of chemotaxis. *Math. Biosci.* 56(3): 217–237, 1981.

[49] Cohen, M. H., and Robertson, A. Wave propagation in the early stages of aggregation of cellular slime molds. *J. Theor. Biol.* 31(1): 101–118, 1971.

[50] ———. Chemotaxis and the early stages of aggregation in cellular slime molds. *J. Theor. Biol.* 31(1): 119–130, 1971.

[51] Collier, J. R., McInerney, D., Schnell, S., Maini, P. K., Gavaghan, D. J., Houston, P., and Stern, C. D. A cell cycle model for somitogenesis: mathematical formulation and numerical simulation. *J. Theor. Biol.* 207(3): 305–316, 2000.

[52] Cooke, J., and Zeeman, E. C. A clock and wavefront model for control of the number of repeated structures during animal morphogenesis. *J. Theor. Biol.* 58(2): 455–476, 1976.

[53] Crick, F. H. C., and Lawrence, P. A. Compartments and polyclones in insect development. *Science* 189(4200): 340–347, 1975.

[54] Dan, K. Cyto-embryology of echinoderms and amphibia. *Int. Rev. Cytol.* 9:321–367, 1960.

[55] David, C. N., and MacWilliams, H. K. Regulation of the self-renewal probability in hydra stem cell clones. *Proc. Natl. Acad. Sci. USA* 75(2): 886–890, 1978.

[56] de Celis, J. F., Tyler, D. M., de Celis, J., and Bray, S. J. Notch signalling mediates segmentation of the drosophila leg. *Development* 125(23): 4617–4626, 1998.

[57] Deakin, M. A. B. Applied catastrophe theory in the social and biological sciences.*Bull. Math. Biol.* 42(5): 647–679, 1980.

[58] Dias, A. S., de Almeida, I., Belmonte, J. M., and Glazier, J. A. Somites without a clock. *Science* 343(6172): 791–795, 2014.

[59] Driever, W., and Nüsslein-Volhard, C. The bicoid protein determines position in the drosophila embryo in a concentration-dependent manner. *Cell* 54(1): 95–104, 1988.

[60] Duistermaat, J. J. Oscillatory integrals, Lagrange immersions, and unfoldings of singularities. *Comm. Pure Appl. Math.* 27(2): 207–281, 1974.

[61] Edelstein, B. B. Dynamics of cell differentiation and associated pattern formation. *J. Theor. Biol.* 37(2): 221–243, 1972.

[62] Eldar, A., Rosin, D., Shilo, B. Z., and Barkai, N. Self-enhanced ligand degradation underlies robustness of morphogen gradients. *Dev. Cell* 5(4): 635–646, 2003.

[63] Ephrussi, A., and St. Johnston, D. Seeing is believing: the bicoid morphogen gradient matures. *Cell* 116(2): 143–152, 2004.

[64] Foty, R. A., Forgacs, G., Pfleger, C. M., and Steinberg, M. S. Liquid properties of embryonic tissues: measurement of interfacial tensions. *Phys. Rev. Lett.* 72(14): 2298–2301, 1994.

[65] Foty, R. A., and Steinberg, M. S. The differential adhesion hypothesis: a direct evaluation. *Dev. Biol.* 278(1): 255–263, 2005.

[66] Garcia-Bellido, A., Ripoll, P., and Morata, G. Developmental compartmentalization of the wing disk of drosophila. *Nature* 245(147): 251–253, 1973.

[67] Gierer, A. Biological features and physical concepts of pattern formation exemplified by hydra. *Cur. Top. Dev. Biol.* 11: 17–59, 1977.

[68] Gierer, A., and Meinhardt, H. A theory of biological pattern formation. *Kybernetik* 12(1): 30–39, 1972.

[69] Giudicelli, F., Özbudak, E. M., Wright, G. J., and Lewis, J. Setting the tempo in development: an investigation of the zebrafish somite clock mechanism. *PLoS Biol.* 5(6): e150, 2007.

[70] Glansdorff, P., and Prigogine, I. On a general evolution criterion in macroscopic physics. *Physica* 30(2): 351–374, 1964.

[71] Glass, L. Classification of biological networks by their qualitative dynamics. *J. Theor. Biol.* 54(1): 85–107, 1975.

[72] Glass, L., and Kauffman, S. A. Co-operative components, spatial localization and oscillatory cellular dynamics. *J. Theor. Biol.* 34(2): 219–237, 1972.

[73] ———. The logical analysis of continuous, non-linear biochemical control networks. *J. Theor. Biol.* 39(1): 103–129, 1973.

[74] Glass, L., and Pasternak, J. S. Stable oscillations in mathematical models of biological control systems. *J. Math. Biology* 6(3): 207–223, 1978.

[75] Gmitro, J. I., and Scriven, L. E. *Intercellular transport*. Academic Press, New York, 1966.

[76] Goel, N., Campbell, R. D., Gordon, R., Rosen, R., Martinez, H., and Yčas. M. Self-sorting of isotropic cells. *J. Theor. Biol.* 28(3): 423–468, 1970.

[77] Goel, N. S., and Rogers, G. Computer simulation of engulfment and other movements of embryonic tissues. *J. Theor. Biol.* 71(1): 103–140, 1978.

[78] Golubitsky, M. An introduction to catastrophe theory and its applications. *SIAM Review* 20(2): 352–387, 1978.

[79] Goodwin, B. C. Model of bacterial growth cycle: statistical dynamics of a system with asymptotic orbital stability. *J. Theor. Biol.* 28(3): 375–391, 1970.

[80] Gordon, R., Goel, N. S., Steinberg. M. S., and Wiseman, L. L. A rheological mechanism sufficient to explain the kinetics of cell sorting. *J. Theor. Biol.* 37(1): 43–73, 1972.

[81] Graner, F., and Glazier, J. A. Simulation of biological cell sorting using a two-dimensional extended Potts model. *Phys. Rev. Lett.* 69(13): 2013–2017, 1992.

[82] Greenberg, J. M., and Hastings, S. P. Spatial patterns for discrete models of diffusion in excitable media. *SIAM J. Appl. Math.* 34(3): 515–523, 1978.

[83] Greenspan, H. P. Deformation of a viscous droplet caused by variable surface-tension. *Stud. Appl. Math.* 57(1): 45–58, 1977.

[84] ______. On the dynamics of cell cleavage. *J. Theor. Biol.* 65(1): 79–99, 1977.

[85] ______. On the motion of a viscous droplet that wets a surface. *J. Fluid Mech.* 84(1): 125–143, 1978.

[86] Gregor, T., Bialek, W., van Steveninck, R. R. D. R., Tank, D. W., and Wieschaus, E. F. Diffusion and scaling during early embryonic pattern formation. *Proc. Natl. Acad. Sci. USA* 102(51): 18403–18407, 2005.

[87] Gregor, T., Wieschaus, E. F., McGregor, A. P., Bialek, W., and Tank, D. W. Stability and nuclear dynamics of the bicoid morphogen gradient. *Cell* 130(1): 141–152, 2001.

[88] Grover, J. W. The enzymatic dissociation and reproducible reaggregation in vitro of 11-day embryonic chick lung. *Dev. Biol.* 3(5): 555–568, 1961.

[89] Gustafson, T., and Wolpert, L. Cellular movement and contact in sea urchin morphogenesis. *Biol. Rev. Camb. Phil. Soc.* 42(3): 442–498, 1967.

[90] Hochberg, D., Zorzano, M. P., and Morán, F. Spatiotemporal patterns driven by autocatalytic internal reaction noise. *J. Chem. Phys.* 122(21): 214701, 2005.

[91] Holtfreter, J. A study of the mechanics of gastrulation. Part I. *J. Exp. Zool.* 94(3): 261–318, 1943.

[92] ______. A study of the mechanics of gastrulation. Part II. *J. Exp. Zool* 95(2): 171-212, 1944.

[93] Hornbruch, A., and Wolpert, L. Cell division in the early growth and morphogenesis of the chick limb. *Nature* 226:764–766, 1970.

[94] Horstmann, D. From 1970 until now: the Keller-Segel model in chemotaxis and its consequences: I. *Jahresber. Deutsch. Math.-Verein.* 105:103–165, 2003.

[95] ______. From 1970 until now: the Keller-Segel model in chemotaxis and its consequences: II. *Jahresber. Deutsch. Math.-Verein.* 106:51–69, 2004.

[96] Hunding, A., Kauffman, S. A., Goodwin, B. C. Drosophila segmentation: supercomputer simulation of prepattern hierarchy. *J. Theor. Biol.* 145(3): 369–384, 1990.

[97] Ishihara, S., and Kaneko, K. Turing pattern with proportion preservation. *J. Theor. Biol.* 238(3): 683–693, 2006.

[98] Jacobson, A. G., and Gordon, R. Changes in the shape of the developing vertebrate nervous system analyzed experimentally, mathematically, and by computer. *Jour. Exp. Zool.* 197(2): 191-246, 1976.

[99] Kalinay, P., and Percus J. K. Extended Fick-Jacobs equation: variational approach. *Phys. Rev. E* 72(6): 061203, 2005.

[100] Kauffman, S. A. Control circuits for determination and transdetermination. *Science* 181(4097): 310–318, 1973.

[101] ______. Measuring a mitotic oscillator: the arc discontinuity. *Bull. Math. Biol.* 36(2): 171–182, 1974.

[102] Kauffman, S. A., Shymko, R. M., and Trabert, K. Control of sequential compartment formation in *Drosophila*. *Science* 199(4326): 259–270, 1978.

[103] Kauffman, S. A., and Wille, J. J. The mitotic oscillator in physarum polycepharum. *J. Theor. Biol.* 55(1): 47–93, 1975.

[104] Kaufman, M., Urbain, J., and Thomas, R. Towards a logical analysis of the immune response. *J. Theor. Biol.* 114(4): 527–561, 1985.

[105] Keller, E. F., and Segel, L. A. Initiation of slime mold aggregation viewed as an instability. *J. Theor. Biol.* 26(3): 399–415, 1970.

[106] ______. Model for chemotaxis. *J. Theor. Biol.* 30(2): 225–234, 1971.

[107] Kerner, E. H. Further considerations on the statistical mechanics of biological associations. *Bull. Math. Biophys.* 21(2): 217–255, 1959.

[108] Lebowitz, J. L., and Rubinow, S. I. A theory for the age and generation time distribution of a microbial population. *J. Math. Bio.* 1(1): 17–36, 1974.

[109] Lewis, J. Autoinhibition with transcriptional delay: a simple mechanism for the zebrafish somitogenesis oscillator. *Curr. Biol.* 13(16): 1398–1408, 2003.

[110] Lewis, J. H. Fate maps and the pattern of cell division: a calculation for the chick wing-bud. *J. Embryol. Exp. Morphol.* 33(2): 419–434, 1975.

[111] Lewis, W. H. Experimental studies on the development of the eye in amphibia. I. On the origin of the lens. Rana palustris. *Amer. J. Anat.* 3(4): 505–536, 1904.

[112] Lindenmayer, A. Mathematical models for cellular interaction in development. *J. Theor. Biol.* 18(3): 280–299, 1968.

[113] ______. Mathematical models for cellular interaction in development. *J. Theor. Biol.* 18(3): 300–315, 1968.

[114] Livshits, M. A., Gurija, G. T., Belintsev, B. N., and Volkenstein, M. V. Positional differentiation as pattern formation in reaction-diffusion systems with permeable boundaries. Bifurcation analysis. *J. Math. Biol.* 11(3): 295–310, 1981.

[115] Macken, C. A., and Perelson, A. S. *Branching processes applied to cell surface aggregation phenomena*. Lecture Notes in Biomathematics, 58. Springer, Berlin-Heidelberg, 1985.

[116] MacWilliams, H. K., Papageorgiou, S. A model of gradient interpretation based on morphogen binding. *J. Theor. Biol.* 72(3): 385–411, 1978.

[117] Maini, P. K. Mathematical models in morphogenesis. *Mathematics inspired by biology*, 151–189. Springer, Berlin-Heidelberg, 1999.

[118] Meinhardt, H. Morphogenesis of lines and nets. *Differentiation* 6(2): 117–123, 1976.

[119] ______. A model of pattern formation in insect embryogenesis. *J. Cell Sci.* 23(1): 117–139, 1977.

[120] ______. A model for pattern formation of hypostome, tentacles and foot in hydra: how to form structures close to each other, how to form them at a distance. *Dev. Biol.* 157(2): 321–333, 1993.

[121] Mittenthal, J. E., and Mazo, R. M. A model for shape generation by strain and cell-cell adhesion in the epithelium of an arthropod leg segment. *J. Theor. Biol.* 100(3): 443–483, 1983.

[122] Murray, J. D., and Oster, G. F. Cell traction models for generating pattern and form in morphogenesis. *J. Math. Biol.* 19(3): 265–279, 1984.

[123] Myasnikova, E., Samsonova, A., Kozlov, K., Samsonova, M., and Reinitz, J. Registration of the expression patterns of drosophila segmentation genes by two independent methods. *Bioinformatics* 17(1): 3–12, 2001.

[124] Nanjundiah, V. Chemotaxis, signal relaying and aggregation morphology. *J. Theor. Biol.* 42(1): 63–105, 1973.

[125] Nardi, J. B., and Stocum, D. L. Surface properties of regenerating limb cells: evidence for gradation along the proximodistal axis. *Differentiation* 25(1-3): 7–31, 1984.

[126] Newman, S. A., and Frisch, H. L. Dynamics of skeletal pattern formation in developing chick limb. *Science* 205(4407): 662–668, 1979.

[127] Ochoa-Espinosa, A., Yu, D., Tsirigos, A., Struffi, P., and Small, S. Anterior-posterior positional information in the absence of a strong bicoid gradient. *Proc. Nat. Acad. Sci* 106(10): 3823–3828, 2009.

[128] Odell, G., Oster, G., Burnside, B., and Alberch, P. A mechanical model for epithelial morphogenesis. *J. Math. Biol.* 9(3): 291–295, 1980.

[129] Percus, J. K. *Combinatorial methods in developmental biology*. Courant Institute of Mathematical Sciences, New York University, New York, 1977.

[130] ———. Tree structures in immunology. *Applications of combinatorics and graph theory to the biological and social sciences*, 259–276. IMA Volumes in Mathematics and Its Applications, 17. Springer, New York, 1989.

[131] Peskin, C. S. *Mathematical aspects of heart physiology*. Courant Institute Lecture Notes. Courant Institute of Mathematical Sciences, New York University, New York, 1975.

[132] Pourquié, O. The segmentation clock: converting embryonic time into spatial pattern. *Science* 301(5631): 328–330, 2003.

[133] Raper, K. B. Developmental patterns in simple slime molds. *Growth* 5:41–76, 1941.

[134] Reinitz, J., and Sharp, D. H. Mechanism of eve stripe formation. *Mech. Dev.* 49(1-2): 133–158, 1995.

[135] Rogers, K. W., and Schier, A. F. Morphogen gradients from generation to interpretation. *Ann. Rev. Cell Dev. Biol.* 27:377–407, 2011.

[136] Rothe, F. Some analytical results about a simple reaction-diffusion system for morphogenesis. *J. Math. Biol.* 7(4): 375–384, 1979.

[137] Rubin, L., and Saunders, J. W., Jr. Ectodermal-mesodermal interactions in the growth of limb buds in the chick embryo: constancy and temporal limits of the ectodermal induction. *Dev. Biology* 28(1): 94–112, 1972.

[138] Segel, L. A., and Jackson, J. L. Dissipative structure: an explanation and an ecological example. *J. Theor. Biol.* 37(3): 345–359, 1972.

[139] Segel, L. A., and Stoeckley, B. Instability of a layer of chemotactic cells, attractant, and degrading enzyme. *J. Theor. Biol.* 37(3): 561–585, 1972.

[140] Sherratt, J. A. "Mathematical modelling of cell adhesion in developmental biology and cancer." Paper presented at Queen's University, Belfast, June 5, 2007.

[141] Steinberg, M. S. Reconstruction of tissues by dissociated cells. *Science* 141(3579): 401–408, 1963.

[142] ———. Adhesion-guided multicellular assembly: A commentary upon the postulates, real and imagined, of the differential adhesion hypothesis, with special attention to computer simulations of cell sorting. *J. Theor. Biol.* 55(2): 431–443, 1975.

[143] Steinberg, M. S., and Garrod, D. R. Observations of the sorting-out of embryonic cells in a monolayer culture. *J. Cell Science* 18(3): 385–403, 1975.

[144] Steinberg, M. S., and Takeichi, M. Experimental specification of cell sorting, tissue spreading, and specific spatial patterning by quantitative differences in cadherin expression. *Proc. Natl. Acad. Sci. USA* 91(1): 206–209, 1994.

[145] Sulsky, D., Childress, S., and Percus, J. K. A model of cell sorting. *J. Theor. Biol.* 106(3): 275–301, 1984.

[146] Summerbell, D., and Lewis, J. H. Time, place and positional value in the chick limb-bud. *J. Embryol. Exp. Morphol.* 33(3): 621–643, 1975.

[147] Summerbell, D., and Wolpert, L. Cell density and cell division in the early morphogenesis of the chick limb. *Nature* 238(88): 24–26, 1972.

[148] Summerbell, D., Lewis, J. H., and Wolpert, L. Positional information in chick limb morphogenesis. *Nature* 244:492–496, 1973.

[149] Takeichi, M. Cadherin cell adhesion receptors as a morphogenetic regulator. *Science* 251(5000): 1451–1455, 1991.

[150] Thom, R. A global dynamical scheme for vertebrate embryology. *Some mathematical questions in biology, IV*, 1–45. Lectures on Mathematics in the Life Sciences, 5. American Mathematical Society, Providence, R.I, 1973.

[151] Thompson, R. N., Yates, C. A., and Baker, R. E. Modelling cell migration and adhesion during development. *Bull. Math. Biol.* 74(12): 2793–2809, 2012.

[152] Tickle, C., Summerbell, D., and Wolpert, L. Positional signalling and specification of digits in chick limb morphogenesis. *Nature* 254(5497): 199–202, 1975.

[153] Till, J. E., McCulloch, E. A., and Siminovich, L. A stochastic model of stem cell proliferation, based on the growth of spleen colony-forming cells. *Proc. Nat. Acad. Sci. USA* 51(1): 29–36, 1964.

[154] Topaz, C. M., Bertozzi, A. L., and Lewis, M. A. A nonlocal continuum model for biological aggregation. *Bull. Math. Biol.* 68(7): 1601–1623, 2006.

[155] Townes, P. L., and Holtfreter, J. Directed movements and selective adhesion of embryonic amphibian cells. *J. Exp. Zool.* 128(1): 53–120, 1955.

[156] Turing, A. M. The chemical basis of morphogenesis. *Phil. Trans. Roy. Soc. Lond.* 237(641): 37–72, 1952.

[157] Tyson, J. J. Classification of instabilities in chemical reaction systems. *J. Chem. Phys.* 62(3): 1010– 1015, 1975.

[158] Wasserman, G. *Stability of unfoldings*. Lecture Notes in Mathematics, 393. Springer, Berlin, 1974.

[159] Wilson, H. V. On some phenomena of coalescence and regeneration in sponges. *J. Exp. Zool.* 5(2): 245–258, 1907.

[160] Wolpert, L., Hicklin, J., and Hornbruch, A. Positional information and pattern regulation in regeneration of hydra. *Symp. Soc. Exp. Biol.* 25: 391–415, 1971.

[161] Wolpert, L., Lewis, J., and Summerbell, D. Morphogenesis of the vertebrate limb. *Cell Patterning* 29: 95–130, 1975.

[162] Zajac, M., Jones, G. L., and Glazier, J. A. Simulating convergent extension by way of anisotropic differential adhesion. *J. Theor. Biol.* 222(2): 247–259, 2003.

Index

Published Titles in This Series

26 **Jerome K. Percus and Stephen Childress,** Mathematical Models in Developmental Biology, 2015
25 **Kurt O. Friedrichs,** Mathematical Methods of Electromagnetic Theory, 2014
24 **Christof Schütte and Marco Sarich,** Metastability and Markov State Models in Molecular Dynamics, 2013
23 **Jerome K. Percus,** Mathematical Methods in Immunology, 2011
22 **Frank C. Hoppensteadt,** Mathematical Methods for Analysis of a Complex Disease, 2011
21 **Frank C. Hoppensteadt,** Quasi-Static State Analysis of Differential, Difference, Integral, and Gradient Systems, 2010
20 **Pierpaolo Esposito, Nassif Ghoussoub, and Yujin Guo,** Mathematical Analysis of Partial Differential Equations Modeling Electrostatic MEMS, 2010
19 **Stephen Childress,** An Introduction to Theoretical Fluid Mechanics, 2009
18 **Percy Deift and Dimitri Gioev,** Random Matrix Theory: Invariant Ensembles and Universality, 2009
17 **Ping Zhang,** Wigner Measure and Semiclassical Limits of Nonlinear Schrödinger Equations, 2008
16 **S. R. S. Varadhan,** Stochastic Processes, 2007
15 **Emil Artin,** Algebra with Galois Theory, 2007
14 **Peter D. Lax,** Hyperbolic Partial Differential Equations, 2006
13 **Oliver Bühler,** A Brief Introduction to Classical, Statistical, and Quantum Mechanics, 2006
12 **Jürgen Moser and Eduard J. Zehnder,** Notes on Dynamical Systems, 2005
11 **V. S. Varadarajan,** Supersymmetry for Mathematicians: An Introduction, 2004
10 **Thierry Cazenave,** Semilinear Schrödinger Equations, 2003
9 **Andrew Majda,** Introduction to PDEs and Waves for the Atmosphere and Ocean, 2003
8 **Fedor Bogomolov and Tihomir Petrov,** Algebraic Curves and One-Dimensional Fields, 2002
7 **S. R. S. Varadhan,** Probability Theory, 2001
6 **Louis Nirenberg,** Topics in Nonlinear Functional Analysis, 2001
5 **Emmanuel Hebey,** Nonlinear Analysis on Manifolds: Sobolev Spaces and Inequalities, 2000
3 **Percy Deift,** Orthogonal Polynomials and Random Matrices: A Riemann-Hilbert Approach, 2000
2 **Jalal Shatah and Michael Struwe,** Geometric Wave Equations, 2000
1 **Qing Han and Fanghua Lin,** Elliptic Partial Differential Equations, Second Edition, 2011